UNITEXT

La Matematica per il 3+2

Volume 160

Editor-in-Chief

Alfio Quarteroni, Politecnico di Milano, Milan, Italy
  École Polytechnique Fédérale de Lausanne (EPFL), Lausanne, Switzerland

Series Editors

Luigi Ambrosio, Scuola Normale Superiore, Pisa, Italy

Paolo Biscari, Politecnico di Milano, Milan, Italy

Ciro Ciliberto, Università di Roma "Tor Vergata", Rome, Italy

Camillo De Lellis, Institute for Advanced Study, Princeton, USA

Victor Panaretos, Institute of Mathematics, École Polytechnique Fédérale de Lausanne (EPFL), Lausanne, Switzerland

Lorenzo Rosasco, DIBRIS, Università degli Studi di Genova, Genova, Italy
  Center for Brains Mind and Machines, Massachusetts Institute of Technology, Cambridge, Massachusetts, US
  Istituto Italiano di Tecnologia, Genova, Italy

The **UNITEXT - La Matematica per il 3+2** series is designed for undergraduate and graduate academic courses, and also includes books addressed to PhD students in mathematics, presented at a sufficiently general and advanced level so that the student or scholar interested in a more specific theme would get the necessary background to explore it.

Originally released in Italian, the series now publishes textbooks in English addressed to students in mathematics worldwide.

Some of the most successful books in the series have evolved through several editions, adapting to the evolution of teaching curricula.

Submissions must include at least 3 sample chapters, a table of contents, and a preface outlining the aims and scope of the book, how the book fits in with the current literature, and which courses the book is suitable for.

For any further information, please contact the Editor at Springer: francesca.bonadei@springer.com

THE SERIES IS INDEXED IN SCOPUS

Sören Bartels

Numerical Mathematics 3x9

Three subject areas in 27 short chapters

Sören Bartels
Department of Applied Mathematics
University of Freiburg
Freiburg, Germany

ISSN 2038-5714   ISSN 2532-3318 (electronic)
UNITEXT
ISSN 2038-5722   ISSN 2038-5757 (electronic)
La Matematica per il 3+2
ISBN 978-3-662-70889-7   ISBN 978-3-662-70890-3 (eBook)
https://doi.org/10.1007/978-3-662-70890-3

This book is a translation of the original German edition "Numerik 3x9," 2nd edition, by Sören Bartels, published by Springer-Verlag GmbH, DE in 2023. The translation was done with the help of an artificial intelligence machine translation tool. A subsequent human revision was done primarily in terms of content, so that the book will read stylistically differently from a conventional translation. Springer Nature works continuously to further the development of tools for the production of books and on the related technologies to support the authors.

Translation from the German language edition: "Numerik 3x9" by Sören Bartels, © Der/die Herausgeber bzw. der/die Autor(en), exklusiv lizenziert an Springer-Verlag GmbH, DE, ein Teil von Springer Nature 2023. Published by Springer Berlin Heidelberg. All Rights Reserved.

© The Editor(s) (if applicable) and The Author(s), under exclusive license to Springer-Verlag GmbH, DE, part of Springer Nature 2025

This work is subject to copyright. All rights are solely and exclusively licensed by the Publisher, whether the whole or part of the material is concerned, specifically the rights of translation, reprinting, reuse of illustrations, recitation, broadcasting, reproduction on microfilms or in any other physical way, and transmission or information storage and retrieval, electronic adaptation, computer software, or by similar or dissimilar methodology now known or hereafter developed.
The use of general descriptive names, registered names, trademarks, service marks, etc. in this publication does not imply, even in the absence of a specific statement, that such names are exempt from the relevant protective laws and regulations and therefore free for general use.
The publisher, the authors and the editors are safe to assume that the advice and information in this book are believed to be true and accurate at the date of publication. Neither the publisher nor the authors or the editors give a warranty, expressed or implied, with respect to the material contained herein or for any errors or omissions that may have been made. The publisher remains neutral with regard to jurisdictional claims in published maps and institutional affiliations.

This Springer imprint is published by the registered company Springer-Verlag GmbH, DE, part of Springer Nature.
The registered company address is: Heidelberger Platz 3, 14197 Berlin, Germany

If disposing of this product, please recycle the paper.

Preface

Numerical mathematics is concerned with the development and analysis of methods for solving mathematical problems. It is a firm component of mathematics education at universities, but often plays a subordinate role, even though all mathematical disciplines result from the goal of solving concrete and practical problems. This important aspect is often neglected in highly developed modern mathematics. On the other hand, numerics is often seen as the time-consuming and sometimes less exciting technical implementation of mathematical concepts, which leads to a false impression of the contents of numerics. In fact, many of the numerical methods go back to scientists like Gauss and Newton, who wanted to quantify and understand profound questions of the natural sciences via explicit calculations. The use of the computer should therefore simplify the handling of numerical methods and not make it more difficult.

In this textbook, the most important ideas and concepts for the algorithmic solution of some basic mathematical problems are discussed and the main difficulties of their practical implementation are examined. In doing so, three questions must always be considered:

- Is it possible to specify a method for the approximate solution of a mathematical problem?
- How do perturbations, for example due to rounding of input data, affect the numerical solution?
- What is the computational complexity of a method to achieve a given accuracy?

These questions are addressed for model problems, such as the solution of systems of linear equations, the calculation of eigenvalues of a matrix, the numerical integration of functions, the approximate solution of nonlinear equations and the approximation of solutions of differential equations.

In developing the course material, I have followed the presentations of various scripts, textbooks and monographs, which are listed at the end of this book. If I have followed a source too closely in the presentation of the material at one point or another, this is to be understood as an appreciation of a particularly successful elaboration. This text makes no claim as to the originality of its contents. Its sole aim is to provide students of mathematics, engineering and the natural sciences an opportunity to familiarise themselves with the basics of numerical mathematics.

The presentation of the classic material is intended to illustrate basic methods of numerics by example. Optimality of the methods or greatest generality of the associated statements was consciously omitted. When solving concrete, possibly practical problems, it is therefore essential to consult the specialised literature, which is also presented in extract form at the end of the book. The application examples listed in the text are intended for motivation and illustration and should not be interpreted as real case studies. For special applications, it is usually necessary to adapt the methods developed for idealised model situations to the special characteristics of the present problem. This book is intended to prepare the reader for this challenge.

The present text results from lectures taught at the Universities of Bonn and Freiburg and is a translation of the second edition of the German version of the textbook. I would like to thank numerous colleagues, assistants and tutors for corrections and suggestions for improvement. I would like to particularly thank Lea Heusler for her careful proofreading of the first edition and Benedikt Albrecht and Nick Seinsche for their important help in the development of the second edition of the text. Furthermore, I would like to thank Yohance Osborne and Vanessa Styles for carefully checking the English translation of the text.

Freiburg, Germany Sören Bartels
December 2024

Contents

Part I Numerical Linear Algebra

1 Basic Concepts 3
   1.1   Problem Statement 3
   1.2   Condition and Stability 4
   1.3   Computational Complexity 8
   1.4   Learning Objectives, Quiz and Application 8

2 Operator Norm and Condition Number 11
   2.1   Vector Norms 11
   2.2   Matrix Norms 12
   2.3   Condition Number 14
   2.4   Learning Objectives, Quiz and Application 15

3 Matrix Factorisations 17
   3.1   Triangular Matrices 17
   3.2 LU Decomposition 18
   3.3   Cholesky Decomposition 20
   3.4   Learning Objectives, Quiz and Application 23

4 Elimination Methods 25
   4.1   Gaussian Elimination Method 25
   4.2   Pivot Strategy 28
   4.3   Learning Objectives, Quiz and Application 30

5 Least Squares Problems 33
   5.1   Gaussian Normal Equation 33
   5.2   Householder Transformations 35
   5.3 QR Decomposition 37
   5.4   Solution of the Least Squares Problem 39
   5.5   Learning Objectives, Quiz and Application 40

6 Singular Value Decomposition and Pseudoinverse 41
   6.1   Singular Value Decomposition 41

	6.2	Pseudoinverse	43
	6.3	Learning Objectives, Quiz and Application	44
7	**The Simplex Method**		45
	7.1	Linear Programs	45
	7.2	The Simplex Step	47
	7.3	Learning Objectives, Quiz and Application	50
8	**Eigenvalue Problems**		51
	8.1	Localisation	51
	8.2	Conditioning	53
	8.3	Power Method	55
	8.4	QR Method	58
	8.5	Jacobi Method	61
	8.6	Learning Objectives, Quiz and Application	64
9	**Iterative Solution Methods**		67
	9.1	Inexact Solution	67
	9.2	Banach's Fixed Point Theorem	67
	9.3	Linear Iteration Methods	69
	9.4	Jacobi and Gauss-Seidel Methods	70
	9.5	Diagonal Dominance and Irreducibility	71
	9.6	Convergence	73
	9.7	Learning Objectives, Quiz and Application	76

Part II Numerical Analysis

10	**General Condition Number and Floating Point Numbers**		79
	10.1	Conditioning	79
	10.2	Floating Point Numbers	81
	10.3	Rounding	83
	10.4	Stability	84
	10.5	Learning Objectives, Quiz and Application	85
11	**Polynomial Interpolation**		87
	11.1	Lagrange Interpolation	87
	11.2	Interpolation Error	88
	11.3	Neville's Algorithm	89
	11.4	Chebyshev Nodes	92
	11.5	Hermite Interpolation	94
	11.6	Learning Objectives, Quiz and Application	96
12	**Interpolation with Splines**		97
	12.1	Splines	97
	12.2	Cubic Splines	99
	12.3	Calculation of Cubic Splines	102
	12.4	Interpolation Error	103
	12.5	Learning Objectives, Quiz and Application	105

13 Discrete Fourier Transform ... 107
- 13.1 Trigonometric Interpolation ... 107
- 13.2 Fourier Bases ... 109
- 13.3 Fast Fourier Transform ... 111
- 13.4 Learning Objectives, Quiz and Application ... 113

14 Numerical Integration ... 115
- 14.1 Quadrature Formulas ... 115
- 14.2 Newton-Cotes Formulas ... 117
- 14.3 Composite Quadrature Formulas ... 118
- 14.4 Gaussian Quadrature ... 120
- 14.5 Extrapolation ... 122
- 14.6 Experimental Convergence Order ... 124
- 14.7 Learning Objectives, Quiz and Application ... 125

15 Nonlinear Problems ... 127
- 15.1 Root Finding and Minimisation Problems ... 127
- 15.2 Approximation of Roots ... 129
- 15.3 One-Dimensional Minimisation ... 132
- 15.4 Multidimensional Minimisation ... 133
- 15.5 Learning Objectives, Quiz and Application ... 135

16 Conjugate Gradient Method ... 137
- 16.1 Quadratic Minimisation ... 137
- 16.2 Conjugate Search Directions ... 138
- 16.3 Calculation of Conjugate Directions ... 139
- 16.4 CG Method ... 141
- 16.5 Convergence of the CG Method ... 142
- 16.6 Learning Objectives, Quiz and Application ... 144

17 Sparse Matrices and Preconditioning ... 147
- 17.1 Sparse Matrices ... 147
- 17.2 Preconditioned CG Method ... 148
- 17.3 Further Preconditioning Matrices ... 150
- 17.4 Learning Objectives, Quiz and Application ... 152

18 Multidimensional Approximation ... 155
- 18.1 Grids and Triangulations ... 155
- 18.2 Approximation on Tensor Product Grids ... 157
- 18.3 Two-Dimensional Fourier Transform ... 159
- 18.4 Approximation on Triangulations ... 159
- 18.5 Learning Objectives, Quiz and Application ... 163

Part III Numerics for Differential Equations

19 Ordinary Differential Equations ... 167
- 19.1 Fundamentals ... 167
- 19.2 The Predator-Prey Model ... 168

19.3	Higher Order Equations	169
19.4	Autonomous Equations	170
19.5	Two-Body Problems	171
19.6	Explicit Solutions	171
19.7	Learning Objectives, Quiz and Application	172

20 Existence, Uniqueness and Stability 173
20.1	Existence and Uniqueness	173
20.2	Gronwall's Lemma	176
20.3	Stability	177
20.4	Learning Objectives, Quiz and Application	179

21 Single-Step Methods 181
21.1	Euler Method	181
21.2	Consistency	183
21.3	Discrete Gronwall Lemma and Convergence	185
21.4	Higher-Order Methods	188
21.5	Learning Objectives, Quiz and Application	189

22 Runge-Kutta Methods 191
22.1	Motivation	191
22.2	Runge-Kutta Methods	192
22.3	Well-Posedness	194
22.4	Consistency	195
22.5	Learning Objectives, Quiz and Application	199

23 Multistep Methods 201
23.1	General Multistep Methods	201
23.2	Consistency	202
23.3	Adams Methods	204
23.4	Predictor-Corrector Method	206
23.5	Learning Objectives, Quiz and Application	207

24 Convergence of Multistep Methods 209
24.1	Difference Equations	209
24.2	Zero-Stability	211
24.3	Convergence	212
24.4	Learning Objectives, Quiz and Application	215

25 Stiff Differential Equations 217
25.1	Stiffness	217
25.2	A-Stability	218
25.3	Gradient Flows	221
25.4	Heat Equation	223
25.5	Learning Objectives, Quiz and Application	224

26 Step Size Control 227
26.1	A Posteriori Error Control	227

	26.2	Adaptive Algorithm	230
	26.3	Control Procedure	230
	26.4	Extrapolation	231
	26.5	Learning Objectives, Quiz and Application	232
27	**Symplectic, Shooting and dG Methods**		233
	27.1	Hamiltonian Systems	233
	27.2	Symplectic Methods	235
	27.3	Shooting Method	239
	27.4	Discontinuous Galerkin Methods	240
	27.5	Learning Objectives, Quiz and Application	241

Part IV Problems and Projects

28	**Problems on Numerical Linear Algebra**		245
	28.1	Basic Concepts	245
	28.2	Operator Norm and Condition Number	246
	28.3	Matrix Factorisations	250
	28.4	Elimination Methods	253
	28.5	Least Squares Problems	255
	28.6	Singular Value Decomposition and Pseudoinverse	258
	28.7	The Simplex Method	260
	28.8	Eigenvalue Problems	262
	28.9	Iterative Solution Methods	266
29	**Problems on Numerical Analysis**		269
	29.1	General Condition Number and Machine Numbers	269
	29.2	Polynomial Interpolation	271
	29.3	Interpolation with Splines	274
	29.4	Discrete Fourier Transform	277
	29.5	Numerical Integration	280
	29.6	Nonlinear Problems	284
	29.7	Conjugate Gradients Method	287
	29.8	Sparse Matrices and Preconditioning	290
	29.9	Multidimensional Approximation	292
30	**Problems on Numerics for Differential Equations**		297
	30.1	Ordinary Differential Equations	297
	30.2	Existence, Uniqueness and Stability	300
	30.3	Single-Step Methods	302
	30.4	Runge-Kutta Methods	306
	30.5	Multistep Methods	308
	30.6	Convergence of Multistep Methods	310
	30.7	Stiff Differential Equations	313
	30.8	Step Size Control	315
	30.9	Symplectic, Shooting and dG Methods	317

Part V Supplementary Material

31 Results from Linear Algebra 323
    31.1 Scalar Product of Vectors 323
    31.2 Determinant of Square Matrices 323
    31.3 Image and Kernel of Linear Mappings 324
    31.4 Eigenvalues and Diagonalisability 325
    31.5 Jordan Normal Form 326

32 Results from Analysis 327
    32.1 Continuous and Differentiable Functions 327
    32.2 Mean Value Theorem and Taylor Polynomials 328
    32.3 Landau Symbols 329
    32.4 Fundamental Theorem of Algebra 329
    32.5 Multidimensional Calculus 330

33 Introduction to C++ 333
    33.1 Structure 333
    33.2 Classes 333
    33.3 Types 334
    33.4 Control Statements 335
    33.5 Logical Expressions and Increments 335
    33.6 Functions 336
    33.7 Pointers 337
    33.8 Dynamic Arrays 337
    33.9 Working with Matrices 338
    33.10 Time Measurement, Saving and Packages 338

34 Introduction to MATLAB 341
    34.1 Structure 341
    34.2 Lists and Arrays 341
    34.3 Matrix Operations 342
    34.4 Manipulation of Arrays 342
    34.5 Elementary Functions 343
    34.6 Loops and Control Statements 343
    34.7 Text and Graphic Output 344
    34.8 Creating New Functions 344
    34.9 Various Commands 345
    34.10 Sparse Matrices 345
    34.11 Examples 345
    34.12 Free Alternative 346

35 Introduction to Python 349
    35.1 Structure 349
    35.2 Elementary Commands 350
    35.3 Types 351
    35.4 Control Statements 351

	35.5	Logical Expressions	352
	35.6	Functions	352
	35.7	Lists...	352
	35.8	Timing, Saving and Plotting......................................	353
36	**Example Programs in MATLAB, C++ and Python**		**355**
	36.1	*LU* Decomposition and Solving Triangular Systems	355
	36.2	Polynomial Interpolation and Neville's Scheme	357
	36.3	Numerical Solution of Ordinary Differential Equations	360

Advanced Topics 367

Bibliography 369

Index 373

Notation

Numbers, Vectors and Matrices

$\mathbb{Z}$	Integers
$\mathbb{N}, \mathbb{N}_0$	Positive and non-negative integers
$\mathbb{R}, \mathbb{C}$	Real and complex numbers
$\mathbb{R}_{\geq 0}, \mathbb{R}_{>0}$	Non-negative and positive real numbers
$[s,t], (s,t)$	Closed and open interval
$\mathbb{R}^n$	n-dimensional Euclidean space
$\mathbb{R}^{n \times m}$	Set of $n \times m$ matrices
$B_r(x), K_r(x)$	Open and closed ball with radius r around x
$A \subset B$	A is a subset of B or $A = B$
$x = (x_i), A = (a_{ij})$	Column vector and matrix
$x^\mathsf{T}, A^\mathsf{T}$	Transposition of a vector or a matrix
$\|\cdot\|$	Norm of a vector or operator norm of a matrix
$x \cdot y = x^\mathsf{T} y$	Scalar product of vectors $x, y \in \mathbb{R}^n$
$x \times y$	Cross product of vectors $x, y \in \mathbb{R}^3$
$x \perp y$	x is orthogonal to y
I_n	$n \times n$ identity matrix
$O(n)$	Orthogonal group
$[x,y]^\mathsf{T}, (x,y)$	Vector with entries x and y
$\begin{bmatrix} x_1 & x_2 \\ y_1 & y_2 \end{bmatrix}$	Matrix with entries x_1, x_2, y_1, y_2
$\mathrm{i} = \sqrt{-1}$	Imaginary unit

Various Symbols

$o(s^p), \mathcal{O}(s^p), \mathcal{O}(n^p)$	Landau symbols
$\lfloor r \rfloor$	Maximum number $k \in \mathbb{Z}$ with $k \leq r$
$|A|$	Cardinality of a finite set

$\mathscr{C}, \widetilde{\mathscr{C}}$	Consistency terms
$N_p(1)$	Level set of the p-norm
$N_g^-(x_0)$	Sublevel set of the function g

Linear Mappings

ker A	Kernel of the linear mapping A
Im A	Image of the linear mapping A
rank A	Rank of the linear mapping A
dim W	Dimension of the vector space W
det A	Determinant of the matrix A
tr A	Trace of the matrix A

Differential Operators

∂_i, ∂_{x_i}, $\frac{\partial}{\partial x_i}$	Partial derivative with respect to the i-th argument
∇f	Gradient of the function f
div F	Divergence of the vector field F
Df, $D^2 f$	Jacobian and Hessian matrix of a function f
y_t	Time derivative of the function y

Function Spaces

$C^k([a,b])$	k-times continuously differentiable functions on $[a,b]$
$\mathscr{P}_m$	Polynomials of degree m
$\mathscr{S}^{m,k}(\mathscr{T}_h)$	Piecewise degree m polynomial, k-times continuously differentiable functions

Prologue: Why Numerics?

Goals and Concepts

Numerical mathematics or simply *numerics* is concerned with the practical implementation of mathematical concepts, for example, to calculate real processes. This can be the infection dynamics of a pandemic, the evaluation and visualisation of a medical computer tomography, the realisation of search algorithms on the internet, the usage behaviour of an internet platform, the training of a neural network, the prediction of the weather, the calculation of ocean currents, the load-bearing capacity of bridges and buildings, the simulation of a crash test or the compression of data for fast transmission of information. As a rule, large amounts of data occur, the implementation is typically done with the help of computers, which leads to additional peculiarities.

Computers can only perform simple arithmetic operations and this only approximately, i.e., with *rounding errors*. Every mathematical task must therefore be reduced to simple problems. The solution of systems of linear equations and the evaluation of explicit calculation rules can be realised very efficiently and robustly. With these two concepts, many tasks such as eigenvalue problems, constrained optimisation tasks, nonlinear equations and data compression problems can be solved approximately.

However, unexpected effects can occur during the development of methods. For example, equivalent formulas can lead to different results when implemented on a computer, different sequences with the same limit can converge at different speeds and rounding errors can accumulate during a calculation. Since rounding errors are anyway unavoidable, it is neither necessary nor sensible to determine exact solutions to problems.

The first part of the book is dedicated to the fast and robust *solution of systems of linear equations* with regular matrices $A \in \mathbb{R}^{n \times n}$, i.e., for a given vector $b \in \mathbb{R}^n$ the determination of $x \in \mathbb{R}^n$ with

$$Ax = b.$$

It is particularly important to understand how disturbances of data affect the solution. Based on this, overdetermined systems of equations or least squares problems, eigenvalue problems and linear optimisation problems are considered.

The core of the second part is the *approximation of functions* with simply representable functions such as piecewise polynomial functions s_h, so that a given accuracy $\varepsilon > 0$ is achieved, i.e.,

$$\|f - s_h\|_{C^0(I)} \leq \varepsilon.$$

This can be used to reduce the calculation of derivatives and integrals to simple problems. Further aspects are the calculation of zeros and minima. In the third part, the numerical *approximation of ordinary differential equations* or initial value problems is examined, which have the general form

$$y'(t) = f(t, y(t)), \quad y(0) = y_0.$$

They form the basis of the simulation of time-dependent problems. Even the simple case $y' = \alpha y$ with solution $y(t) = y_0 e^{\alpha t}$ leads to insights that can be transferred to large classes of problems. With these methods, trajectories of bodies, Hamiltonian systems for the description of solar systems and one-dimensional boundary value problems can be numerically approximated.

Difficulties and Ideas

We consider some typical and partly surprising phenomena of the direct algorithmic implementation of mathematical concepts.

Rounding Errors

Since binary computers can only represent finitely many numbers, rounding errors are inevitable. Even if modern computers calculate with high accuracy, this can easily lead to difficulties. For example, if a party receives $n_P = 2\,099\,580$ out of a total of $n_G = 42 \cdot 10^6$ votes cast, a computer delivers the share

```
n_P/n_G = 0.0500
```

thus supposedly 5.00% of the votes. However, a legal 5% hurdle in an election does not provide for rounding and a more accurate representation of the quotient shows the result

$$\frac{n_P}{n_G} = 0.04999000,$$

```
1 % machine_precision.m
2 x = 1;
3 while 1+x > 1
4     x = x/2;
5 end
6 disp(2*x);
```

```
1 >> machine_precision
2    2.2204e-16
```

Fig. 1 Determination of machine accuracy (left) and result of calculation (right)

so the party did not receive the required votes. Here, a misleading result is created by rounding in the visual representation of the number. Another error occurs due to rounding-related arithmetic operations of the computer. The relative calculation accuracy of a computer can be determined by halving the number $x = 1$ until the expression $1 + x$ is no longer distinguished from 1 by the computer, see Fig. 1. A typical accuracy is at $1 \cdot 10^{-16}$, so one can assume 15 correct decimal places. Instead of setting the calculation accuracy in relation to a single vote, the 5% hurdle can be checked more easily with the inequality $n_P/n_G \geq 1/20$ or $20 n_P \geq n_G$.

Convergence Speed

The number $\sqrt{2}$ can be constructed by successively determining the decimal places. Starting from $r_0 = 1$, decimal places are added to obtain numbers r_k with k decimal places, which are maximised subject to the constraint $r_k^2 < 2$. In the first step, $r_1 = 1.4$ is set, since $(1.5)^2 > 2$ holds. In the k-th step, $r_{k-1}^2 < 2$ and

$$r_k = r_{k-1} + \ell \cdot 10^{-k}$$

where $\ell \in \{0, 1, \ldots, 9\}$ is chosen to maximise r_k subject to the constraint $r_k^2 < 2$. Thus, in each step, one obtains another correct decimal place and accordingly, for the error

$$\delta_k = \left|\sqrt{2} - r_k\right| < 10^{-k}.$$

The error is reduced by the factor $q = 1/10$ in each step. With a trick, one obtains approximations where the number of correct decimal places doubles in each step. For this, we consider more generally the calculation of $\sqrt{a}$ for a positive number $a > 0$. The equation $x^2 = a$ is obviously equivalent to

$$x^2 = \frac{1}{2}x^2 + \frac{1}{2}a \quad \Longleftrightarrow \quad x = \frac{1}{2}\left(x + \frac{a}{x}\right).$$

The second identity characterises the solution as a fixed point x^* of a function $x \mapsto \Phi(x)$ and this observation can be used to define the *fixed point iteration*

$$x_{k+1} = \Phi(x_k) = \frac{1}{2}\left(x_k + \frac{a}{x_k}\right)$$

with a suitable starting value $x_0 > 0$, which is referred to as *Heron's method*. Here, one can prove so-called quadratic convergence of the errors $e_k = |\sqrt{a} - x_k|$, i.e.

$$e_{k+1} \leq c e_k^2$$

or the doubling of correct decimal places in each step, provided $c e_k^2 < 1$ holds. An implementation can be found in Fig. 2. The convergence speed of a fixed point iteration can be quantified with a Taylor approximation. If $\Phi'(x_*) = 0$, then

$$x_{k+1} - x_* = \Phi(x_k) - \Phi(x_*) = \frac{1}{2}\Phi''(\xi)(x_k - x_*)^2,$$

which implies *local, quadratic convergence*. If $\Phi'(x_*) \neq 0$, then the *local, linear convergence* $e_{k+1} \leq q e_k$ follows analogously, if $|\Phi'(x)| \leq q < 1$ for all $x \in B_\varepsilon(x_*)$. Typical courses of corresponding fixed point iterations are depicted in Fig. 3.

```
1 % heron.m
2 a = 2.0; delta = 1.0e-15;
3 x = a/2; e = abs(x-sqrt(a));
4 while e > delta
5     x = (x+a/x)/2;
6     e = abs(x-sqrt(a));
7     disp([x,e]);
8 end
```

```
1 >> format shortE
2 >> heron
3    1.5000e+00    8.5786e-02
4    1.4167e+00    2.4531e-03
5    1.4142e+00    2.1239e-06
6    1.4142e+00    1.5947e-12
7    1.4142e+00    2.2204e-16
```

Fig. 2 Calculation of the square root according to Heron (left) and results of the calculation (right)

Fig. 3 Linear (left) and quadratic (right) convergence of fixed point iterations

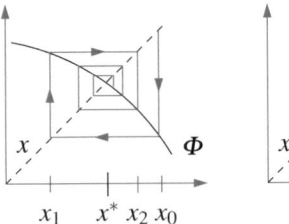
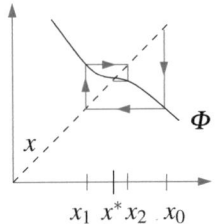

Instabilities

Rounding errors can become very noticeable in problems with specific bad properties. As an example, we consider the approximation of the number π by computing the area of the unit circle. For this purpose, the circle is approximated as shown in Fig. 4 with n congruent triangles, whose heights are denoted by k_n so that the area $A = \pi$ is approximated by $A_n = nk_n/2$. Such an approach was already used by Archimedes in the third century BC.

We have that $k_n = \sin(2\pi/n)$, however, only basic operations and the square root should be used. Using the identity $\sin\alpha = 2\sin(\alpha/2)\cos(\alpha/2)$ and the pq-formula results in the *recursion formula*

$$2k_{2n}^2 = 1 - \sqrt{1 - k_n^2}.$$

From $\sin(\pi/6) = 1/2$ one obtains the starting value $k_{12} = 1/2$ and can thus determine a sequence of heights. The results generated with the program shown in Fig. 5 and listed in Fig. 4 show that the approximations of π initially improve, then stagnate and finally become completely useless. However, if one uses a binomial

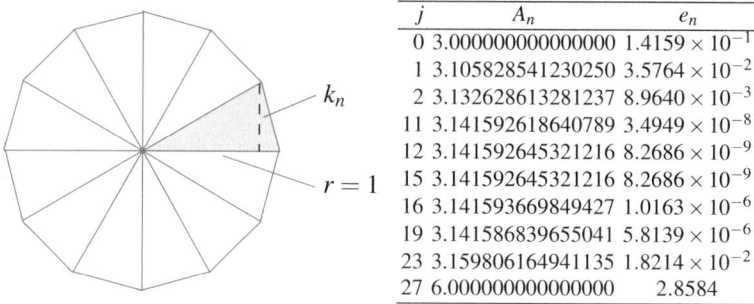

j	A_n	e_n
0	3.000000000000000	1.4159×10^{-1}
1	3.105828541230250	3.5764×10^{-2}
2	3.132628613281237	8.9640×10^{-3}
11	3.141592618640789	3.4949×10^{-8}
12	3.141592645321216	8.2686×10^{-9}
15	3.141592645321216	8.2686×10^{-9}
16	3.141593669849427	1.0163×10^{-6}
19	3.141586839655041	5.8139×10^{-6}
23	3.159806164941135	1.8214×10^{-2}
27	6.000000000000000	2.8584

Fig. 4 Approximation of the unit circle area with n triangles (left) and numerically determined areas A_n as well as errors $e_n = |A_n - \pi|$ with $n = 2^j \cdot 12$ (right)

```
1 % pi_approx.m
2 n = 12; k = 0.5; J = 30;
3 for j = 1:J
4     n = 2*n;
5     k = sqrt((1-sqrt(1-k^2))/2);
6     A = n*k/2; e = abs(pi-A);
7     disp([j,A,e]);
8 end
```

```
1 % pi_approx_mod.m
2 n = 12; k = 0.5; J = 30;
3 for j = 1:J
4     n = 2*n;
5     k = k/sqrt(2*(1+sqrt(1-k^2)));
6     A = n*k/2; e = abs(pi-A);
7     disp([j,A,e]);
8 end
```

Fig. 5 Approximation of the circle number π with direct (left) and modified (right) calculation of the heights k_n

formula, one obtains the equivalent representation

$$2k_{2n}^2 = \left(1 - \sqrt{1 - k_n^2}\right)\frac{1 + \sqrt{1 - k_n^2}}{1 + \sqrt{1 - k_n^2}} = \frac{k_n^2}{1 + \sqrt{1 - k_n^2}}.$$

This formula allows π to be approximated to machine precision. It will be shown that generally the subtraction of nearly equal numbers should be avoided.

Computational Effort

The calculation of the determinant of a square matrix $A \in \mathbb{R}^{n \times n}$ can be performed using the Laplace expansion theorem. With the recursion formula

$$\det A = \sum_{j=1}^{n}(-1)^{1+j} a_{1j} \det \widehat{A}_{1j},$$

where $\widehat{A}_{1j}$ is the submatrix obtained by deleting the first row and j-th column, the calculation can be reduced to computing determinants of smaller matrices until finally matrices with only one entry appear, Fig. 6 shows a practical implementation. The computational effort grows dramatically, when transitioning from $n = 8$ to $n = 10$ the computing time increases by a factor of $90 = 9 \cdot 10$ and for matrices of dimension $n \geq 12$ the method is hardly feasible in a reasonable time. Practically and theoretically, it is seen that $n!$ operations are necessary. Alternatively, the Gaussian elimination method provides a factorisation $A = LU$ with triangular matrices L and U, where for the diagonal entries of L, $\ell_{ii} = 1$ may be required. Thus, with the

```
1 % det_laplace.m
2 function val = det_laplace(A)
3 n = size(A,1); val = 0;
4 if n == 1
5     val = A(1,1);
6 else
7     for j = 1:n
8         I = 2:n;
9         J = [1:j-1,j+1:n];
10        hat_A_1j = A(I,J);
11        val = val+(-1)^(1+j) ...
12            *A(1,j) ...
13            *laplace(hat_A_1j);
14    end
15 end
```

```
1 % det_laplace_hilb.m
2 for n = 4:2:10
3     A = hilb(n);
4     tic; d = det_laplace(A); toc
5 end
```

```
1 >> det_laplace_hilb
2 Elapsed time is 0.000814 seconds.
3 Elapsed time is 0.002340 seconds.
4 Elapsed time is 0.108463 seconds.
5 Elapsed time is 9.220774 seconds.
```

Fig. 6 Calculation of the determinant with the Laplace expansion theorem (left) and run times for matrix sizes $n = 4, 6, 8, 10$ (right)

Prologue: Why Numerics?

rules for the determinant, it follows that

$$\det A = \det L \det U = (\ell_{11}\ell_{22}\ldots\ell_{nn})(u_{11}u_{22}\ldots u_{nn}) = u_{11}u_{22}\ldots u_{nn},$$

where we have used that the determinant of a triangular matrix is given by the product of the diagonal entries or eigenvalues. If the factorisation is given, the determinant can be computed with $n-1$ operations. A check of the elimination method shows that the factorisation can be found with n^3 operations, here required row swapping may be included in the operations.

Robustness to Disturbances

Rounding errors can be considered as disturbances and so one can abstractly assess whether a problem can be approximated or numerically solved at all, regardless of specific algorithms. For illustration we consider the determination of the roots of a polynomial. Specifically, we choose

$$p(x) = (x-a)^n - 0$$

with a given number a, which is then the n-fold root of the polynomial. We now disturb the term 0 and subtract a small number $\varepsilon > 0$ instead, i.e. we consider the polynomial

$$p_\varepsilon(x) = (x-a)^n - \varepsilon.$$

The complex roots shown in Fig. 7 are given by $\tilde{x}_k = a + s_k \varepsilon^{1/n}$ with the n-th roots of unity $s_k = e^{i2\pi k/n}$, $k = 1, 2, \ldots, n$, which are evenly distributed on the boundary of the unit circle in the complex plane, in the case $n = 2$ they are $s_1 = -1$ and $s_2 = 1$. The error between the correct roots $x_k = a$ and those of the disturbed polynomial is $e_k = |x_k - \tilde{x}_k| = \varepsilon^{1/n}$ and this becomes smaller as ε becomes smaller. However, the problem is that the ratio of the output error to the input error, i.e.

$$\frac{\max_{k=1,\ldots,n} |x_k - \tilde{x}_k|}{\|p - p_\varepsilon\|_{C^0(\mathbb{R})}} = \frac{\varepsilon^{1/n}}{\varepsilon} = \varepsilon^{(1-n)/n}$$

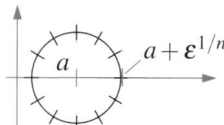

Fig. 7 The (complex) roots of the disturbed polynomial $p_\varepsilon(x) = (x-a)^n - \varepsilon$ are located on the circle around a with radius $r = \varepsilon^{1/n}$

is unbounded for $\varepsilon \to 0$ and $n \geq 2$. Small disturbances in the data of the problem thus have a disproportionate effect on the result. Therefore, the determination of roots of polynomials is referred to as an *ill-conditioned problem*. Directly connected with this is the poor conditioning of the determination of eigenvalues of a matrix. A real-life ill-conditioned problem is the vertical positioning of a pen.

Inexact Solving

The Gaussian elimination method for solving a system of linear equations leads to a cost of n^3 computational operations. However, an exact solution in terms of computer arithmetic is rarely necessary, as not only rounding errors influence the result, but the data can also not be considered exact due to measurement and model errors. This observation leads to the idea that by merely approximating the solution of the linear system, the computational effort can be significantly reduced. One approach is based on the decomposition of the matrix A into its diagonal part D and the rest $R = A - D$. Provided D is regular, the equation $Ax = b$ is thus equivalent to the equations

$$Dx = b - Rx \quad \Longleftrightarrow \quad x = D^{-1}(b - Rx).$$

The second equation can be interpreted as a fixed point equation $x = \Phi(x)$ and leads to the *iteration*

$$x_{k+1} = D^{-1}(b - Rx_k)$$

with a starting vector $x_0 \in \mathbb{R}^n$. In some cases, good approximations are obtained in a few steps. The evaluation of the right-hand side generally requires a cost of n^2 computational operations, but in many cases A or R have many vanishing entries and the cost is only a moderate multiple cn of n. If the iteration converges quickly, the cost of solving the system is reduced from n^3 to $\tilde{c}n$, which is enormous for typical sizes of n in the range $[10^2, 10^7]$. To exploit this aspect, the definition of A must be modified, as in the program shown in Fig. 8, to avoid unnecessary multiplications with zero. Corresponding runtimes are displayed in Fig. 9. A better convergence behaviour is achieved with the iteration $x_{k+1} = (D+U)^{-1}(b - Lx_k)$, where U and L are the submatrices of A above and below the diagonal, respectively, and in each step a system of equations with triangular matrix $D+U$ must be solved. Rounding errors are not a problem here, as convergent fixed point iterations have a self-stabilising effect in the sense that each iterate can be considered as a new starting value.

```
1  % jacobi_iteration.m
2  n = 10^2; b = ones(n,1);
3  e = ones(n,1); e_s = ones(n-1,1);
4  A = diag(4*e,0)-diag(e_s,1)-diag(e_s,-1);
5  % A = spdiags([-e,4*e,-e],[-1,0,1],n,n);
6  D = diag(A); D_inv = D.^(-1); R = A-diag(D);
7  x = zeros(n,1); tol = 1.0e-3; ctr = 0;
8  while norm(A*x-b) > tol
9      x = D_inv.*(b-R*x); ctr = ctr+1; disp(ctr);
10 end
```

Fig. 8 Solving a system of equations with the Jacobi iteration, the alternative definition of the band matrix A avoids unnecessary multiplications with zero entries

$$A = \begin{bmatrix} 4 & -1 & & & \\ -1 & \ddots & \ddots & & \\ & \ddots & \ddots & -1 \\ & & -1 & 4 \end{bmatrix}$$

n	A fully populated	A sparsely populated
10^2	0.005273 s	0.047754 s
10^3	0.028120 s	0.009399 s
10^4	1.042249 s	0.023457 s
10^5	—	0.106429 s
10^6	—	0.512903 s

Fig. 9 If unnecessary multiplications are avoided with band matrices, the iterative method leads to low memory requirements and short computation times even for very large matrices

Approximation with Polynomials

A theorem by Weierstraß states that any continuous function on a compact interval can be approximated arbitrarily well by polynomials. However, these results do not show how to find the polynomials or what degree of polynomial is needed to achieve a given accuracy. To calculate such polynomials, pairwise different points $x_0, x_1, \ldots, x_n$ in the interval $[a, b]$ can be chosen together with a polynomial p defined by the requirement

$$p(x_i) = f(x_i), \quad i = 0, 1, \ldots, n.$$

To fulfil these $n + 1$ *interpolation conditions*, the polynomial must have at least degree n. From the fundamental theorem of algebra, it follows that a polynomial of this degree is uniquely defined. With a basis $(p_j)_{j=0,\ldots,n}$, such as the monomials $p_j(x) = x^j$, the coefficient vector $c \in \mathbb{R}^{n+1}$ of p results from the system of equations $Ac = f$ with $A_{ij} = p_j(x_i)$ and $f_i = f(x_i)$, $i, j = 0, 1, \ldots, n$. However, for certain functions f and uniformly distributed points $x_0, x_1, \ldots, x_n$ it is observed that the polynomials do not converge uniformly for increasing numbers n, see Figs. 10 and 11. Using Rolle's theorem, it can be seen that the distances between the support points should be chosen smaller at the edges, which is optimally realised by so-called *Chebyshev nodes*. In addition to this effect, it should be noted, that the monomial basis leads to a matrix A with unfavourable properties with respect to small disturbances.

```matlab
% interpolation.m
f = @(x) 1./(1+25*x.^2);
delta = 0.01; X = (-1:delta:1); Y = f(X); n = 11;
x_eq = zeros(n+1,1); y_eq = zeros(n+1,1);
x_ch= zeros(n+1,1); y_ch = zeros(n+1,1);
dx = 2/n; dtheta = pi/(2*(n+1));
for k = 1:n+1
    x_eq(k) = -1+(k-1)*dx; y_eq(k) = f(x_eq(k));
    x_ch(k) = cos((2*k-1)*dtheta); y_ch(k) = f(x_ch(k));
end
p_eq = polyfit(x_eq,y_eq,n); p_ch = polyfit(x_ch,y_ch,n);
plot(X,Y,'--',x_eq,y_eq,'o',X,polyval(p_eq,X));
title('equidistant'); pause
plot(X,Y,'--',x_ch,y_ch,'o',X,polyval(p_ch,X));
title('chebyshev');
```

Fig. 10 Calculation and representation of an interpolation polynomial with evenly and unevenly distributed support points

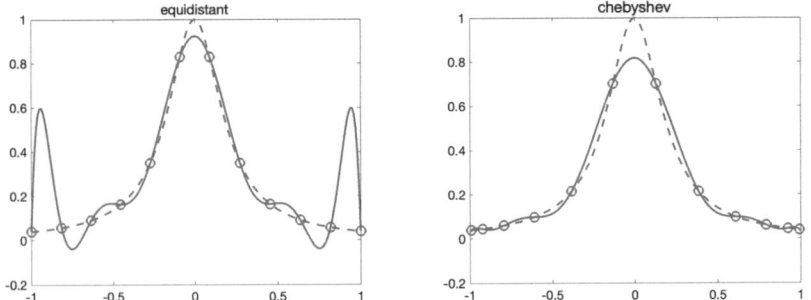

Fig. 11 Polynomial interpolation with equidistant support points (left) and Chebyshev nodes (right)

Choice of Suitable Bases

Every vector $x \in \mathbb{R}^n$ can be represented with respect to the canonical basis $e_1, e_2, \ldots, e_n$ such that

$$x = \sum_{k=1}^{n} \alpha_k e_k,$$

where the coefficients α_k correspond to the components of the vector. If the vector x has special properties, for example, it is given as a sampled audio signal at times $t_1, t_2, \ldots, t_n$, it makes sense to choose a basis $v_1, v_2, \ldots, v_n$ that takes these properties into account. In this case, many coefficients in the linear combination

Fig. 12 Functions can often be represented as a sum of sine oscillations

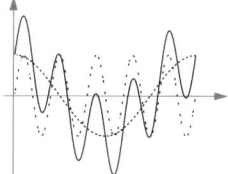

have the property of being very small in magnitude or negligible, and hence we have

$$x = \sum_{k=1}^{n} \beta_k v_k \approx \sum_{\ell=1}^{m} \beta_{k_\ell} v_{k_\ell}, \quad m \ll n.$$

For example, if $n = 10^4$, the vector x can often be well represented with $m \sim 10^2$ relevant pieces of information. This is referred to as *data compression*, which is the basis of the digital age. The mathematical challenge lies in the efficient implementation of the basis change. If the vectors $(v_k)_{k=1,\ldots,n}$ are chosen as fundamental oscillations, then the *fast Fourier transformation* allows for an almost optimal basis change. An example is shown in Fig. 12.

Large Intermediate Results

Swapping rows is only necessary in the Gaussian elimination process when so-called pivot elements, with which the elimination of entries below the diagonal is performed, are identically zero. To avoid instabilities or strong effects of rounding errors, swapping rows should also be performed when pivot elements are small compared to other entries. Otherwise, intermediate values that are large in magnitude can occur, as can be checked using the example

$$\begin{bmatrix} \varepsilon & 1 \\ 1 & 1 \end{bmatrix} \begin{bmatrix} x_1 \\ x_2 \end{bmatrix} = \begin{bmatrix} 1 \\ 2 \end{bmatrix}$$

with solution $(x_1, x_2) \approx (1, 1)$ for $0 \leq \varepsilon \ll 1$. The fact that intermediate results can lead to large computational errors is shown by the perturbation calculation for the sum $s = y_1 + y_2 + \cdots + y_n$ with exact summands y_i and disturbed values $\widetilde{y}_i = (1 + \sigma_i \varepsilon_i) y_i$, with $\sigma_i \in \{\pm 1\}$ and $\varepsilon_i \geq 0$, so that for the disturbed sum $\widetilde{s}$ we have

$$\widetilde{s} = \sum_{i=1}^{n} \widetilde{y}_i = \sum_{i=1}^{n} (1 + \sigma_i \varepsilon_i) y_i = s + \sum_{i=1}^{n} \sigma_i \varepsilon_i y_i.$$

```
1  % intermediate_vals.m
2  x = 1.0; s = 1.0;
3  eps_p = 10^(-5); k = 10^3;
4  x_p = (1+eps_p)*x;
5  s_p = sqrt(x_p)+k*exp(x_p-1)...
6       +k*sin(3*pi*x_p/2);
7  e_rel_s = abs((s_p-s)/s);
8  e_rel_x = abs(eps_p);
9  kappa_rel = e_rel_s/e_rel_x;
10 disp(kappa_rel);
```

```
1  >> intermediate_vals
2     1.0006e+03
```

Fig. 13 If intermediate results or summands are larger in magnitude than the result, a large amplification of relative errors can occur

For the relative error in the result $|s - \tilde{s}|/|s|$ it follows from the triangle inequality and the relative errors $|\tilde{y}_i - y_i|/|y_i| = \varepsilon_i$ of the data, that

$$\varepsilon_s = \frac{|s - \tilde{s}|}{|s|} \leq \frac{1}{|s|} \sum_{i=1}^{n} |\varepsilon_i||y_i| \leq \left(\frac{\sum_{i=1}^{n} |y_i|}{|s|}\right) \max_{i=1,\ldots,n} \varepsilon_i = \kappa \varepsilon_y.$$

So, a large amplification of the relative error can occur if $|s|$ is small compared to the absolute summands $|y_i|$. The first inequality is an equality when the disturbances have the same sign as the summands, and the second inequality is an equality when all disturbances are of equal size. The program shown in Fig. 13 calculates the value

$$s = \sqrt{x} + k \exp(x-1) + k \sin(x 3\pi/2)$$

for a disturbance $\tilde{x} = (1 + \varepsilon_p)x$ of $x = 1$, which leads to an amplification of the relative errors by the factor $\kappa \approx k$ and thus confirms the result.

A special case of the above estimate is the subtraction of nearly equal numbers, which corresponds to the case $y_1 \approx -y_2$ and leads to so-called *cancellation effects*. For example, if two rods have the lengths $\ell_1 = 101.51$ and $\ell_2 = 100.49$ in centimetres and these were approximately measured with $\tilde{\ell}_1 = 102.00$ and $\tilde{\ell}_2 = 100.00$, then the relative errors ε_i, $i = 1, 2$, are less than 0.5%; however, the relative error of the differences $\delta = 1.02$ and $\tilde{\delta} = 2.00$ is almost 100.0%, which corresponds to an error amplification of $\kappa \approx 200$.

Descent Methods

To determine a (local) minimum of a differentiable function $g : \mathbb{R}^n \to \mathbb{R}$, it makes sense, as when descending in a mountain landscape, to gradually reduce the function values. In order to reach a minimum as quickly as possible, the direction with locally the greatest reduction of the function value should be chosen for the next iteration step. This is given by the negative gradient of the function. Starting from an initial

Fig. 14 Illustration of the descent method for determining a minimum of a function

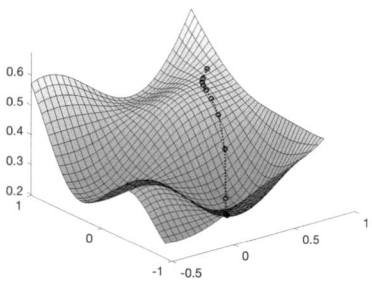

value $x_0 \in \mathbb{R}^n$, a sequence of iterates x_k, $k = 0, 1, \ldots$, is thus determined by the rule

$$x_{k+1} = x_k - \alpha_k \nabla g(x_k).$$

The *step size* α_k should be chosen sensibly so that one does not actually get a value of the function which is larger than before. Figure 14 shows a path resulting from a descent method on the graph of the function. In addition to optimising the step sizes, for a class of minimisation problems also the optimisation of the search directions is of interest. For quadratic minimisation problems of the form $g(x) = (1/2)\|Ax - b\|^2$ it can be ensured that the descent directions are orthogonal to each other in a suitable sense and one obtains the minimum with a maximum of n steps.

Implicit and Explicit Methods

Approximate solutions of the initial value problem $y' = f(t, y)$, $y(0) = y_0$, are obtained by approximating the derivative of y by a secant slope

$$y'(t) \approx \frac{y(t + \tau) - y(t)}{\tau}$$

which in the case of this right-sided difference quotient with step size $\tau > 0$ and time steps $t_k = k\tau$ leads to the Euler method

$$y_{k+1} = y_k + \tau f(t_k, y_k)$$

with starting value y_0. Thus one obtains a sequence of approximations $(y_k)_{k=0,\ldots,K}$ by simple successive or *explicit* evaluation of the right-hand side. However, experiments in the case $f(t, y) = \alpha y$ with $\alpha < 0$ and exact, bounded solution $y(t) = y_0 e^{\alpha t}$ show that the approximations only remain bounded for sufficiently small step sizes, see Fig. 15. This is improved, if instead of the right-sided difference quotient a left-sided one is taken, which leads to the method

```
1  % euler_expl.m
2  alpha = -2; y_0 = 1; T = 10;
3  f = @(t,s) alpha*s;
4  K = 8; tau = T/K;
5  y = zeros(K+1,1); y(1) = y_0;
6  for k = 1:K
7      t_k = (k-1)*tau;
8      y(k+1) = y(k)+tau*f(t_k,y(k));
9  end
10 plot(tau*(0:K),y,'o-');
11 title('expl. Euler');
```

Fig. 15 The explicit Euler method is a simple to implement method (left), which however can lead to unbounded approximations (right)

```
1  % euler_impl.m
2  alpha = -2; y_0 = 1; T = 10;
3  K = 16; tau = T/K;
4  y = zeros(K+1,1); y(1) = y_0;
5  for k = 2:K+1
6      t_k = (k-1)*tau;
7      % y(k) = y(k-1)+tau*f(t_k,y(k));
8      y(k) = (1-alpha*tau)^(-1)*y(k-1);
9  end
10 plot(tau*[0:K],y,'o-');
11 title('impl. Euler');
```

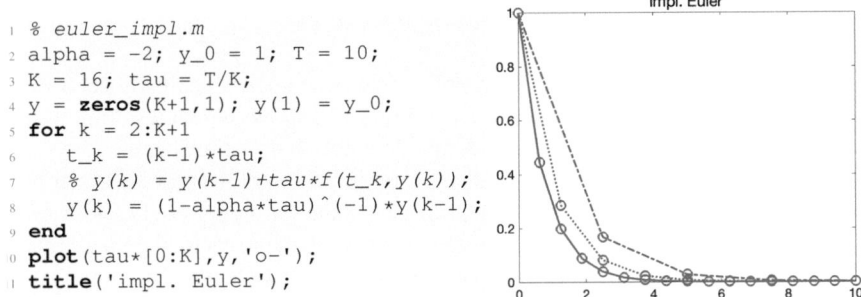

Fig. 16 The implicit Euler method has better stability properties than the explicit method

$$y_k = y_{k-1} + \tau f(t_k, y_k)$$

with starting value y_0, see Fig. 16. The price for the better stability properties of the method is however the required solution of a possibly nonlinear system of equations in each iteration step. This method is therefore referred to as an *implicit method*.

Multi-Term Recursion

To obtain better approximations of derivatives, it is obvious to use more than just two time points, for example one could use the following combination of three values y_{k+2}, y_{k+1}, y_k to approximate the derivative $y'(t_r)$

$$y'(t_r) \approx \alpha_0 y_k + \alpha_1 y_{k-1} + \alpha_2 y_{k-2}.$$

Possible coefficients α_ℓ, $\ell = 0, 1, 2$, result from Taylor approximations; however, not all values found using this method result in good choices. Criteria to choose the coefficients can be found by considering the trivial differential equation $y'(t) = 0$ with starting value y_0 and constant solution $y(t) = y_0$. A sensible method should guarantee that in this case, approximations remain bounded. These fulfil for given initial values y_0, y_1 the three-term recursion

$$\alpha_2 y_{k+2} + \alpha_1 y_{k+1} + \alpha_0 y_k = 0$$

or in matrix representation with $\gamma_\ell = -\alpha_\ell/\alpha_2$, $\ell = 0, 1$, the relation

$$\begin{bmatrix} y_{k+1} \\ y_{k+2} \end{bmatrix} = \begin{bmatrix} 0 & 1 \\ \gamma_0 & \gamma_1 \end{bmatrix} \begin{bmatrix} y_k \\ y_{k+1} \end{bmatrix} = A \begin{bmatrix} y_k \\ y_{k+1} \end{bmatrix}.$$

A transformation of the iteration matrix A into Jordan normal form yields

(i) $J = T^{-1} A T = \begin{bmatrix} \lambda_1 & 0 \\ 0 & \lambda_2 \end{bmatrix}$, (ii) $J = T^{-1} A T = \begin{bmatrix} \lambda_1 & 1 \\ 0 & \lambda_1 \end{bmatrix}$,

with an orthogonal matrix T and geometrically simple or multiple eigenvalues $\lambda_1, \lambda_2 \in \mathbb{C}$. Relevant for the stability of the numerical method is now whether the matrix J is non-expansive, i.e., whether $\|Jz\|_* \leq \|z\|_*$ for all $z \in \mathbb{C}^2$ with a suitable vector norm $\|\cdot\|_*$. In the first case, this is given if $|\lambda_1|, |\lambda_2| \leq 1$, and in the second, if $|\lambda_1| < 1$. Examples of unstable and stable three-term recursions are given by the coefficients $(\alpha_2, \alpha_1, \alpha_0) = (1, 4, -5)$ and $(\alpha_2, \alpha_1, \alpha_0) = (3, -4, 1)$, respectively. The results of an unstable iteration are shown in Fig. 17.

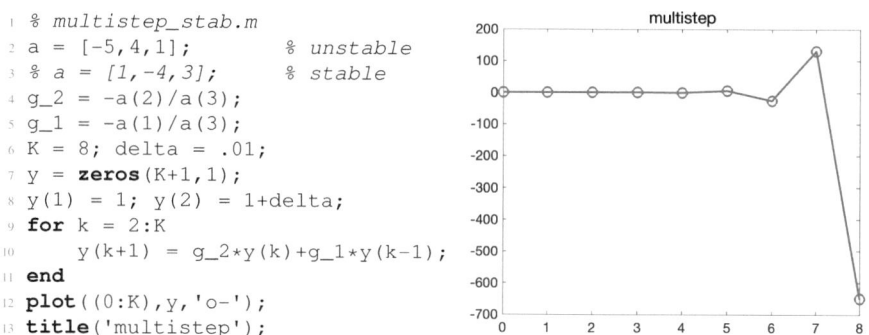

```
1  % multistep_stab.m
2  a = [-5,4,1];        % unstable
3  % a = [1,-4,3];      % stable
4  g_2 = -a(2)/a(3);
5  g_1 = -a(1)/a(3);
6  K = 8; delta = .01;
7  y = zeros(K+1,1);
8  y(1) = 1; y(2) = 1+delta;
9  for k = 2:K
10     y(k+1) = g_2*y(k)+g_1*y(k-1);
11 end
12 plot((0:K),y,'o-');
13 title('multistep');
```

Fig. 17 Multistep methods (left) can lead to oscillating, rapidly growing approximations (right)

Sources of Error

In the numerical solution of a mathematical problem, numerous, mostly unavoidable errors arise.

- *Model error:* This refers to the error of the often simplified representation of a real problem by mathematical equations as well as measurement errors in determining specific problem properties.
- *Method error:* The algorithm used to solve a problem leads to errors caused by approximations of continuous quantities such as derivatives or termination criteria in iterative methods.
- *Rounding error:* All arithmetic operations of the computer must be considered as error-prone.

It turns out that *relative errors* are better suited for evaluating a method than *absolute errors*. The *conditioning* of a problem is understood to be the (independent of the numerical method) susceptibility of the problem to disturbances, the *stability* of a method is the error amplification caused by the calculation steps, and *convergence* is the reduction of the method error when approximations are improved and termination criteria are reduced.

Approach of Numerics

The development of numerical methods for the approximate solution of a mathematical problem should consider the following aspects:

- Observe and understand unexpected phenomena
- Develop methods that avoid problems
- Find suitable fixed-point equations
- Identify dominant sources of error
- Use meaningful convergence concepts
- Utilise special properties of problem classes
- Construct problem-adapted bases
- Critically discuss conditions for convergence

Problems

The following tasks can be worked on experimentally or theoretically.

(a) Determine for $\ell = 1, 2, \ldots, 10$ the smallest machine number x such that the comparison $10^\ell + x > 10^\ell$ is evaluated by the computer as correct. Interpret your results.

(b) Check the condition $\Phi'(x^*) = 0$ for quadratic convergence of the Heron method and construct suitable initial values for the approximation of $\sqrt{a}$, $a > 0$.

(c) Approximate the derivative of the exponential function at the point $x = 1$ by secant slopes $(f(x+h) - f(x))/h$ and $(f(x+h) - f(x-h))/(2h)$ with different step sizes $h > 0$ and comment on your results.

(d) Provide a stable method for the approximate calculation of $\exp(x)$ for any $x \in \mathbb{R}$.

(e) How long does it take to calculate the solution of a linear system of equations of size $n = 10^\ell$, $\ell = 3, 4, 5, 6$, with the Gaussian elimination method, if the computer can perform 10^9 operations per second, and how much storage space is required?

(f) Test the Jacobi method $x_{k+1} = D^{-1}(b - Rx_k)$ and the Gauss-Seidel method $x_{k+1} = (D+U)^{-1}(b - Lx_k)$ for band matrices $A \in \mathbb{R}^{n \times n}$ with off-diagonal entries -1 and main diagonal entries $a_{ii} = 2$ or $a_{ii} = 4$ for $i = 1, 2, \ldots, n$.

(g) Test the polynomial interpolation with equidistant support points and Chebyshev nodes in the case $f(x) = \cos(x)$. Calculate extrema of some derivatives of the functions $f(x) = \cos(x)$ and $f(x) = (1 + 25x^2)^{-1}$ in the interval $[-1, 1]$.

(g) Determine a basis of the polynomial space of maximum degree 3, so that a polynomial q with the properties $q(0) = v_0$, $q(1) = v_1$ and $q'(0) = v_2$, $q'(1) = v_3$ can be represented with the coefficients v_0, v_1, v_2, v_3.

(h) Experimentally determine error amplifications of Gaussian elimination without pivot search for the system of equations $Ax = b$, where $a_{11} = \varepsilon$, $a_{12} = a_{21} = a_{22} = 1$ and $b_1 = 1 + \varepsilon$ as well as $b_2 = 2$ for some $\varepsilon > 0$.

(i) In the case of termination of the descent method with termination criterion $\|\nabla g(x_k)\| \leq \varepsilon$ with a given number $0 < \varepsilon \ll 1$, is there always an approximation of a global minimum?

(j) Discuss sources of error in the calculation of the trajectory of a body with the equation resulting from Newton's laws $x(t) = x_0 + tv_0 + (t^2/2)(0, -g)$ with $g = 9.81 \text{m/s}^2$ and given $x_0, v_0 \in \mathbb{R}^2$.

(k) What special property is observable in the implicit Euler method for the equation $y'(t) = 1$, $y(0) = y_0$, on a large time interval $[0, T]$?

(l) Investigate the stability of the three-term recursion $y_{k+2} = -2y_{k+1} + y_k$ and test this with the initial values $y_0 = 1$ and $y_1 = \sqrt{2} - 1$.

Part I
Numerical Linear Algebra

Chapter 1
Basic Concepts

1.1 Problem Statement

Numerical mathematics deals with the practical calculation of mathematical objects such as

$$\int_0^1 e^{-x^2}\,dx, \quad \min_{x\in[0,1]} F(x), \quad f(x)=0, \quad Ax=b, \quad Ax=\lambda x, \quad y'=f(t,y).$$

Abstractly, this can be formulated as the evaluation of a mapping.

Definition 1.1 A *mathematical operation* consists in the evaluation of a mapping $\phi: X \to Y$ at $x \in X$.

For example, $\phi(x) = f^{-1}(x)$, $\phi(x) = A^{-1}x$ or $\phi(x) = \sin(x)$. Many of the objects listed above are not defined by closed formulas and can possibly only be determined *approximately*. Moreover, only a finite number of so-called *machine numbers* are available on computers, so not every real number can be entered exactly and elementary arithmetic operations like 1/3 can only be determined approximately. This leads to *rounding errors*. Other sources of error are *model errors*, which occur in the simplified mathematical description of a real process, and *data errors*, which can be caused by measurements. Many of these inaccuracies are unavoidable and therefore it is usually neither necessary nor sensible to solve a mathematical problem exactly. By *approximate solving* the *computational effort* can often be significantly reduced. The calculation of the determinant of a matrix $A \in \mathbb{R}^{n\times n}$ using the Laplace expansion theorem, for example, leads to $n!$ arithmetic operations, which for large dimensions n is hardly feasible in a reasonable time. However, it is often possible to construct at least approximately a factorisation $A \approx LR$ with triangular matrices $L, R \in \mathbb{R}^{n\times n}$ with the help of which the determinant $\det A \approx \det L \det R$ can be determined with an effort comparable to

n. Solving a system of linear equations $Ax = b$ is closely related to this. In practice, the inverse matrix A^{-1} is usually not determined explicitly, but the system is directly or iteratively solved. The expression $x = A^{-1}b$ therefore stands in numerics mostly for the solution of the linear system $Ax = b$ and less often for the multiplication of b with A^{-1}. More generally, the following typical questions are discussed in numerics:

- Computability of problems (algorithmics)
- Influence of perturbations (conditioning and stability)
- Error between calculated and exact solution (convergence)
- Computational effort of methods (complexity)

An important goal is to achieve a good compromise between accuracy and effort of a method. This is investigated for the following problems:

- Systems of linear equations
- Eigenvalue problems
- Interpolation of functions
- Integration of functions
- Root finding and optimisation
- Initial value problems

1.2 Condition and Stability

We consider an example that illustrates the effects of perturbations on the solution of a problem.

Example 1.1 For each $\varepsilon \in \mathbb{R} \setminus \{0\}$, the unique solution of the linear system

$$\begin{bmatrix} 1 & 1 \\ 1 & 1+\varepsilon \end{bmatrix} x = \begin{bmatrix} 2 \\ 2 \end{bmatrix}$$

is given by $x = [2, 0]^T$. We assume that ε is very small and perturb the right side in the second component, that is, we consider

$$\begin{bmatrix} 1 & 1 \\ 1 & 1+\varepsilon \end{bmatrix} \tilde{x} = \begin{bmatrix} 2 \\ 2+\varepsilon \end{bmatrix}.$$

The unique solution is given by $\tilde{x} = [1, 1]^T$. Although the perturbation in the right hand side is arbitrarily small, the solutions x and $\tilde{x}$ differ greatly.

The effects of perturbations on the solution of a problem lead to the concept of conditioning.

1.2 Condition and Stability

Definition 1.2 A mathematical operation $\phi(x)$ is *ill conditioned (at the point x)*, if small relative perturbations ε_x of the data cause large relative errors ε_ϕ in the solution that is, if a perturbation $\tilde{x}$ exists with

$$\varepsilon_\phi = \frac{|\phi(\tilde{x}) - \phi(x)|}{|\phi(x)|} \gg \frac{|\tilde{x} - x|}{|x|} = \varepsilon_x,$$

where $x \neq 0$ and $\phi(x) \neq 0$ hold. Otherwise, the operation is called *well conditioned* and there exists a moderate constant $c_{\text{cond}} \geq 0$ with $\varepsilon_\phi \leq c_{\text{cond}} \varepsilon_x$.

The relation $a \gg b$ means that a is significantly larger than b, for example $a \geq 100b$. What is considered significantly larger or as a moderate constant is generally problem-dependent. To show that the multiplication of two numbers is well conditioned, we consider the componentwise relative error of the arguments.

Proposition 1.1 *The operation $\phi(x, y) = xy$ is well conditioned in the sense that for $x, y \in \mathbb{R}$ with $x, y \neq 0$ (and consequently $\phi(x, y) \neq 0$) and perturbations $\tilde{x}, \tilde{y} \in \mathbb{R}$ the relative errors*

$$\varepsilon_\phi = \frac{|\phi(\tilde{x}, \tilde{y}) - \phi(x, y)|}{|\phi(x, y)|}, \quad \varepsilon_x = \frac{|\tilde{x} - x|}{|x|}, \quad \varepsilon_y = \frac{|\tilde{y} - y|}{|y|}$$

fulfill the estimate

$$\varepsilon_\phi \leq \varepsilon_x + \varepsilon_y + \varepsilon_x \varepsilon_y.$$

If ε_x and ε_y are small, then the relative error ε_ϕ is also small.

Proof We have

$$\varepsilon_\phi = \frac{|\tilde{x}\tilde{y} - xy|}{|xy|} = \frac{|(\tilde{x} - x)\tilde{y} + x(\tilde{y} - y)|}{|xy|} \leq \frac{|\tilde{x} - x|}{|x|} \frac{|\tilde{y} - y + y|}{|y|} + \frac{|\tilde{y} - y|}{|y|}$$

and the triangle inequality $|\tilde{y} - y + y| \leq |\tilde{y} - y| + |y|$ implies the claim. □

Remark 1.1 Other well conditioned operations are the addition of two positive or two negative numbers and the inversion of non-zero numbers. Ill conditioned, on the other hand, is the subtraction of nearly equal numbers, as will be shown below.

Obviously, an operation needs to be well conditioned to be able to meaningfully solve a given problem numerically, as rounding errors otherwise could cause large errors.

Definition 1.3 A *procedure* or *algorithm* for the (approximate) solution of an operation ϕ is a mapping $\tilde{\phi} : X \to Y$, which is defined by the execution of elementary, possibly rounding error-prone operations, where in the simplest case

$$\tilde{\phi} = f_J \circ f_{J-1} \circ \cdots \circ f_1.$$

Example 1.2

(i) The operation $\phi(x) = x^4$ can be realised by $\widetilde{\phi} = f \circ f$, with the multiplication provided by the computer $f(x) = x \boxdot x$.
(ii) The root $\phi(x) = \sqrt{x}$ of a number $x > 0$ is given according to Heron as the limit of the sequence $z_{n+1} = (z_n + x/z_n)/2$ with $z_0 > 0$. Thus, $\widetilde{\phi}$ can be defined as the J-fold application of this iteration rule with initialisation $z_0 = 1$.

For an operation, different methods are usually conceivable, but even if the operation is well conditioned, not all methods lead to good results, as rounding errors can have different effects during the execution of a method.

Example 1.3 The operation defined by the function

$$\phi(x) = \frac{1}{x} - \frac{1}{x+1} = \frac{1}{x(x+1)}$$

is well conditioned for large values $|x|$, because for a perturbation $\widetilde{x} = (1 + \varepsilon_x)x$ with a small number ε_x we get

$$\phi(x) - \phi(\widetilde{x}) = \frac{(1+\varepsilon_x)x\big((1+\varepsilon_x)x + 1\big) - x(x+1)}{(1+\varepsilon_x)x\big((1+\varepsilon_x)x + 1\big)x(x+1)} \approx \frac{2x^2}{x^4}\varepsilon_x.$$

This implies that the relative error satisfies $\varepsilon_\phi \leq 4\varepsilon_x$, provided $|x| \geq 1$ holds. The numerical realisation can be done via the methods

$$\widetilde{\phi}_1(x) = \left(\frac{1}{x}\right) - \left(\frac{1}{x+1}\right), \quad \widetilde{\phi}_2(x) = \frac{1}{\big(x(x+1)\big)}$$

where the brackets determine the order of execution of operations. Numerical experiments show that $\widetilde{\phi}_1$ and $\widetilde{\phi}_2$ differ greatly for large numbers x.

Definition 1.4 An algorithm $\widetilde{\phi}$ is called *unstable*, if there is a perturbation $\widetilde{x}$ of x, such that the relative error $\varepsilon_{\widetilde{\phi}}$ caused by rounding errors and perturbations is significantly larger than the error ε_ϕ caused only by the perturbation, i.e. if $\phi(x) \neq 0$ and

$$\varepsilon_{\widetilde{\phi}} = \frac{|\widetilde{\phi}(\widetilde{x}) - \phi(x)|}{|\phi(x)|} \gg \frac{|\phi(\widetilde{x}) - \phi(x)|}{|\phi(x)|} = \varepsilon_\phi.$$

An algorithm is called *stable*, if it is not unstable, and in this case there exists a moderate constant $c_{\text{stab}} \geq 0$ with $\varepsilon_{\widetilde{\phi}} \leq c_{\text{stab}} \varepsilon_\phi$.

Remark 1.2 A necessity for the stability of an algorithm is that each individual computational step is a well conditioned operation.

The above algorithm $\widetilde{\phi}_1$ is unstable due to so-called *cancellation effects*, which occur when subtracting nearly equal-sized numbers.

1.2 Condition and Stability

Example 1.4 For $x = 0.677354$ and $y = 0.677335$ the value $\phi(x, y) = x - y = 0.000019 = 0.19 \cdot 10^{-4}$. For the perturbation $\tilde{x} = (1 + \varepsilon_x)x$ with $\varepsilon_x = 1.0 \cdot 10^{-4}$ it follows

$$\varepsilon_\phi = \frac{|\phi(\tilde{x}, y) - \phi(x, y)|}{|\phi(x, y)|} = \frac{x}{x-y}\varepsilon_x = \frac{0.677354 \cdot 10^{-4}}{0.19 \cdot 10^{-4}} \approx 3.565021$$

The perturbation of 0.01% thus causes a relative error of over 350%, corresponding to an amplification factor $\kappa = |x|/|x - y| \approx 35000$.

The subtraction of nearly equal-sized numbers is therefore an ill conditioned operation. Cancellation phenomena occur independently of the size of the substracted numbers if these are nearly equal and are often a result of intermediate values that are large in magnitude.

Proposition 1.2

(i) Let $\phi(x, y) = x - y \neq 0$ and let $\tilde{x}, \tilde{y}$ be perturbations of x, y. Then, the relative errors $\varepsilon_x = |x - \tilde{x}|/|x|$ and $\varepsilon_y = |y - \tilde{y}|/|y|$ are amplified by the inverse of the exact difference $\delta = x - y$ and the sum of the absolute values of x and y, i.e.,

$$\varepsilon_\phi = \frac{|(x - \tilde{x}) + (y - \tilde{y})|}{|\delta|} \leq |\delta|^{-1}(|x| + |y|)\max\{\varepsilon_x, \varepsilon_y\}.$$

(ii) If $s = y_1 + y_2 + \cdots + y_n$ and $\tilde{s} = \tilde{y}_1 + \tilde{y}_2 + \cdots + \tilde{y}_n$, with relative errors $\varepsilon_i = |\tilde{y}_i - y_i|/|y_i|$, then the relative error $\varepsilon_s = |\tilde{s} - s|/|s|$ for the perturbed sum satisfies

$$\varepsilon_s \leq \left(\frac{1}{|s|}\sum_{i=1}^{n}|y_i|\right)\max_{j=1,\ldots,n}\varepsilon_j,$$

i.e., a strong error amplification occurs if $|y_1| + |y_2| + \cdots + |y_n| \gg |s|$.

Proof

(i) The first estimate follows from an application of the triangle inequality.
(ii) With factors $\sigma_i \in \{\pm 1\}$ so that $\sigma_i \varepsilon_i = (\tilde{y}_i - y_i)/y_i$ the perturbed sum is given by

$$\tilde{s} = \sum_{i=1}^{n}(1 + \sigma_i \varepsilon_i)y_i = s + \sum_{i=1}^{n}\sigma_i \varepsilon_i y_i.$$

The triangle inequality implies the error bound which is an equality if, e.g., $\sigma_i \varepsilon_i y_i \geq 0$ for $i = 1, 2, \ldots, n$.

□

Remark 1.3 The error caused by rounding and approximate solving in the numerical solution of an operation can be estimated using the conditioning of the operation and the stability of the method, because

$$\frac{|\phi(x) - \widetilde{\phi}(\widetilde{x})|}{|\phi(x)|} \leq c_{\text{stab}} \frac{|\phi(x) - \phi(\widetilde{x})|}{|\phi(x)|} \leq c_{\text{stab}} c_{\text{cond}} \frac{|x - \widetilde{x}|}{|x|}.$$

1.3 Computational Complexity

In addition to the stability of a numerical algorithm, the computational complexity is an important quantity.

Definition 1.5 For an operation $\phi : \mathbb{R}^n \to \mathbb{R}^m$ and a corresponding algorithm $\widetilde{\phi} : \mathbb{R}^n \to \mathbb{R}^m$, the *(computational) complexity* is the number of required elementary operations in evaluating $\widetilde{\phi}$.

An exact determination of the complexity is usually not necessary and instead the dependence on the problem size n is examined. The so-called *Landau notation* is helpful in this regard.

Definition 1.6 The sequence $(a_n)_{n \in \mathbb{N}}$ is *(asymptotically) of the order* of the sequence $(b_n)_{n \in \mathbb{N}}$, if numbers $c > 0$ and $N \in \mathbb{N}$ exist, so that $|a_n| \leq c|b_n|$ for all $n \geq N$. In this case, we use the *Landau notation* $a_n = \mathcal{O}(b_n)$.

For the complexity a_n of an algorithm, it is relevant whether this is of a polynomial order n^p.

Example 1.5

(i) The multiplication of a vector $x \in \mathbb{R}^n$ with a fixed number $a \in \mathbb{R}$ leads to an complexity of order $\mathcal{O}(n)$.
(ii) The Gaussian algorithm for solving a linear system has a complexity of $\mathcal{O}(n^3)$, while Cramer's rule with a calculation of the determinant according to Laplace's expansion theorem leads to a complexity of order $\mathcal{O}(n!)$.

1.4 Learning Objectives, Quiz and Application

You should be able to explain the concept of the conditioning of a mathematical operation and illustrate it with examples. Furthermore, you should be able to define the stability of an algorithm and describe possible problems such as cancellation effects. You should be able to explain the Landau notation and determine the complexity of basic matrix operations.

1.4 Learning Objectives, Quiz and Application

Quiz 1.1 Decide for each of the following statements whether it is true or false. You should be able to justify your answer.

We have that $n^p = \mathcal{O}(\ln(1+n))$ for every $0 < p \leq 1$.	
A stable algorithm is well conditioned.	
In practice, cancellation effects are rather unlikely.	
The sequential execution of two well conditioned operations is well conditioned.	
If a system of linear equations is well conditioned for one right-hand side, it is well conditioned for any right-hand side.	

Application 1.1 A union of n countries decides to introduce a common currency. The conversion rates imply fixed exchange rates between the national currencies, which are denoted by m_{ij}. We have that $m_{ji} = m_{ij}^{-1}$. For the practical implementation, approximations $\widetilde{m}_{ij}$ should be suitably chosen.

(i) What is a sufficient tolerance for the relative errors $\varepsilon_{ij} = (\widetilde{m}_{ij} - m_{ij})/m_{ij}$, so that a maximum relative deviation of 0.01% results from exchanging five times at random?

(ii) Alternatively, the conversion rates can be rounded so that, for example, six significant decimal places are retained, which means approximately $\widetilde{m}_{ij} = 0.00123456$ or $\widetilde{m}_{ij} = 12.3456$ if $m_{ij} = 0.00123456789$ or $m_{ij} = 12.3456789$ respectively. Is this approach more sensible?

Chapter 2
Operator Norm and Condition Number

2.1 Vector Norms

In order to be able to specify the concepts of conditioning and stability, distances between points in $\mathbb{R}^n$ or lengths of vectors must be measurable.

Definition 2.1 A *norm* on $\mathbb{R}^n$ is a mapping $\|\cdot\| : \mathbb{R}^n \to \mathbb{R}_{\geq 0}$ with the following properties:

(i) $\|x\| = 0 \implies x = 0$ for all $x \in \mathbb{R}^n$ (definiteness);
(ii) $\|x + y\| \leq \|x\| + \|y\|$ for all $x, y \in \mathbb{R}^n$ (triangle inequality);
(iii) $\|\lambda x\| = |\lambda|\|x\|$ for all $\lambda \in \mathbb{R}$ and $x \in \mathbb{R}^n$ (homogeneity).

Example 2.1 The ℓ^p-norms are for $1 \leq p \leq \infty$ and $x = [x_1, \ldots, x_n]^\mathsf{T} \in \mathbb{R}^n$ defined by

$$\|x\|_p = \begin{cases} \left(\sum_{j=1}^n |x_j|^p\right)^{1/p}, & p < \infty, \\ \max_{j=1,\ldots,n} |x_j|, & p = \infty. \end{cases}$$

The norm $\|\cdot\|_2$ is called *Euclidean norm* and satisfies $\|x\|_2^2 = x \cdot x = x^\mathsf{T} x$.

Remarks 2.1

(i) The ℓ^p-norms are equivalent in the sense that for all $1 \leq p, q \leq \infty$ a constant $c_{pq} \geq 1$ exists, so that for all $x \in \mathbb{R}^n$ we have

$$c_{pq}^{-1} \|x\|_p \leq \|x\|_q \leq c_{pq} \|x\|_p.$$

The constant c_{pq} depends on p, q and n.

Fig. 2.1 Level sets $N_p(1)$ of different ℓ^p-norms in $\mathbb{R}^2$

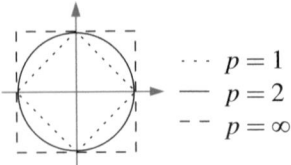

(ii) The ℓ^p-norms differ by their level sets

$$N_p(1) = \{x \in \mathbb{R}^n : \|x\|_p = 1\},$$

see Fig. 2.1.

2.2 Matrix Norms

In the following, we always identify matrices $A \in \mathbb{R}^{m \times n}$ with linear mappings $A : \mathbb{R}^n \to \mathbb{R}^m$, for which the respective canonical bases are chosen. A linear mapping is also referred to as a linear operator.

Definition 2.2 For norms $\|\cdot\|_{\mathbb{R}^m}$ and $\|\cdot\|_{\mathbb{R}^n}$ on $\mathbb{R}^m$ and $\mathbb{R}^n$ respectively, the *(induced) operator norm* for all $A \in \mathbb{R}^{m \times n}$ is defined by

$$\|A\|_{op} = \sup_{x \in \mathbb{R}^n, \|x\|_{\mathbb{R}^n} = 1} \|Ax\|_{\mathbb{R}^m}.$$

The operator norm measures how strongly level sets are deformed.

Example 2.2 A symmetric matrix $A \in \mathbb{R}^{2 \times 2}$ maps the circular level set $N_2(1)$ to an ellipse contained in the circle with radius $\|A\|_2$.

The operator norm defines a norm with the following properties.

Lemma 2.1 *For fixed norms $\|\cdot\|$ on $\mathbb{R}^n$ and $\mathbb{R}^m$, let $\|\cdot\|_{op}$ be the induced operator norm on $\mathbb{R}^{m \times n}$. Then:*

(i) $\|\cdot\|_{op}$ *defines a norm on* $\mathbb{R}^{m \times n}$;
(ii) $\|A\|_{op} = \sup_{x \in \mathbb{R}^n, \|x\| = 1} \|Ax\| = \inf\{c \geq 0 : \forall x \in \mathbb{R}^n \, \|Ax\| \leq c\|x\|\}$;
(iii) *for $A \neq 0$ and $x \in \mathbb{R}^n$ with $\|x\| \leq 1$ and $\|Ax\| = \|A\|_{op}$ it follows that $\|x\| = 1$;*
(iv) *the infimum and the supremum in (ii) are attained.*

Proof Exercise. □

Remark 2.2 From (ii) it follows that $\|Ax\| \leq \|A\|_{op}\|x\|$ for all $x \in \mathbb{R}^n$.

2.2 Matrix Norms

For some ℓ^p-norms, the induced operator norms can be explicitly given. The entries of a matrix $A \in \mathbb{R}^{m \times n}$ are denoted by a_{ij}, $1 \le i \le m$, $1 \le j \le n$.

Examples 2.3

(i) The ℓ^1-norm on $\mathbb{R}^m$ and $\mathbb{R}^n$ induces the *maximum absolute column sum*

$$\|A\|_1 = \max_{j=1,\ldots,n} \sum_{i=1}^{m} |a_{ij}|.$$

(ii) The ℓ^∞-norm on $\mathbb{R}^m$ and $\mathbb{R}^n$ induces the *maximum absolute row sum*

$$\|A\|_\infty = \max_{i=1,\ldots,m} \sum_{j=1}^{n} |a_{ij}|.$$

(iii) The ℓ^2-norm on $\mathbb{R}^m$ and $\mathbb{R}^n$ induces the *spectral norm*

$$\|A\|_2 = \sqrt{\varrho(A^\mathsf{T} A)} = \left(\max\{|\lambda| : \lambda \text{ is an eigenvalue of } A^\mathsf{T} A\}\right)^{1/2}.$$

The number $\varrho(A^\mathsf{T} A)$ is called *spectral radius* of $A^\mathsf{T} A$.

Some further properties of the operator norm are the following.

Lemma 2.2 *Let norms on $\mathbb{R}^\ell$, $\mathbb{R}^m$ and $\mathbb{R}^n$ be fixed and the induced operator norms be denoted by $\|\cdot\|$.*

(i) *For $A \in \mathbb{R}^{\ell \times m}$ and $B \in \mathbb{R}^{m \times n}$ we have $\|AB\| \le \|A\| \|B\|$.*
(ii) *The identity matrix $I_n \in \mathbb{R}^{n \times n}$ satisfies $\|I_n\| = 1$.*
(iii) *Every induced operator norm on $\mathbb{R}^{n \times n}$ satisfies $\|A\|_{op} \ge |\lambda|$ for all matrices $A \in \mathbb{R}^{n \times n}$ and every eigenvalue λ of A.*

Proof According to the previous lemma, $\|ABx\| \le \|A\| \|Bx\| \le \|A\| \|B\| \|x\|$ and this implies $\|AB\| \le \|A\| \|B\|$. The other statements follow directly from the definition of the operator norm. □

The Euclidean norm can be defined in a straightforward way on $\mathbb{R}^{m \times n}$, but it is not an induced operator norm.

Example 2.4 The *Frobenius norm* of a matrix $A \in \mathbb{R}^{m \times n}$ is defined by $\|A\|_F = \left(\sum_{i=1}^{m} \sum_{j=1}^{n} a_{ij}^2\right)^{1/2}$. It is not an induced operator norm for $n > 1$, since $\|I_n\|_F = \sqrt{n}$ holds. Also the scaled Frobenius norm $n^{-1/2} \|A\|_F$ is not an induced operator norm, because this violates the property $\|A\|_{op} \ge |\lambda|$ for every eigenvalue λ of A.

2.3 Condition Number

With the help of the concept of the operator norm, the conditioning of a system of linear equations can be specified.

Proposition 2.1 *Let $\|\cdot\|$ be an operator norm on $\mathbb{R}^{n\times n}$. Let $A \in \mathbb{R}^{n\times n}$ be regular and let $x, \tilde{x}, b, \tilde{b} \in \mathbb{R}^n \setminus \{0\}$, such that*

$$Ax = b, \quad A\tilde{x} = \tilde{b}.$$

Then it follows that

$$\frac{\|x - \tilde{x}\|}{\|x\|} \leq \|A\| \|A^{-1}\| \frac{\|b - \tilde{b}\|}{\|b\|}$$

Proof We have that $\|x - \tilde{x}\| = \|A^{-1}(b - \tilde{b})\| \leq \|A^{-1}\| \|b - \tilde{b}\|$ and $\|b\| = \|Ax\| \leq \|A\| \|x\|$ or $\|x\| \geq \|A\|^{-1} \|b\|$. From this it follows

$$\frac{\|x - \tilde{x}\|}{\|x\|} \leq \frac{\|A^{-1}\| \|b - \tilde{b}\|}{\|x\|} \leq \frac{\|A^{-1}\| \|b - \tilde{b}\|}{\|A\|^{-1} \|b\|},$$

thus the claimed estimate. $\square$

The product $\|A\| \|A^{-1}\|$ controls the amplification of the relative error when solving a system of linear equations.

Definition 2.3 The *condition number* of a regular matrix $A \in \mathbb{R}^{n\times n}$ with respect to the operator norm induced by the norm $\|\cdot\|$ on $\mathbb{R}^n$ is defined by

$$\operatorname{cond}_{\|\cdot\|}(A) = \|A\| \|A^{-1}\|.$$

In the case of an ℓ^p-norm, we write cond_p instead of $\operatorname{cond}_{\|\cdot\|_p}$.

Remarks 2.3

(i) The condition number of a matrix is always bounded below by 1, since for every operator norm $1 = \|AA^{-1}\| \leq \|A\| \|A^{-1}\| = \operatorname{cond}_{\|\cdot\|}(A)$.
(ii) If A is symmetric with eigenvalues $\lambda_1, \ldots, \lambda_n$, then

$$\operatorname{cond}_2(A) = \frac{\max_{j=1,\ldots,n} |\lambda_j|}{\min_{k=1,\ldots,n} |\lambda_k|}.$$

Let us consider the condition number of the matrix from the earlier Example 1.1, in which perturbations of the right side caused large errors.

Example 2.5 The matrix $A = \begin{bmatrix} 1 & 1 \\ 1 & 1+\varepsilon \end{bmatrix}$ has the eigenvalues $\lambda_{1,2} = 1 + \varepsilon/2 \pm (1 + \varepsilon^2/4)^{1/2}$. With the Taylor approximation $(1+x)^{1/2} \approx 1 + x/2$ it follows that for small numbers ε we have $\lambda_1 \approx 2 + \varepsilon/2$ and $\lambda_2 \approx \varepsilon/2$. Thus, $\text{cond}_2(A) \approx 4\varepsilon^{-1}$, which explains the sensitive behaviour of corresponding systems of equations to perturbations.

Geometrically interpreted, the condition number measures the distortion defined by the linear mapping A, but is independent of uniform scalings.

Example 2.6 For a symmetric matrix $A \in \mathbb{R}^{2\times 2}$, $\text{cond}_2(A)$ describes the ratio of the radii of the ellipse $A(N_2(1))$.

2.4 Learning Objectives, Quiz and Application

You should be familiar with various characterisations of the operator norm as well as some concrete examples. You should be able to define the condition number and explain its significance for the approximate solution of systems of linear equations.

Quiz 2.1 Decide for each of the following statements whether it is true or false. You should be able to justify your answer.

If a system of linear equations is well conditioned with respect to an operator norm, it is also well conditioned with respect to any other operator norm.	
For $A, B \in \mathbb{R}^{n\times n}$ and $\lambda, \mu \in \mathbb{R}$ and an arbitrary operator norm $\|\cdot\|$ holds $\|\lambda A + \mu B\| \leq \lambda \|A\| + \mu \|B\|$.	
For $A = \begin{bmatrix} -2 & 4 \\ 0 & 5 \end{bmatrix}$, $\|A\|_\infty = 6$ and $\|A\|_1 = 9$ hold.	
For $A \in \mathbb{R}^{m\times n}$ and $B \in \mathbb{R}^{n\times p}$, $\ker AB = \ker A$ holds.	
If λ is an eigenvalue of A, then $\|A\| \leq |\lambda|$ holds for every operator norm.	

Application 2.1 The routes of two airplanes flying in a plane are given by $t \mapsto x^i + tv^i$ with $x^i, v^i \in \mathbb{R}^2$ for $i = 1, 2$, where $\|v^i\|_2 = 350\,\text{km/h}$ applies. Calculate the point where the routes of the airplanes intersect and the respective times when the airplanes arrive at this point. How large can measurement errors in determining the initial positions x^i, $i = 1, 2$, be at most, so that the error in the calculation of the intersection point is less than 5 km?

Chapter 3
Matrix Factorisations

3.1 Triangular Matrices

Systems of linear equations can be solved in a canonical way when they are defined by a triangular matrix. This motivates the factorisation of matrices using triangular matrices. In this chapter, we follow the presentation in [10].

Definition 3.1 A matrix $L \in \mathbb{R}^{n \times n}$ is called *lower triangular matrix*, if $\ell_{ij} = 0$ for $i < j$. A matrix $U \in \mathbb{R}^{n \times n}$ is called *upper triangular matrix*, if U^T is a lower triangular matrix. A triangular matrix $D \in \mathbb{R}^{n \times n}$ is called *normalised*, if $d_{ii} = 1$ for $i = 1, 2, \dots, n$.

Linear systems with a regular triangular matrix can be solved using backward or forward substitution. The diagonal elements of a regular triangular matrix U are non-zero because $0 \neq \det U = u_{11} u_{22} \dots u_{nn}$.

Algorithm 3.1 (Backward Substitution) *Let* $U \in \mathbb{R}^{n \times n}$ *be a regular upper triangular matrix and* $b \in \mathbb{R}^n$. *Compute* $x \in \mathbb{R}^n$ *by:*

$$\text{for } i = n : -1 : 1; \quad x_i = \Big(b_i - \sum_{j=i+1}^{n} u_{ij} x_j\Big)/u_{ii}; \quad \text{end}$$

Remark 3.1 In the i-th step, $n - i$ multiplications and subtractions as well as one division are performed, so that the total effort of the backward substitution is given by

$$\sum_{i=1}^{n}(1 + 2(n-i)) = n + 2\sum_{k=1}^{n-1} k = n + (n-1)n = n^2.$$

The sets of regular lower and upper triangular matrices are groups.

Lemma 3.1 *Let $U, V \in \mathbb{R}^{n \times n}$ be upper triangular matrices. Then UV is an upper triangular matrix and if U is regular, then U^{-1} is also an upper triangular matrix with diagonal entries u_{ii}^{-1}, $i = 1, 2, \ldots, n$.*

Proof Exercise. □

3.2 LU Decomposition

If a factorisation $A = LU$ of a regular matrix $A \in \mathbb{R}^{n \times n}$ into a lower (*lower*) and an upper (*upper*) triangular matrix $L \in \mathbb{R}^{n \times n}$ and $U \in \mathbb{R}^{n \times n}$ is given, then the linear system $Ax = b$ can be solved in two steps:

(i) Solve $Ly = b$. (ii) Solve $Ux = y$.

This implies that $Ax = (LU)x = L(Ux) = Ly = b$. Perturbations are amplified in the first step with $\mathrm{cond}(L)$ and in the second with $\mathrm{cond}(U)$, so in total with $\mathrm{cond}(L) \mathrm{cond}(U)$. The method is therefore only stable if $\mathrm{cond}(L) \mathrm{cond}(U) \approx \mathrm{cond}(A)$ holds. This is generally not the case.

Example 3.1 For $A = \begin{bmatrix} \varepsilon & 1 \\ 1 & 0 \end{bmatrix}$ with $0 < \varepsilon \ll 1$ we have $A^{-1} = \begin{bmatrix} 0 & 1 \\ 1 & -\varepsilon \end{bmatrix}$ and we have $\|A\|_\infty = \|A^{-1}\|_\infty = 1 + \varepsilon$ so $\mathrm{cond}_\infty(A) = (1 + \varepsilon)^2 \approx 1$. A factorisation is given by

$$L = \begin{bmatrix} 1 & 0 \\ \varepsilon^{-1} & 1 \end{bmatrix}, \quad U = \begin{bmatrix} \varepsilon & 1 \\ 0 & -\varepsilon^{-1} \end{bmatrix}.$$

We have $\|L\|_\infty = \|L^{-1}\|_\infty = 1 + \varepsilon^{-1}$ and $\|U\|_\infty = \varepsilon^{-1}$, $\|U^{-1}\|_\infty = 1 + \varepsilon^{-1}$, thus

$$\mathrm{cond}_\infty(L) = (1 + \varepsilon^{-1})^2 \approx \varepsilon^{-2}, \quad \mathrm{cond}_\infty(U) = (1 + \varepsilon^{-1})/\varepsilon \approx \varepsilon^{-2}.$$

Definition 3.2 A factorisation $A = LU$ with lower triangular matrix $L \in \mathbb{R}^{n \times n}$ and upper triangular matrix $U \in \mathbb{R}^{n \times n}$ is called *LU decomposition* of A. It is called *normalised*, if L is normalised, that is, only ones are on the diagonal of L.

Proposition 3.1 *For a regular matrix $A \in \mathbb{R}^{n \times n}$ the following statements are equivalent:*

(i) *There exists a uniquely determined normalised LU decomposition of A.*
(ii) *All upper left submatrices $A_k = (a_{ij})_{1 \le i, j \le k} \in \mathbb{R}^{k \times k}$, $k = 1, 2, \ldots, n$, of A are regular.*

3.2 LU Decomposition

Proof (i) $\implies$ (ii). If $A = LU$ and A is regular, then L and U are also regular, because $0 \neq \det(A) = \det(L)\det(U)$. Furthermore, all submatrices L_k and U_k are regular, since for example $\det(L) = \ell_{11}\ell_{22}\ldots\ell_{nn}$ holds. Since for each submatrix A_k the decomposition $A_k = L_k U_k$ holds, the regularity of A_k follows.

(ii) $\implies$ (i). For $n = 1$ the implication is clear and assume it is proven for $n - 1$. Then there exists a uniquely determined normalised LU decomposition $A_{n-1} = L_{n-1}U_{n-1}$. Let the vectors $[b^\mathsf{T}, a_{nn}]$ and $[c^\mathsf{T}, a_{nn}]$ be the last column and row of A respectively. To prove the statement for n it suffices to show that uniquely determined vectors $\ell, u \in \mathbb{R}^{n-1}$ and $r \in \mathbb{R}$ exist with

$$\begin{bmatrix} A_{n-1} & b \\ c^\mathsf{T} & a_{nn} \end{bmatrix} = \begin{bmatrix} L_{n-1} & 0 \\ \ell^\mathsf{T} & 1 \end{bmatrix} \begin{bmatrix} U_{n-1} & u \\ 0 & r \end{bmatrix} = \begin{bmatrix} L_{n-1}U_{n-1} & L_{n-1}u \\ (U_{n-1}^\mathsf{T}\ell)^\mathsf{T} & \ell^\mathsf{T} u + r \end{bmatrix}.$$

Because $A_{n-1} = L_{n-1}U_{n-1}$ this is equivalent to

$$b = L_{n-1}u, \quad c = U_{n-1}^\mathsf{T}\ell, \quad a_{nn} = \ell^\mathsf{T} u + r.$$

Since L_{n-1} and U_{n-1} are regular, uniquely determined solutions u and ℓ exist, which then uniquely determine r. $\square$

Examples 3.2

(i) If A is positive definite, that is $Ax \cdot x > 0$ for all $x \in \mathbb{R}^n \setminus \{0\}$, or strictly diagonally dominant, that is $\sum_{j=1,\ldots,n, j\neq i} |a_{ij}| < |a_{ii}|$ for $i = 1, 2, \ldots, n$, then A has an LU decomposition.

(ii) The matrix $A = \begin{bmatrix} 0 & 1 \\ 1 & 0 \end{bmatrix}$ does not have an LU decomposition.

The LU decomposition of a matrix can be determined easily.

Lemma 3.2 *If $A = LU$ is a normalised LU decomposition of A, it follows for $1 \leq i, k \leq n$*

$$a_{ik} = u_{ik} + \sum_{j=1}^{i-1} \ell_{ij}u_{jk}, \quad a_{ki} = \ell_{ki}u_{ii} + \sum_{j=1}^{i-1} \ell_{kj}u_{ji}.$$

Proof Because $\ell_{ij} = 0$ for $j > i$ and $\ell_{jj} = 1$ we have

$$a_{ik} = \sum_{j=1}^{n} \ell_{ij}u_{jk} = \sum_{j=1}^{i} \ell_{ij}u_{jk} = u_{ik} + \sum_{j=1}^{i-1} \ell_{ij}u_{jk}$$

and because $u_{ji} = 0$ for $j > i$ we have

$$a_{ki} = \sum_{j=1}^{n} \ell_{kj} u_{ji} = \sum_{j=1}^{i} \ell_{kj} u_{ji} = \ell_{ki} u_{ii} + \sum_{j=1}^{i-1} \ell_{kj} u_{ji}.$$

This proves the assertion. □

The formulas of the lemma can be solved for u_{ik} for $i \leq k$ and since $u_{ii} \neq 0$ for ℓ_{ki} for $k > i$. In the following algorithm, the rows of U and columns of L are determined alternately.

Algorithm 3.2 (*LU Decomposition*) *The matrix $A \in \mathbb{R}^{n \times n}$ has a normalised LU decomposition. The non-trivial entries of L and U are given by:*

for $i = 1 : n$
$\quad$ for $k = i : n;\quad u_{ik} = a_{ik} - \sum_{j=1}^{i-1} \ell_{ij} u_{jk};\quad$ end
$\quad$ for $k = i+1 : n;\quad \ell_{ki} = \bigl(a_{ki} - \sum_{j=1}^{i-1} \ell_{kj} u_{ji}\bigr)/u_{ii};\quad$ end
end

Remarks 3.2

(i) The calculation of u_{ik} requires $i - 1$ multiplications and subtractions, for ℓ_{ki} an additional division is required, so that in the i-th step

$$(n - i + 1)2(i - 1) + (n - i)(2(i - 1) + 1) = (4n + 5)i - 4i^2 - (3n + 2)$$

operations are carried out. By summing over $i = 1, 2, \ldots, n$ the total computational effort $2n^3/3 + \mathcal{O}(n^2)$ is obtained.

(ii) The entries of A can be successively overwritten by the non-trivial entries of L and U, so no additional storage space is necessary.

3.3 Cholesky Decomposition

If $A \in \mathbb{R}^{n \times n}$ is symmetric, only $n(n + 1)/2$ many entries of A are relevant and it is canonical to look for a factorisation $A = LL^\mathsf{T}$ with a lower triangular matrix $L \in \mathbb{R}^{n \times n}$. What is necessary for this is that A is symmetric and positive semi-definite, because the factorisation implies that

$$A^\mathsf{T} = (LL^\mathsf{T})^\mathsf{T} = LL^\mathsf{T} = A,$$
$$x^\mathsf{T} A x = x^\mathsf{T}(LL^\mathsf{T})x = (L^\mathsf{T} x)^\mathsf{T}(L^\mathsf{T} x) = \|L^\mathsf{T} x\|_2^2 \geq 0.$$

3.3 Cholesky Decomposition

If A or L is regular, it follows that A must be positive definite. In this case, the conditions for the existence of the Cholesky decomposition are also sufficient and imply their uniqueness.

Definition 3.3 The matrix $A \in \mathbb{R}^{n \times n}$ is called *positive definite* if for all $x \in \mathbb{R}^n \setminus \{0\}$ we have that $x^T A x > 0$. If only $x^T A x \geq 0$ holds for all $x \in \mathbb{R}^n$, then A is called *positive semidefinite*.

Lemma 3.3 *Let A be symmetric and positive definite. Then $\det A > 0$ and all submatrices $A_k = (a_{ij})_{1 \leq i, j \leq k}$ are positive definite.*

Proof Exercise. □

Definition 3.4 A factorisation $A = LL^T$ with a lower triangular matrix L is called *Cholesky decomposition* of A.

Proposition 3.2 *Let $A \in \mathbb{R}^{n \times n}$ be symmetric and positive definite. Then there exists a uniquely determined lower triangular matrix $L \in \mathbb{R}^{n \times n}$ with $A = LL^T$ and $\ell_{ii} > 0$ for $i = 1, 2, \ldots, n$.*

Proof If $n = 1$, then $a_{11} > 0$ and the construction follows by choice of $\ell_{11} = \sqrt{a_{11}}$. The submatrix $A_{n-1} = (a_{ij})_{1 \leq i, j \leq n-1}$ is positive definite and symmetric. Let us suppose that we have the factorisation $A_{n-1} = L_{n-1} L_{n-1}^T$ with the desired properties. Let $[b^T, a_{nn}]$ be the last row of A. Then a vector $c \in \mathbb{R}^{n-1}$ and a number $\alpha > 0$ are to be constructed such that

$$\begin{bmatrix} A_{n-1} & b \\ b^T & a_{nn} \end{bmatrix} = \begin{bmatrix} L_{n-1} & 0 \\ c^T & \alpha \end{bmatrix} \begin{bmatrix} L_{n-1}^T & c \\ 0 & \alpha \end{bmatrix} = \begin{bmatrix} L_{n-1} L_{n-1}^T & L_{n-1} c \\ (L_{n-1} c)^T & \alpha^2 + c^T c \end{bmatrix}$$

holds. Because $A_{n-1} = L_{n-1} L_{n-1}^T$ this is equivalent to the equations $L_{n-1} c = b$ and $c^T c + \alpha^2 = a_{nn}$. Since L_{n-1} has positive diagonal entries, L_{n-1} is regular and c is uniquely determined. To be able to solve the second equation with a real number $\alpha > 0$, we must prove $\alpha^2 = a_{nn} - c^T c > 0$. We have

$$\det A = \det \begin{bmatrix} L_{n-1} & 0 \\ c^T & \alpha \end{bmatrix} \det \begin{bmatrix} L_{n-1}^T & c \\ 0 & \alpha \end{bmatrix} = \alpha^2 (\det L_{n-1})^2.$$

Since $\det A > 0$ and $\det L_{n-1} > 0$ it follows $\alpha^2 > 0$, that is, there exists a unique $\alpha > 0$, which completes the factorisation. □

The factorisations can again be determined successively.

Lemma 3.4 *If $A = LL^T$, then it follows*

$$a_{ik} = \begin{cases} \ell_{ik} \ell_{kk} + \sum_{j=1}^{k-1} \ell_{ij} \ell_{kj} & \text{for } i > k, \\ \ell_{kk}^2 + \sum_{j=1}^{k-1} \ell_{kj}^2 & \text{for } i = k. \end{cases}$$

Proof Since $\ell_{kj} = 0$ for $j > k$, we have

$$a_{ik} = \sum_{j=1}^{n} \ell_{ij}\ell_{kj} = \sum_{j=1}^{k} \ell_{ij}\ell_{kj}$$

and this implies the claim. □

These identities can be solved for ℓ_{kk} and ℓ_{ik}.

Algorithm 3.3 (LL^T Decomposition) Let $A \in \mathbb{R}^{n \times n}$ be symmetric and positive definite. The non-trivial entries of L are given by:

for $k = 1 : n$
$$\ell_{kk} = \left(a_{kk} - \sum_{j=1}^{k-1} \ell_{kj}^2\right)^{1/2}$$
for $i = k+1 : n$; $\quad \ell_{ik} = \left(a_{ik} - \sum_{j=1}^{k-1} \ell_{ij}\ell_{kj}\right)/\ell_{kk};\quad$ end

end

Remark 3.3 The algorithm calculates the Cholesky decomposition with $n^3/3 + \mathcal{O}(n^2)$ operations.

Example 3.2 The matrix $A = \begin{bmatrix} a & b \\ b & c \end{bmatrix}$ is positive definite if $a > 0$ and $ca - b^2 > 0$ hold. In this case, one obtains $A = LL^\mathsf{T}$ with

$$L = \begin{bmatrix} a^{1/2} & 0 \\ b/a^{1/2} & (c - b^2/a)^{1/2} \end{bmatrix}.$$

The solution of a linear system can be determined using the Cholesky decomposition as follows:

(i) Solve $Ly = b$. (ii) Solve $L^\mathsf{T} x = y$.

To show that this defines a stable algorithm, we use that the spectral norm of a matrix $M \in \mathbb{R}^{n \times n}$ is given by

$$\|M\|_2^2 = \rho(M^\mathsf{T} M) = \max\{|\lambda| : \lambda \text{ is an eigenvalue of } M^\mathsf{T} M\}.$$

If M is symmetric, then $\|M\|_2 = \rho(M)$.

Proposition 3.3 If $A \in \mathbb{R}^{n \times n}$ is symmetric and positive definite, then for the Cholesky decomposition $A = LL^\mathsf{T}$, we have that

$$\operatorname{cond}_2(L) = \operatorname{cond}_2(L^\mathsf{T}) = \bigl(\operatorname{cond}_2(A)\bigr)^{1/2}.$$

Proof The symmetric but generally different matrices $L^\mathsf{T} L$ and LL^T have the same eigenvalues, because since L is regular, we have for all $x \in \mathbb{R}^n$ and $\lambda \in \mathbb{R}$

$$L^\mathsf{T} L x = \lambda x \quad \Longleftrightarrow \quad LL^\mathsf{T}(Lx) = \lambda(Lx).$$

With $\rho(LL^\mathsf{T}) = \rho(L^\mathsf{T} L)$ it follows $\|L\|_2 = \|L^\mathsf{T}\|_2$ and similarly $\|L^{-1}\|_2 = \|L^{-\mathsf{T}}\|_2$. This implies $\mathrm{cond}_2(L) = \mathrm{cond}_2(L^\mathsf{T})$. With $LL^\mathsf{T} = A$ and since A is symmetric, we have

$$\|L\|_2^2 = \|L^\mathsf{T}\|_2^2 = \rho(LL^\mathsf{T}) = \rho(A) = \|A\|_2$$

and

$$\|L^{-1}\|_2^2 = \rho(L^{-\mathsf{T}} L^{-1}) = \rho((LL^\mathsf{T})^{-1}) = \rho(A^{-1}) = \|A^{-1}\|_2.$$

With these identities, it follows overall

$$\mathrm{cond}_2(L) = \|L\|_2 \|L^{-1}\|_2 = \|A\|_2^{1/2} \|A^{-1}\|_2^{1/2} = \bigl(\mathrm{cond}_2(A)\bigr)^{1/2}.$$

This proves the claim. □

3.4 Learning Objectives, Quiz and Application

You should be able to define the LU and Cholesky factorisations, name sufficient and necessary conditions for their existence and derive algorithms for practical computations. You should be able to explain the effort and stability properties of solving linear systems using these factorisations.

Quiz 3.1 Decide for each of the following statements whether it is true or false. You should be able to justify your answer.

If $x^\mathsf{T} A x < 0$ holds for all $x \in \mathbb{R}^n$, then A has an LU decomposition.	
If A has an LU decomposition and A is symmetric, then $U = L^\mathsf{T}$.	
If A is invertible with Cholesky decomposition $A = LL^\mathsf{T}$, then $L^{-\mathsf{T}} L$ defines a Cholesky decomposition of A^{-1}.	
If a Cholesky decomposition $A = LL^\mathsf{T}$ is given, then the linear system $Ax = b$ can be solved with the effort $\mathcal{O}(n^2)$.	
If A is symmetric and invertible, then A is positive definite.	

Application 3.1 For the evaluation of financial derivatives such as options, the simulation of multidimensional Brownian motions is required. For this, n-dimensional random variables are needed that follow a correlated normal distribution, that is

$X \sim N(\mu, \Sigma)$ with an expected value $\mu \in \mathbb{R}^n$ and a symmetric, positive definite covariance matrix $\Sigma \in \mathbb{R}^{n \times n}$. This means that

$$\Sigma_{ij} = E[(X_i - \mu_i)(X_j - \mu_j)]$$

and $\mu_i = E(X_i)$ for $i, j = 1, 2, \ldots, n$. If Y is a standard normally distributed vector random variable, that is $Y \sim N(0, I_n)$, and $\Sigma = LL^\mathsf{T}$ is the Cholesky decomposition of Σ, then by means of $X = \mu + LY$ a random variable with $X \sim N(\mu, \Sigma)$ is obtained. In MATLAB a realisation of X can be generated using pseudo-random variables by X=mu+L*randn(n,1). Use $n = 3$,

$$\Sigma = \begin{bmatrix} 1 & 1 & 0 \\ 1 & 5 & 1 \\ 0 & 1 & 5 \end{bmatrix}, \quad \mu = \begin{bmatrix} -5 \\ 0 \\ 5 \end{bmatrix},$$

generate 1000 realisations of the variable X and display the histograms of the components X_i using the command hist in the range $[-10, 10]$ for $i = 1, 2, 3$.

Chapter 4
Elimination Methods

4.1 Gaussian Elimination Method

Systems of linear equations appear in various areas of applications. They allow us to determine (approximately) internal quantities from certain external, measurable quantities, which are often not directly accessible.

Example 4.1 Can the total value of coins in a jar be determined by their weight and volume?

The Gaussian method successively transforms a linear system into an equivalent system with an upper triangular matrix. We follow the presentation in [10] in this chapter.

Algorithm 4.1 (Gaussian Elimination) *Let $A \in \mathbb{R}^{n \times n}$ and $b \in \mathbb{R}^n$.*

(1) *Set $A^{(1)} = A$ and $b^{(1)} = b$ and $k = 1$.*
(2) *The matrix $A^{(k)}$ satisfies $a_{ij}^{(k)} = 0$ for $1 \le j \le k-1$ and $i \ge j+1$ and with $\ell_{ik} = a_{ik}^{(k)}/a_{kk}^{(k)}$ for $i = k+1, \ldots, n$ the normalised lower triangular matrix $L^{(k)} \in \mathbb{R}^{n \times n}$ is defined as follows:*

$$A^{(k)} = \begin{bmatrix} a_{11}^{(1)} & \cdots & & \cdots & a_{1n}^{(1)} \\ & \ddots & & & \vdots \\ & & a_{kk}^{(k)} & \cdots & a_{kn}^{(k)} \\ & & \vdots & & \vdots \\ & & a_{nk}^{(k)} & \cdots & a_{nn}^{(k)} \end{bmatrix}, \quad L^{(k)} = \begin{bmatrix} 1 & & & & \\ & \ddots & & & \\ & & 1 & & \\ & & -\ell_{k+1,k} & & \\ & & \vdots & \ddots & \\ & & -\ell_{nk} & & 1 \end{bmatrix}.$$

Then for $A^{(k+1)} = L^{(k)} A^{(k)}$, it holds that $a_{ij}^{(k+1)} = 0$ for $1 \leq j \leq k$ and $i \geq j+1$, that is

$$A^{(k+1)} = \begin{bmatrix} a_{11}^{(1)} & \cdots & & & \cdots & a_{1n}^{(1)} \\ & \ddots & & & & \vdots \\ & & a_{kk}^{(k)} & \cdots & \cdots & a_{kn}^{(k)} \\ & & & a_{k+1,k+1}^{(k+1)} & \cdots & a_{k+1,n}^{(k+1)} \\ & & & \vdots & & \vdots \\ & & & a_{n,k+1}^{(k+1)} & \cdots & a_{nn}^{(k+1)} \end{bmatrix}.$$

Also set $b^{(k+1)} = L^{(k)} b^{(k)}$.
(3) Stop if $k+1 = n$; otherwise increase $k \to k+1$ and repeat step (2).

Proposition 4.1 *If $A \in \mathbb{R}^{n \times n}$ is regular, then the Gaussian method is feasible if and only if A has an LU decomposition. The method then provides the normalised LU decomposition with $U = A^{(n)}$ and $L = \left(L^{(n-1)} \ldots L^{(1)}\right)^{-1}$. The modified right-hand side $y = b^{(n)}$ is given by $y = L^{-1} b$ and the solution of the linear system $Ax = b$ is the solution of the system $Ux = y$.*

Proof

(i) Assume that the matrix A has an LU decomposition. The Gaussian method is implementable, provided $a_{kk}^{(k)} \neq 0$ holds at every step. We consider the left, upper $k \times k$ submatrix $A_k^{(k)}$ of $A^{(k)} = L^{(k-1)} \ldots L^{(1)} A$, that is

$$A_k^{(k)} = \begin{bmatrix} a_{11}^{(1)} & \cdots & & \cdots & a_{1n}^{(1)} \\ & a_{22}^{(2)} & & & a_{2n}^{(2)} \\ & & \ddots & & \vdots \\ & & & & a_{kk}^{(k)} \end{bmatrix}.$$

With the left, upper submatrices $L_k^{(j)}$ of $L^{(j)}$ and A_k of A then

$$A_k^{(k)} = L_k^{(k-1)} \ldots L_k^{(1)} A_k.$$

Since the normalised triangular matrices $L_k^{(j)}$ are regular, $A_k^{(k)}$ is regular exactly when A_k is regular. This is given by the result on the existence of the LU decomposition. Thus, it follows that $0 \neq \det A_k^{(k)} = a_{11}^{(1)} a_{22}^{(2)} \ldots a_{kk}^{(k)}$ so $a_{kk}^{(k)} \neq 0$ and the procedure is well-defined.

(ii) Conversely, if the Gaussian method is feasible, then $U = A^{(n)} = L^{(n-1)} \ldots L^{(1)} A$ is an upper triangular matrix and it suffices to show that $L = \left(L^{(n-1)} \ldots L^{(1)}\right)^{-1}$ is a normalised lower triangular matrix. With the k-th

4.1 Gaussian Elimination Method

canonical basis vector $e_k \in \mathbb{R}^n$ and

$$\ell_k = [0, \ldots, 0, \ell_{k+1,k}, \ldots, \ell_{nk}]^\mathsf{T}$$

$L^{(k)} = I_n - \ell_k e_k^\mathsf{T}$. With $e_k^\mathsf{T} \ell_k = 0$ it follows

$$L^{(k)}(I_n + \ell_k e_k^\mathsf{T}) = (I_n - \ell_k e_k^\mathsf{T})(I_n + \ell_k e_k^\mathsf{T}) = I_n - \ell_k e_k^\mathsf{T} + \ell_k e_k^\mathsf{T} - \ell_k e_k^\mathsf{T} \ell_k e_k^\mathsf{T} = I_n,$$

that is $(L^{(k)})^{-1} = I_n + \ell_k e_k^\mathsf{T}$. With complete induction it follows

$$L = \left(L^{(n-1)} \ldots L^{(1)}\right)^{-1} = \left(L^{(1)}\right)^{-1} \ldots \left(L^{(n-1)}\right)^{-1} = I_n + \sum_{j=1}^{n-1} \ell_j e_j^\mathsf{T}$$

or

$$L = \begin{bmatrix} 1 & & & \\ \ell_{21} & 1 & & \\ \vdots & \ddots & \ddots & \\ \ell_{n1} & \ldots & \ell_{n,n-1} & 1 \end{bmatrix}.$$

This shows that $A = LU$ is the normalised LU decomposition of A.

$\square$

Remark 4.1 The proof shows that no additional calculations are required to determine L.

For the implementation of the Gaussian method, the matrices $L^{(k)}$ do not need to be explicitly set up.

Algorithm 4.2 (Gaussian Method) *Let* $A \in \mathbb{R}^{n \times n}$ *be an LU-decomposable matrix and* $b \in \mathbb{R}^n$. *Calculate the LU decomposition and the vector* $y = L^{-1}b$ *by:*

for $k = 1 : n - 1$
    for $i = k+1 : n$; $\ell_{ik} = a_{ik}^{(k)}/a_{kk}^{(k)}$; $b_i^{(k+1)} = b_i^{(k)} - \ell_{ik} b_k^{(k)}$;
        for $j = k+1 : n$; $a_{ij}^{(k+1)} = a_{ij}^{(k)} - \ell_{ik} a_{kj}^{(k)}$;   end;
    end
end

Remark 4.2 The algorithm provides the non-trivial entries of the LU decomposition of the matrix A and the modified right-hand side y with $(2/3)n^3 + \mathcal{O}(n^2)$ computational steps. The entries of U are given by $u_{ij} = a_{ij}^{(i)}$. The matrix A can be overwritten with the calculated quantities.

4.2 Pivot Strategy

The Gaussian elimination defined above is not feasible for every matrix and can lead to instabilities.

Example 4.2 The system of linear equations

$$\begin{bmatrix} \varepsilon & 1 \\ 1 & 1 \end{bmatrix} \begin{bmatrix} x_1 \\ x_2 \end{bmatrix} = \begin{bmatrix} 1 \\ 2 \end{bmatrix}$$

is well conditioned for $0 \leq \varepsilon < 1/2$ and has the solution $x_1 = 1/(1-\varepsilon)$ and $x_2 = (1-2\varepsilon)/(1-\varepsilon)$. The first step of the back substitution in the Gaussian elimination initially provides an approximation for x_2, which for very small numbers ε considering rounding is given by $\tilde{x}_2 = 1$. If this result is used to calculate $\tilde{x}_1$ in the equation $\varepsilon \tilde{x}_1 + \tilde{x}_2 = 1$, then the result is $\tilde{x}_1 = 0$, which is not a good approximation of the correct value x_1. However, considering the equivalent system of equations resulting from a row swap

$$\begin{bmatrix} 1 & 1 \\ \varepsilon & 1 \end{bmatrix} \begin{bmatrix} x_1 \\ x_2 \end{bmatrix} = \begin{bmatrix} 2 \\ 1 \end{bmatrix},$$

no instabilities occur in the Gaussian elimination.

The avoidance of instabilities in Gaussian elimination is achieved through a *pivot search*. For this, the above procedure in the k-loop before the i-loop is extended as follows:

- determine $p \in \{k, \ldots, n\}$ with $|a_{pk}^{(k)}| = \max_{i=k,\ldots,n} |a_{ik}^{(k)}|$;
- swap the rows p and k in $[A^{(k)}|b^{(k)}]$ and obtain $[\tilde{A}^{(k)}|\tilde{b}^{(k)}]$;
- eliminate entries in $[\tilde{A}^{(k)}|\tilde{b}^{(k)}]$ and obtain $[A^{(k+1)}|b^{(k+1)}]$.

Practically, the swapping is not actually performed, but corresponding indices are renamed by defining a vector $\pi \in \mathbb{N}^n$ that describes the swaps:

- initialise π with $\pi = [1, \ldots, n]$;
- if the rows k and p are to be swapped, then swap $\pi(k)$ and $\pi(p)$.

The swapping of rows can also be represented with a *permutation matrix*. It holds that $\tilde{A}^{(k)} = P^{(k)} A^{(k)}$, where $P^{(k)}$ is obtained by swapping the rows p and k in the identity matrix I_n.

Remark 4.3 Instead of the column pivot search, a *total pivot search* can be performed, in which case columns of the remaining matrix are also swapped. However, this leads to a high effort.

4.2 Pivot Strategy

Proposition 4.2 *If $A \in \mathbb{R}^{n \times n}$ is regular and $b \in \mathbb{R}^n$, then the Gaussian method with pivot search is feasible. It provides the normalised LU decomposition $PA = LU$ with $|\ell_{ij}| \le 1$ for all $1 \le i, j \le n$ as well as the modified right-hand side $b^{(n)} = L^{-1} Pb$. Here, $P = P^{(n-1)} \ldots P^{(1)}$.*

Proof The method is not feasible exactly when in the k-th step with $1 \le k \le n-1$, it holds that $|a_{pk}^{(k)}| = \max_{i=k,\ldots,n} |a_{ik}^{(k)}| = 0$, that is, the k-th column of the matrix $A^{(k)}$ has only vanishing entries from the diagonal element onwards,

$$A^{(k)} = \begin{bmatrix} a_{11}^{(1)} & \cdots & & & \cdots & a_{1n}^{(1)} \\ & \ddots & & & & \vdots \\ & & a_{k-1,k-1}^{(k-1)} & a_{k-1,k}^{(k-1)} & \cdots & a_{k-1,n}^{(k-1)} \\ & & 0 & a_{k,k+1}^{(k)} & \cdots & a_{kn}^{(k)} \\ & & \vdots & \vdots & \cdots & \vdots \\ & & 0 & a_{n,k+1}^{(k)} & \cdots & a_{nn}^{(k)} \end{bmatrix},$$

thus the first k columns of $A^{(k)}$ are linearly dependent and consequently $A^{(k)}$ is not regular. But then A cannot be regular either, because $A^{(k)}$ arises from A through regular transformations. This is a contradiction and it follows $\max_{i=k,\ldots,n} |a_{ik}^{(k)}| > 0$. For the coefficients of L, $\ell_{ik} = a_{ik}^{(k)}/a_{pk}^{(k)}$ holds and after choosing $a_{pk}^{(k)}$ it follows that $|\ell_{ik}| \le 1$. To derive the decomposition $PA = LU$, we note with $(P^{(k)})^{-1} = P^{(k)}$, that

$$A^{(1)} = A,$$
$$A^{(2)} = L^{(1)} P^{(1)} A^{(1)} = L^{(1)} P^{(1)} A,$$
$$A^{(3)} = L^{(2)} P^{(2)} A^{(2)} = L^{(2)} P^{(2)} L^{(1)} P^{(1)} A = L^{(2)} \big[P^{(2)} L^{(1)} P^{(2)}\big] \big[P^{(2)} P^{(1)}\big] A,$$
$$A^{(4)} = L^{(3)} P^{(3)} A^{(3)}$$
$$= L^{(3)} \big[P^{(3)} L^{(2)} P^{(3)}\big] \big[P^{(3)} P^{(2)} L^{(1)} P^{(2)} P^{(3)}\big] \big[P^{(3)} P^{(2)} P^{(1)}\big] A$$

and corresponding identities for $A^{(5)}, \ldots, A^{(n)}$. With

$$\widehat{L}^{(k)} = P^{(n-1)} P^{(n-2)} \ldots P^{(k+1)} L^{(k)} P^{(k+1)} \ldots P^{(n-2)} P^{(n-1)}$$

it holds

$$A^{(n)} = \widehat{L}^{(n-1)} \ldots \widehat{L}^{(1)} PA.$$

The matrix $A^{(n)} = U$ is an upper triangular matrix and with $L^{(k)} = I_n - \ell_k e_k^T$ it follows that

$$\widehat{L}^{(k)} = \begin{bmatrix} 1 & & & & \\ & \ddots & & & \\ & & 1 & & \\ & & -\widehat{\ell}_{k+1,k} & & \\ & & \vdots & \ddots & \\ & & -\widehat{\ell}_{nk} & & 1 \end{bmatrix}$$

and $L = \left(\widehat{L}^{(n-1)} \ldots \widehat{L}^{(1)}\right)^{-1}$ is a normalised lower triangular matrix. □

Remarks 4.4

(i) To solve $Ax = b$ using an LU decomposition $PA = LU$, one solves the systems of equations $Ly = Pb$ and $Ux = y$. In the modified Gaussian method, $y = b^{(n)} = L^{-1}Pb$ and one solves $Ux = b^{(n)}$.

(ii) In an implementation, the vector π must be created to obtain U and L from the overwritten matrix A, which means additional storage space is needed. The effort for the Gaussian method with pivot search also amounts to $2n^3/3 + \mathcal{O}(n^2)$ operations.

4.3 Learning Objectives, Quiz and Application

You should be able to motivate and apply the Gaussian elimination method and explain its relationships to LU decomposition. You should be able to illustrate the importance of pivot strategies.

Quiz 4.1 Decide for each of the following statements whether it is true or false. You should be able to justify your answer.

With the Gaussian elimination method, the inverse A^{-1} of an LU decomposable matrix A can be determined with the effort $\mathcal{O}(n^4)$.	
If $A \in \mathbb{R}^{n \times n}$ is positive definite, no pivot search is necessary to perform the Gaussian elimination method.	
If $L^{(1)}, L^{(2)}, \ldots, L^{(n-1)}$ are the elimination matrices in the Gaussian elimination method for a system of equations with system matrix A, then the factor L in the LU decomposition of A is given by $L = L^{(1)} L^{(2)} \ldots L^{(n-1)}$.	
The pivot search prevents the occurrence of cancellation effects.	
Permutation matrices are obtained by row swaps in the identity matrix.	

4.3 Learning Objectives, Quiz and Application

Application 4.1 The combustion of glucose is described by the chemical reaction equation

$$x_1 \, C_6H_{12}O_6 + x_2 \, O_2 \longrightarrow x_3 \, CO_2 + x_4 \, H_2O$$

A minimal integer solution $x = [x_1, x_2, \ldots, x_4]^\mathsf{T} \neq 0$ is to be determined, such that the same number of atoms of each involved substance is on the left and right side. How can the Gaussian elimination method be modified to construct a solution?

Chapter 5
Least Squares Problems

5.1 Gaussian Normal Equation

In many applications, *overdetermined systems of equations* occur, that is for $A \in \mathbb{R}^{m \times n}$ with $m \geq n$ and $b \in \mathbb{R}^m$, a vector $x \in \mathbb{R}^n$ is sought such that

$$Ax \approx b.$$

The problem is generally not exactly solvable, since more conditions than variables can occur.

Example 5.1 For measurement data (t_i, y_i), $i = 1, 2, \ldots, m$, a number $c \in \mathbb{R}$ is sought with $y_i \approx c t_i$. The number c then describes the slope of a straight line, which approximates the pairs of points as well as possible, see Fig. 5.1.

Definition 5.1 Given $A \in \mathbb{R}^{m \times n}$ and $b \in \mathbb{R}^m$, the *least squares problem* is defined as:

$$\text{Minimise } x \mapsto \|Ax - b\|_2^2$$

For $x \in \mathbb{R}^n$, the vector $r = b - Ax$ is the *residual* of x.

The method is referred to as *method of least squares*, due to the presence of the Euclidean norm.

Proposition 5.1 *The solutions of the least squares problem are exactly the solutions of the* Gaussian normal equation

$$A^\mathsf{T} A x = A^\mathsf{T} b,$$

© The Author(s), under exclusive license to Springer-Verlag GmbH, DE, part of Springer Nature 2025
S. Bartels, *Numerical Mathematics 3x9*, La Matematica per il 3+2 160, https://doi.org/10.1007/978-3-662-70890-3_5

Fig. 5.1 In linear regression problems, an approximation of given measured values by a straight line is sought

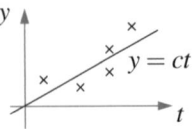

Fig. 5.2 Abstract solution of the least squares problem using the decomposition $b = Ax + r$ with $r \in \ker A^\mathsf{T}$

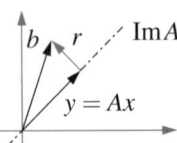

and, in particular, a solution $x \in \mathbb{R}^n$ exists. If $z \in \mathbb{R}^n$ is another solution, then $Ax = Az$ and the associated residuals agree.

Proof According to results of linear algebra, it holds

$$\mathbb{R}^m = \operatorname{Im} A + \ker A^\mathsf{T}$$

and this decomposition is orthogonal. A proof follows from considering the set $(\operatorname{Im} A)^\perp$. Thus, for $b \in \mathbb{R}^m$ uniquely determined vectors $y \in \operatorname{Im} A$ and $r \in \ker A^\mathsf{T}$ exist with $y \cdot r = 0$ and $b = y + r$. Furthermore, an $x \in \mathbb{R}^n$ exists with $y = Ax$, see Fig. 5.2. In all it follows

$$A^\mathsf{T} b = A^\mathsf{T} y + A^\mathsf{T} r = A^\mathsf{T} Ax + 0 = A^\mathsf{T} Ax,$$

that is, x solves the normal equation. To show that x is also a solution of the least squares problem, let $z \in \mathbb{R}^n$ be arbitrary. With $r = b - Ax$ and $A^\mathsf{T} r = 0$ it follows

$$\begin{aligned}
\|b - Az\|_2^2 &= \|(b - Ax) + A(x - z)\|_2^2 \\
&= \|b - Ax\|_2^2 + 2r \cdot A(x - z) + \|A(x - z)\|_2^2 \\
&= \|b - Ax\|_2^2 + 2(A^\mathsf{T} r) \cdot (x - z) + \|A(x - z)\|_2^2 \\
&= \|b - Ax\|_2^2 + \|A(x - z)\|_2^2 \\
&\geq \|b - Ax\|_2^2.
\end{aligned}$$

So, x is a minimum point and thus a solution of the least squares problem. Equality holds exactly when $A(x - z) = 0$ so $x - z \in \ker A = \ker A^\mathsf{T} A$ holds, which is to say z is a minimum point and fulfils the normal equation. In particular, it follows that $A(x - z) = 0$ holds, when $z \in \mathbb{R}^n$ is another solution. □

Remark 5.1 The identity $A^\mathsf{T} r = 0$ states that r is perpendicular or normal to the columns of A.

Lemma 5.1 *The matrix $A^\mathsf{T} A$ is symmetric and positive semi-definite. It is positive definite if and only if $\ker A = \{0\}$, that is, when A is injective or the column vectors of A are linearly independent, i.e. $\operatorname{rank} A = n$ if $m \geq n$. In this case, the solution to the normal equation is unique.*

Proof We have $(A^\mathsf{T} A)^\mathsf{T} = A^\mathsf{T} A$ and

$$x^\mathsf{T}(A^\mathsf{T} A)x = (Ax)^\mathsf{T}(Ax) = \|Ax\|_2^2 \geq 0$$

with equality if and only if $Ax = 0$. This implies the assertion, since positive definite matrices are regular. □

Remark 5.2 The condition number of $A^\mathsf{T} A$ is generally larger than that of A, because for $m = n$ and a regular matrix $A \in \mathbb{R}^{n \times n}$ we have

$$\operatorname{cond}_2(A^\mathsf{T} A) = \|A^\mathsf{T} A\|_2 \|(A^\mathsf{T} A)^{-1}\|_2 = \frac{\lambda_{max}(A^\mathsf{T} A)}{\lambda_{min}(A^\mathsf{T} A)} = \operatorname{cond}_2(A)^2,$$

so $\operatorname{cond}_2(A^\mathsf{T} A) \geq \operatorname{cond}_2(A)$, since $\operatorname{cond}_2(A) \geq 1$.

Because of this observation, least squares problems are not solved using the normal equation.

5.2 Householder Transformations

Since the Euclidean norm is invariant under rotations, we have

$$\|\widetilde{Q}(Ax - b)\|_2 = \|Ax - b\|_2$$

for every rotation $\widetilde{Q}$ and more generally for orthogonal matrices. We will try to construct an orthogonal matrix Q such that QA has a generalised upper triangular shape, which allows a simple solution to the least squares problem.

Definition 5.2 The matrix $Q \in \mathbb{R}^{\ell \times \ell}$ is called *orthogonal* if $Q^\mathsf{T} Q = I_\ell$. The set of orthogonal matrices is denoted by $O(\ell)$.

Lemma 5.2 *For all $P, Q \in O(\ell)$, $PQ \in O(\ell)$, $Q^{-1} = Q^\mathsf{T} \in O(\ell)$, $\|Qx\|_2 = \|x\|_2$ for all $x \in \mathbb{R}^\ell$ and $\operatorname{cond}_2(Q) = 1$.*

Proof For all $x \in \mathbb{R}^\ell$ we have

$$\|Qx\|_2^2 = (Qx)^\mathsf{T}(Qx) = x^\mathsf{T}(Q^\mathsf{T} Q)x = x^\mathsf{T} x = \|x\|_2^2$$

Fig. 5.3 Householder transformations define reflections on a plane

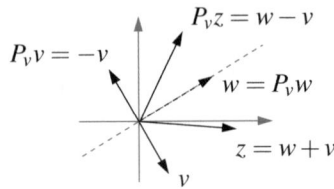

and thus $\|Q\|_2 = 1$. From the identities

$$QQ^\mathsf{T} = (QQ^\mathsf{T})^\mathsf{T} = I_\ell^\mathsf{T} = I_\ell,$$
$$Q^{-\mathsf{T}}Q^{-1} = (QQ^\mathsf{T})^{-1} = (Q^\mathsf{T}Q)^{-\mathsf{T}} = I_\ell^{-\mathsf{T}} = I_\ell,$$
$$(PQ)^\mathsf{T}(PQ) = Q^\mathsf{T}(P^\mathsf{T}P)Q = Q^\mathsf{T}Q = I_\ell$$

it follows that $PQ \in O(\ell)$ and $Q^{-1} = Q^\mathsf{T} \in O(\ell)$. These properties imply $\mathrm{cond}_2(Q) = \|Q\|_2 \|Q^{-1}\|_2 = 1$. □

Definition 5.3 For $v \in \mathbb{R}^\ell$ with $\|v\|_2 = 1$ the matrix $P_v = I_\ell - 2vv^\mathsf{T}$ is called a *Householder transformation*.

Householder transformations realise reflections on the plane perpendicular to v, see Fig. 5.3.

Lemma 5.3 *Every Householder transformation $P_v = I_\ell - 2vv^\mathsf{T}$ is symmetric and orthogonal. It holds that $P_v v = -v$ and $P_v w = w$ for all $w \in \mathbb{R}^\ell$ with $w \cdot v = 0$.*

Proof Exercise. □

Every vector $x \in \mathbb{R}^\ell \setminus \{0\}$ can be mapped to a multiple of the canonical basis vector $e_1 \in \mathbb{R}^\ell$ with a Householder transformation in $\mathbb{R}^\ell$.

Lemma 5.4 *Let $x \in \mathbb{R}^\ell \setminus \{0\}$ and $x \notin \mathrm{span}\{e_1\}$ and define, with $\sigma = \mathrm{sign}(x_1)$ if $x_1 \neq 0$ and $\sigma = 1$ otherwise, the vector $v \in \mathbb{R}^\ell$ by*

$$v = \frac{x + \sigma \|x\|_2 e_1}{\|x + \sigma \|x\|_2 e_1\|_2}.$$

Then it holds

$$P_v x = (I_\ell - 2vv^\mathsf{T}) x = -\sigma \|x\|_2 e_1.$$

Proof The matrix P_v remains unchanged when x is replaced by $\tilde{x} = \sigma x/\|x\|_2$, and from $P_v \tilde{x} = -e_1$ it follows that $P_v x = \sigma \|x\|_2 P_v \tilde{x} = -\sigma \|x\|_2 e_1$. Therefore, it is sufficient to consider the case $\|x\|_2 = 1$ and $\sigma = 1$. Since $x \notin \mathrm{span}\{e_1\}$, v is

5.3 QR Decomposition

well-defined and it holds $\|v\|_2 = 1$. With $\tilde{v} = x + e_1$ it holds due to $\|e_1\|_2^2 = \|x\|_2^2$, that

$$2\tilde{v}^T x = 2(x + e_1)^T x = \|x + e_1\|_2^2 = \|\tilde{v}\|_2^2$$

and thus due to $v = \tilde{v}/\|\tilde{v}\|_2$

$$P_v x = (I_\ell - 2vv^T)x = x - 2v\frac{\tilde{v}^T x}{\|\tilde{v}\|_2} = x - v\|\tilde{v}\|_2 = x - \tilde{v} = -e_1.$$

This proves the lemma. □

Remark 5.3 The introduction of σ avoids cancellation effects.

5.3 QR Decomposition

With the help of Householder transformations, we will step by step transform the first columns of submatrices of A to multiples of canonical basis vectors e_1 of corresponding length and thus generate an upper triangular structure.

Proposition 5.2 Let $A \in \mathbb{R}^{m \times n}$ with $m \geq n$ and rank $A = n$. Then there exist $Q \in O(m)$ and a generalised upper triangular matrix $R \in \mathbb{R}^{m \times n}$, that is $r_{ij} = 0$ for $i > j$, such that

$$A = QR = Q \begin{bmatrix} r_{11} & r_{12} & \cdots & r_{1n} \\ & r_{22} & \cdots & r_{2n} \\ & & \ddots & \vdots \\ & & & r_{nn} \\ & & & \\ & & & \end{bmatrix}.$$

Furthermore, $|r_{ii}| > 0$ for all $1 \leq i \leq n$. The factorisation is called QR decomposition.

Proof In the first step, we set $A_1 = A$. Let $x = a_1 \in \mathbb{R}^m$ be the first column of A_1. If x is a multiple of e_1, we set $Q_1 = I_m$. Otherwise, we define $Q_1 = P_v$ as in the previous lemma. It follows $Q_1 a_1 = r_{11} e_1$ with $|r_{11}| = \|Q_1 a_1\|_2 = \|a_1\|_2 > 0$ and thus

$$Q_1 A_1 = \begin{bmatrix} r_{11} & r_1^T \\ & A_2 \end{bmatrix}$$

with a matrix $A_2 \in \mathbb{R}^{(m-1)\times(n-1)}$ and a vector $r_1 \in \mathbb{R}^{n-1}$. In the second step, let $a_2 \in \mathbb{R}^{m-1}$ be the first column of A_2 and $\tilde{Q}_2 \in \mathbb{R}^{(m-1)\times(m-1)} \in O(m-1)$ be the identity matrix I_{m-1} or a Householder transformation $\tilde{Q}_2 = P_{\tilde{v}}$, such that $\tilde{Q}_2 a_2 = r_{22} e_1 \in \mathbb{R}^{m-1}$ with $|r_{22}| = \|\tilde{Q}_2 a_2\| = \|a_2\|_2 > 0$. This implies

$$\tilde{Q}_2 A_2 = \begin{bmatrix} r_{22} & r_2^\mathsf{T} \\ & A_3 \end{bmatrix}$$

with a matrix $A_3 \in \mathbb{R}^{(m-2)\times(n-2)}$ and a vector $r_2 \in \mathbb{R}^{n-2}$ and it follows

$$Q_2 Q_1 A = \begin{bmatrix} 1 & \\ & \tilde{Q}_2 \end{bmatrix} Q_1 A = \begin{bmatrix} r_{11} & r_1^\mathsf{T} \\ & \tilde{Q}_2 A_2 \end{bmatrix} = \begin{bmatrix} r_{11} & [\ r_1^\mathsf{T}\ ] \\ & r_{22} & r_2^\mathsf{T} \\ & & A_3 \end{bmatrix}.$$

The first two rows remain unchanged in the following steps. The matrix Q_2 is orthogonal, and, in particular, it is the Householder transformation for the vector $v = [0, \tilde{v}]^\mathsf{T}$, where $\tilde{v} = 0$ in the case $\tilde{Q}_2 = I_{m-1}$. After n steps, we obtain the factorisation

$$Q_n Q_{n-1} \ldots Q_1 A = R.$$

Since each Householder transformation is orthogonal and symmetric, it follows that $Q_j^{-1} = Q_j^\mathsf{T} = Q_j$ for $j = 1, 2, \ldots, n$. This results in

$$(Q_n Q_{n-1} \ldots Q_1)^{-1} = Q_1^{-1} Q_2^{-1} \ldots Q_n^{-1} = Q_1^\mathsf{T} Q_2^\mathsf{T} \ldots Q_n^\mathsf{T} = Q_1 Q_2 \ldots Q_n$$

and with $Q = Q_1 Q_2 \ldots Q_n$ the claimed factorisation $A = QR$ follows. The entries r_{ii}, $i = 1, 2, \ldots, n$, of R satisfy $|r_{ii}| = \|a_i\|_2 \neq 0$, since A would otherwise not have full rank. □

Remarks 5.4

(i) In the case $m = n$, the factorisation is uniquely determined up to the sign of the diagonal entries of R, because if $A = QR = Q'R'$, it follows that $E = (Q')^{-1} Q = R' R^{-1}$ is an upper triangular matrix in $O(n)$. Since E^{-1} is an upper triangular matrix, the identity $E^\mathsf{T} = E^{-1}$ can only hold, if E is a diagonal matrix with diagonal elements in $\{\pm 1\}$ and thus it follows $Q = Q'E$ and $R = ER'$.

(ii) The Householder transformations are not realised via matrix-matrix multiplications, because with $w = A^\mathsf{T} v$

$$P_v A = (I_m - 2vv^\mathsf{T})A = A - 2v(v^\mathsf{T} A) = A - 2vw^\mathsf{T}.$$

(iii) The vectors v_i, $i = 1, 2, \ldots, n$, which define the Householder transformations can be stored in the lower triangular part of A, where $v_i = 0$ if $Q_i = I_m$.

Setting $\widehat{v}_i = [0, v_i] \in \mathbb{R}^m$, $i = 1, 2, \ldots, n$, it also holds

$$Q = \prod_{i=1}^{n}(I_m - 2\widehat{v}_i \widehat{v}_i^T).$$

Algorithm 5.1 (QR Decomposition) *Let $A \in \mathbb{R}^{m \times n}$ with rank $A = n$. Initialise $A_1 = A$ and $i = 1$.*

(1) *Let $a_i \in \mathbb{R}^{m-i+1}$ be the first column of the right lower block $A_i \in \mathbb{R}^{(m-i+1) \times (n-i+1)}$ of A.*
(2) *If $a_i = e_1$, then continue with (5).*
(3) *Define $\widetilde{v} = a_i + \sigma \|a_i\|_2 e_1$ and $v = \widetilde{v}/\|\widetilde{v}\|_2$.*
(4) *Replace the block A_i with $A_i - vw^T$ where $w = 2A_i^T v$.*
(5) *Stop if $i = n$; otherwise increase $i \to i + 1$ and repeat step (1).*

Remark 5.5 In the i-th iteration step,

- $4(m - i + 1) + 4$ operations are required to calculate v,
- $(m - i + 2)(n - i + 1)$ operations are required to calculate w,
- $(m - i)(n - i + 1)$ operations are required to calculate $A_i - vw^T$

are needed. In total, the effort to calculate the factorisation is thus $2mn^2 - (2/3)n^3 + \mathcal{O}(mn)$. In the case $m = n$, the calculation is twice as expensive as that of the LU decomposition.

5.4 Solution of the Least Squares Problem

We use the QR decomposition to construct a stable method for the least squares problem.

Proposition 5.3 *Let $A \in \mathbb{R}^{m \times n}$ with $m \geq n$ and rank $A = n$. With the QR decomposition $A = QR$ and*

$$Q^T b = \begin{bmatrix} c \\ d \end{bmatrix}, \quad Q^T A = R = \begin{bmatrix} \widehat{R} \\ 0 \end{bmatrix}$$

with $c \in \mathbb{R}^n$, $d \in \mathbb{R}^{m-n}$ and an upper triangular matrix $\widehat{R} \in \mathbb{R}^{n \times n}$, the solution of the least squares problem defined by A and b is given by $\widehat{R}x = c$.

Proof With $\|Qz\|_2 = \|z\|_2$ for all $z \in \mathbb{R}^m$ and $Q^T Q = I_m$ it follows that

$$\|b - Ax\|_2^2 = \|Q(Q^T b - Q^T Ax)\|_2^2 = \left\| \begin{bmatrix} c \\ d \end{bmatrix} - \begin{bmatrix} \widehat{R} \\ 0 \end{bmatrix} x \right\|_2^2 = \|\widehat{R}x - c\|_2^2 + \|d\|_2^2.$$

Fig. 5.4 Transformation of the least squares problem using the QR decomposition of A

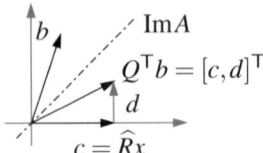

Since rank $A = n$, $\widehat{R}$ is regular. The right-hand side is obviously minimal for $x = \widehat{R}^{-1}c$. □

A geometric interpretation of the use of QR decomposition for the solution of the least squares problem is shown in Fig. 5.4.

Remark 5.6 From $Q \in O(n)$ it follows for regular matrices $A \in \mathbb{R}^{n \times n}$, that $\text{cond}_2(R) = \text{cond}_2(A)$. The QR decomposition thus defines a stable algorithm for solving least squares problems.

5.5 Learning Objectives, Quiz and Application

You should be familiar with applications that lead to least squares problems, and you should be able to derive the Gaussian normal equation and its most important properties. You should be able to explain the construction of the QR decomposition of a matrix and describe its significance in solving least squares problems.

Quiz 5.1 Decide for each of the following statements whether it is true or false. You should be able to justify your answer.

The least squares problem always has a solution.	
If rank $A = n \leq m$, then $A^T A$ is invertible.	
A Householder transformation is defined by $I_n - 2(v^T v)^{-2} vv^T$, provided $v \in \mathbb{R}^n \setminus \{0\}$ holds.	
If Q is orthogonal, then both the row and column vectors of Q are pairwise orthogonal.	
For every vector norm $\|\cdot\|$ on $\mathbb{R}^n$, every orthogonal matrix $Q \in O(n)$ and every vector $x \in \mathbb{R}^n$, $\|Qx\| = \|x\|$ holds.	

Application 5.1 Theoretical considerations of two physical processes lead to the assumption that the quantities y and t are related in the form $y(t) = c_0 + c_1 t + c_2 t^2 + c_3 t^3$ and the quantities z and v follow the relation $z(v) = c/v$. Experiments yield measurement data (t_i, y_i) and (v_i, z_i) for $i = 1, 2, \ldots, m$. Formulate least squares problems for the approximate determination of $c_0, c_1, \ldots, c_3$ and c and set up the corresponding Gaussian normal equations. How can the validity of the assumption about the relationship be assessed after calculating the coefficients?

Chapter 6
Singular Value Decomposition and Pseudoinverse

6.1 Singular Value Decomposition

The symmetric and positive semidefinite matrix $A^\mathsf{T} A \in \mathbb{R}^{n \times n}$ for $A \in \mathbb{R}^{m \times n}$ plays an important role in the least squares problem. It is diagonalisable and there exists an orthonormal basis consisting of eigenvectors $v_1, v_2, \ldots, v_n$ with associated eigenvalues

$$\lambda_1 \geq \lambda_2 \geq \cdots \geq \lambda_p > \lambda_{p+1} = \cdots = \lambda_n = 0$$

with $0 \leq p \leq n$, where eigenvalues are listed multiple times if necessary according to their multiplicity. For $i = 1, 2, \ldots, p$ we define $u_i = \lambda_i^{-1/2} A v_i$. For $1 \leq i, j \leq p$ then

$$u_i^\mathsf{T} u_j = \lambda_i^{-1/2} \lambda_j^{-1/2} (A v_i)^\mathsf{T} (A v_j) = (\lambda_i \lambda_j)^{-1/2} v_i^\mathsf{T} (A^\mathsf{T} A v_j)$$
$$= (\lambda_i \lambda_j)^{-1/2} v_i^\mathsf{T} (\lambda_j v_j) = \lambda_j (\lambda_i \lambda_j)^{-1/2} v_i^\mathsf{T} v_j = \delta_{ij}.$$

The vectors $(u_1, u_2, \ldots, u_p)$ thus form an orthonormal set of vectors in $\mathbb{R}^m$. We supplement it with vectors $(u_{p+1}, u_{p+2}, \ldots, u_m)$ to an orthonormal basis $(u_1, u_2, \ldots, u_m)$ of $\mathbb{R}^m$. We have

$$A^\mathsf{T} u_i = \lambda_i^{-1/2} A^\mathsf{T} A v_i = \lambda_i^{1/2} v_i, \quad i = 1, 2, \ldots, p.$$

From $\ker A^T A = \ker A$ and $n = \dim \ker A + \dim \operatorname{Im} A$ we deduce the identity $\operatorname{Im} A = \operatorname{span}\{u_1, u_2, \ldots, u_p\}$ and thus $\{u_{p+1}, \ldots, u_m\} = (\operatorname{Im} A)^\perp = \ker A^T$ respectively

$$A^T u_i = 0, \quad i = p+1, \ldots, m.$$

Using $\sigma_i = \lambda_i^{1/2}$, $i = 1, 2, \ldots, p$, we obtain the following proposition.

Proposition 6.1 *Let $A \in \mathbb{R}^{m \times n}$. Then there exist numbers $\sigma_1 \geq \sigma_2 \geq \cdots \geq \sigma_p > 0$ and orthonormal bases $(u_i)_{i=1,\ldots,m}$ of $\mathbb{R}^m$ and $(v_j)_{j=1,\ldots,n}$ of $\mathbb{R}^n$ with the properties*

$$Av_i = \sigma_i u_i, \quad A^T u_i = \sigma_i v_i, \quad i = 1, 2, \ldots, p,$$
$$Av_j = 0, \quad A^T u_k = 0, \quad j = p+1, \ldots, n, \ k = p+1, \ldots, m.$$

The numbers σ_i^2, $i = 1, 2, \ldots, p$, are exactly the non-zero eigenvalues of $A^T A$ and are called singular values *of A. For*

$$U = [u_1, \ldots, u_m] \in \mathbb{R}^{m \times m}, \quad V = [v_1, \ldots, v_n] \in \mathbb{R}^{n \times n}$$

we have that $U \in O(m)$ and $V \in O(n)$ and with

$$\Sigma = \begin{bmatrix} \sigma_1 & & & & 0 & \ldots & 0 \\ & \ddots & & & \vdots & & \vdots \\ & & \sigma_p & 0 & \ldots & 0 \\ 0 & \ldots & 0 & 0 & \ldots & 0 \\ \vdots & & \vdots & \vdots & & \vdots \\ 0 & \ldots & 0 & 0 & \ldots & 0 \end{bmatrix} \in \mathbb{R}^{m \times n}$$

it follows

$$A = U \Sigma V^T = \sum_{i=1}^p \sigma_i u_i v_i^T, \quad A^T = V \Sigma^T U^T = \sum_{i=1}^p \sigma_i v_i u_i^T.$$

The factorisation is called singular value decomposition (SVD).

Proof The statements follow from the construction and the application of the factorisations to the orthonormal bases. □

6.2 Pseudoinverse

With the help of the singular value decomposition, the concept of the inverse matrix can be generalised to non-regular and non-square matrices.

Definition 6.1 If $A = U \Sigma V^T$ is the singular value decomposition of A and $\Sigma^+ \in \mathbb{R}^{n \times m}$ is defined by

$$\Sigma^+ = \begin{bmatrix} \sigma_1^{-1} & & & 0 \\ & \ddots & & \vdots \\ & & \sigma_p^{-1} & 0 \\ 0 & \cdots & 0 & 0 \end{bmatrix} \in \mathbb{R}^{n \times m},$$

then $A^+ = V \Sigma^+ U^T = \sum_{i=1}^{p} \sigma_i^{-1} v_i u_i^T \in \mathbb{R}^{n \times m}$ is called *pseudoinverse* or *Moore–Penrose-Inverse* of A.

Remarks 6.1

(i) By construction of A^+ we have $\ker A^+ = \text{span}\{u_{p+1}, \ldots, u_m\} = \ker A^T$ and $\text{Im } A^+ = \text{span}\{v_1, v_2, \ldots, v_p\} = \text{Im } A^T$.
(ii) The matrix A^+ is the uniquely determined solution in $X \in \mathbb{R}^{n \times m}$ of the algebraic equations

$$AXA = A, \quad XAX = X, \quad (AX)^T = AX, \quad (XA)^T = XA.$$

For example, because $U^T U = I_m$ and $V^T V = I_n$, we have that

$$A^+ A A^+ = (V \Sigma^+ U^T)(U \Sigma V^T)(V \Sigma^+ U^T) = V \Sigma^+ \Sigma \Sigma^+ U^T$$
$$= V \Sigma^+ U^T = A^+.$$

With the pseudoinverse, the least squares problem can be solved.

Proposition 6.2 *The vector $A^+ b$ is, among all solutions of the least squares problem, the one with the minimal norm.*

Proof With $A^+ A A^+ = A^+$ and Remark 6.1 (i) it follows

$$AA^+ b - b \in \ker A^+ = \ker A^T,$$

that is $A^T A (A^+ b) = A^T b$ or, that $A^+ b$ is a solution of the Gaussian normal equation. If $z \in \mathbb{R}^n$ is another solution, then because $\ker A^T A = \ker A$, we have that

$$A^T A (A^+ b - z) = 0 \quad \Longleftrightarrow \quad A(A^+ b - z) = 0.$$

With $w = A^+b - z \in \ker A$ it follows from $A^+b \in \operatorname{Im} A^+ = (\ker A)^\perp$, that $(A^+b) \cdot w = 0$ and for $z = A^+b - w$ we get

$$\|z\|_2^2 = \|A^+b\|_2^2 + \|w\|_2^2 \geq \|A^+b\|_2^2.$$

Thus, A^+b is a solution with minimal norm. □

Remark 6.2 If rank $A = n \leq m$, then it follows from $A^+b = (A^T A)^{-1} A^T b$ for all $b \in \mathbb{R}^m$, that $A^+ = (A^T A)^{-1} A^T$ and in particular $A^+ = A^{-1}$ if $n = m$.

6.3 Learning Objectives, Quiz and Application

You should be able to describe the ideas for the construction of the singular value decomposition of a matrix and to concretise the definition of the pseudoinverse as well as its relation to least squares problems.

Quiz 6.1 Decide for each of the following statements whether they are true or false. You should be able to justify your answer.

The squares of the singular values of a matrix are the eigenvalues of AA^T.	
For the first singular value σ_1 of A, $\sigma_1 = \|A\|_2$ holds.	
If $A \in \mathbb{R}^{n \times n}$ is symmetric, then the singular value decomposition defines a diagonalisation of A.	
A solution to the least squares problem is defined by the solution of the system of linear equations $A^+ x = b$.	
There exists a solution $z \in \mathbb{R}^n$ to the least squares problem with the property $\|A^+b\|_2 = \|z\|_2$.	

Application 6.1 The matrix $A \in \mathbb{R}^{m \times n}$ describes certain data such as the greyscale of the pixels of an image. To compress the data, the singular value decomposition $A = \sum_{i=1}^{p} \sigma_i u_i v_i^T$ is first determined. For $\varepsilon > 0$ and $i = 1, 2, \ldots, p$, let

$$\tilde{\sigma}_i = \begin{cases} \sigma_i, & \text{if } \sigma_i \geq \varepsilon, \\ 0, & \text{if } \sigma_i < \varepsilon, \end{cases}$$

and

$$\tilde{A} = \sum_{i=1}^{p} \tilde{\sigma}_i u_i v_i^T.$$

Show that

$$\|A - \tilde{A}\|_{\mathscr{F}} \leq p\varepsilon$$

and rank $\tilde{A} \leq \operatorname{rank} A$.

Chapter 7
The Simplex Method

7.1 Linear Programs

In applications such as the minimisation of production costs, linear optimisation problems with linear inequality constraints arise. To formulate such problems succinctly, we use the notation $a \le b$ for vectors $a, b \in \mathbb{R}^m$, if $a_i \le b_i$ for $i = 1, 2, \ldots, m$. In this chapter, we follow the presentation in [12].

Definition 7.1 A *linear program* is an optimisation problem

$$\text{Minimise } g(y) = p^\mathsf{T} y \text{ subject to the constraint } Uy \le d$$

with given $p \in \mathbb{R}^\ell$, $U \in \mathbb{R}^{q \times \ell}$ and $d \in \mathbb{R}^q$. A linear program is in *standard form*, if it can be written in the form

$$\text{Minimise } f(x) = c^\mathsf{T} x \text{ subject to the constraint } Ax = b, \ x \ge 0$$

with given $c \in \mathbb{R}^n$, $A \in \mathbb{R}^{m \times n}$ and $b \in \mathbb{R}^m$.

Remark 7.1 By introducing additional variables, any linear program can be transformed into standard form. Here, one decomposes $y_i = v_i - w_i$ with $v_i, w_i \ge 0$ and writes an inequality $Uy \le d$ as an equation $Uy + z = d$ with $z \ge 0$. The new variable is then the vector $x = [v, w, z]$.

Definition 7.2 The *feasible set* of a linear program in standard form is $M = \{x \in \mathbb{R}^n : Ax = b, \ x \ge 0\}$.

Remark 7.2 The feasible set is convex and can be empty, a single element, and bounded or unbounded, see Fig. 7.1.

Fig. 7.1 The feasible set M of a linear program is the intersection of convex sets

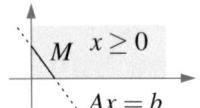

Definition 7.3 A point $x \in M$ is called a *corner*, if it cannot be written as a proper convex combination in M, that is for all $z, y \in M$ and $\lambda \in (0, 1)$ with $x = \lambda z + (1 - \lambda)y$ it follows $x = y = z$.

We will always assume that M is non-empty. Without proof, we use the following results.

Proposition 7.1 *Assume that the feasible set M is non-empty and bounded.*

(i) *The set M has finitely many corners $y^1, y^2, \ldots, y^L \in M$ and these span M, that is $M = \{x = \sum_{i=1}^{L} \lambda_i y^i : \lambda_i \in [0, 1], \sum_{i=1}^{L} \lambda_i = 1\}$.*
(ii) *The linear program has a solution and the minimum is attained at a corner of M.*

Remark 7.3 If M is unbounded, the problem can be solvable or unsolvable.

To solve a linear program, it is thus sufficient to consider corners. The set of possible solutions is thus reduced to finitely many points.

Definition 7.4 The *index set* I_x of a corner $x \in M$ consists of the indices of the non-zero components

$$I_x = \{i \in \{1, 2, \ldots, n\} : x_i > 0\}$$

and let $J_x = \{1, 2, \ldots, n\} \setminus I_x$. The sets I_x and J_x are considered ordered and for a vector $z \in \mathbb{R}^n$ and the matrix $A \in \mathbb{R}^{m \times n}$ with column vectors $(a_i : i = 1, 2, \ldots, n)$ we denote

$$z_{I_x} = (z_i : i \in I_x), \quad z_{J_x} = (z_j : j \in J_x),$$
$$A_{I_x} = (a_i : i \in I_x), \quad A_{J_x} = (a_j : j \in J_x).$$

When it is clear from the context which corner is meant, the index x is omitted from I_x and J_x. For $z \in \mathbb{R}^n$ we then have

$$Az = A_I z_I + A_J z_J.$$

Proposition 7.2 *The following statements apply to the corners of M:*

(i) *A point $x \in M$ is a corner if and only if the column vectors $(a_i : i \in I_x)$ are linearly independent.*
(ii) *Each corner $x \in M$ is uniquely determined by its index set.*

7.2 The Simplex Step

Proof

(i) If x is not a corner, then there exist $y, z \in M \setminus \{x\}$ and $\lambda \in (0, 1)$ with $x = \lambda y + (1 - \lambda)z$. We have $(y - z)_J = 0$ with $J = \{1, 2, \ldots, n\} \setminus I$ and $I = I_x$, since from $x_i = 0$ and $y, z \geq 0$ also $y_i = z_i = 0$ follows. This implies $0 = b - b = A(y - z) = A_I(y - z)_I$ and because $(y - z)_I \neq 0$ the linear dependence of the columns of the matrix A_I follows. Conversely, if the columns of A_I are linearly dependent, then there exists $\tilde{y} \neq 0$ with $A_I \tilde{y} = 0$ and $\tilde{y}$ can be completed by zeros to $y \in \mathbb{R}^n$ with $y_I = \tilde{y}$. Since $x_i > 0$ for all $i \in I$, then with $\varepsilon > 0$ sufficiently small, we have that $x \pm \varepsilon y \geq 0$. Furthermore, $Ay = 0$ and thus $A(x \pm \varepsilon y) = b$, so $x \pm \varepsilon y \in M$. With $\lambda = 1/2$ then $x = \lambda(x + \varepsilon y) + (1 - \lambda)(x - \varepsilon y)$ is a true convex combination and x is not a corner.

(ii) If x is a corner, then according to (i) the column vectors of A_I are linearly independent and from $b = Ax = A_I x_I$ it follows that x_I is uniquely determined.

□

The number of equality constraints $Ax = b$ defined by $A \in \mathbb{R}^{m \times n}$ in a linear program in standard form is usually less than the number of unknowns, that is, $m \leq n$.

Definition 7.5 A corner $x \in M$ is called *degenerate* if $|I_x| < m$ holds. Otherwise, it is called *non-degenerate*.

Remark 7.4 If $x \in M$ is a non-degenerate corner and $\text{rank } A = m$, that is, the constraints are linearly independent, then $|I_x| = m$ and the matrix $A_I \in \mathbb{R}^{m \times m}$ is invertible.

Example 7.1 The corner $x = [0, 1]^T$ with index set $I_x = \{2\}$ is non-degenerate for $A = [0, 1]$ and degenerate for $A = \begin{bmatrix} 0 & 1 \\ 0 & 2 \end{bmatrix}$.

7.2 The Simplex Step

To solve a linear program, it is sufficient to consider the corners of the feasible set. Starting from a corner, a new one is constructed so that the function value is reduced. Let $\text{rank } A = m$ and M be non-empty. We proceed as follows:

(1) Let $x \in M$ be a corner and if it is degenerate, let I_x be supplemented to an m-element set I so that A_I is regular. Let $J = \{1, 2, \ldots, n\} \setminus I$.
(2) For all $z \in M$ it follows from $b = Az = A_I z_I + A_J z_J$, that

$$z_I = A_I^{-1} b - A_I^{-1} A_J z_J. \tag{7.1}$$

Thus, the components with respect to I are uniquely determined by those with respect to J; in particular, due to $x_J = 0$, it follows that $x_I = A_I^{-1}b$. For the function value $f(z) = c^T z$ it follows

$$c^T z = c_I^T z_I + c_J^T z_J = c_I^T(A_I^{-1}b - A_I^{-1}A_J z_J) + c_J^T z_J$$
$$= c_I^T x_I + (c_J^T - c_I^T A_I^{-1} A_J) z_J = c^T x + (c_J - A_J^T A_I^{-T} c_I)^T z_J.$$

With $u_J = c_J - A_J^T A_I^{-T} c_I$ we deduce that

$$f(z) = f(x) + u_J^T z_J. \tag{7.2}$$

Due to $z \geq 0$, a reduction of the function value is only possible if u_J is negative in one component. Otherwise, x is already the solution to the problem.

(3) Let $u_r < 0$ for an $r \in J$. We consider the vector z defined via

$$z_j = 0, \; j \in J \setminus \{r\}, \quad z_r = t$$

with a number $t \geq 0$ to be chosen. We use (7.1) to supplement z_J to a vector $z \in \mathbb{R}^n$ that satisfies $Az = b$, that is we set

$$z_I = A_I^{-1}b - A_I^{-1}A_J z_J = x_I - t A_I^{-1} a_r.$$

Here, a_r is the r-th column of A. With (7.2) it follows

$$f(z) = f(x) + t u_r \leq f(x).$$

When choosing t, it is still necessary to ensure $z \geq 0$.

(4) Let $d = A_I^{-1} a_r$, so that $z_I = x_I - td$. If $d \leq 0$, then $z \geq 0$ for any choice of $t \geq 0$ and $f(z) \to -\infty$ for $t \to \infty$, that is M is unbounded and the problem is not solvable.

(5) Let $d_i > 0$ for an $i \in I$. The condition $z \geq 0$ is fulfilled, as long as

$$z_i = x_i - t d_i \geq 0$$

holds. To maximally reduce the function value $f(z)$ and at the same time ensure $z \geq 0$, we choose

$$t = \min_{i \in I, d_i > 0} \frac{x_i}{d_i} = \frac{x_s}{d_s}.$$

This implies in particular $z_s = 0$. If x is non-degenerate, then due to $x_i > 0$ for all $i \in I$, it follows that $t > 0$ and the function value is genuinely reduced.

(6) We show that $z \in M$ is a corner. We have $I_z \subset I^{new} = (I_x \setminus \{s\}) \cup \{r\}$ and according to Proposition 7.2 it suffices to show, that the vectors $(a_i : i \in I^{new})$

7.2 The Simplex Step

are linearly independent. Let $\gamma_i \in \mathbb{R}$, $i \in I^{new}$, such that

$$0 = \sum_{i \in I^{new}} \gamma_i a_i = \sum_{i \in I \setminus \{s\}} \gamma_i a_i + \gamma_r a_r$$

holds. With $d = A_I^{-1} a_r$ or $a_r = A_I d$ it follows

$$0 = \sum_{i \in I \setminus \{s\}} \gamma_i a_i + \gamma_r \sum_{i \in I} d_i a_i = \sum_{i \in I \setminus \{s\}} (\gamma_i + \gamma_r d_i) a_i + \gamma_r d_s a_s.$$

Since the vectors $(a_i : i \in I)$ are linearly independent, it follows $\gamma_i + \gamma_r d_i = 0$ for $i \in I \setminus \{s\}$ and $\gamma_r d_s = 0$. Because $d_s \neq 0$ this implies $\gamma_i = 0$ for all $i \in I^{new}$. Thus z is a corner.

We have proven the following result.

Proposition 7.3 *Let* rank $A = m$ *and* $x \in M$ *be a corner. With* $I_x \subset I$, *such that* $A_I \in \mathbb{R}^{m \times m}$ *is regular, let* $J = \{1, 2, \ldots, n\} \setminus I$. *If* $u = c_J - A_J^T A_I^{-T} c_I \geq 0$, *then* x *is the solution of the problem. If* $u_r < 0$ *for some* $r \in J$, *then define* $d = A_I^{-1} a_r$. *If* $d \leq 0$, *then the problem is unsolvable. If* $d_s > 0$ *with* $s \in I$, *such that* $t = \min_{i \in I, d_i > 0} \frac{x_i}{d_i} = \frac{x_s}{d_s}$, *then, by setting*

$$x_i^{new} = \begin{cases} x_i - t d_i, & i \in I \setminus \{s\}, \\ t, & i = r, \\ 0, & i \in (J \setminus \{r\}) \cup \{s\}, \end{cases}$$

a corner x^{new} *of* M *is defined with the property*

$$f(x^{new}) \leq f(x).$$

If x *is non-degenerate, then the inequality is strict.*

Remarks 7.5

(i) The *simplex method* consists in the repeated application of the simplex step, until the case $d \leq 0$ for unsolvability occurs, the sufficient termination criterion $u \geq 0$ for a corner is fulfilled or a corner is passed through a second time. Since there are only finitely many corners, the method always terminates.

(ii) So-called *cycles* can occur, i.e. one returns to a corner already visited without a reduction occurring or the minimum being reached. However, this is not observed for practically relevant problems.

(iii) The newly constructed corner x^{new} can be degenerate, even if x is non-degenerate.

(iv) There are $\binom{n}{m}$ many corners, so in the worst case $\mathcal{O}(n!)$ many corners would have to be passed through to reach the minimum. In practice, only polynomial effort with respect to n is observed, but there are examples where 2^n many simplex steps are required.

(v) The algorithmic construction of a corner for the initialization of the method is by no means trivial.

7.3 Learning Objectives, Quiz and Application

You should be able to explain geometric properties of linear programs and the essential ideas of the simplex step.

Quiz 7.1 Decide for each of the following statements whether it is true or false. You should be able to justify your answer.

Every linear program in normal form has a solution.	
The simplex step realizes a proper reduction of the function to be minimized.	
The point $x = 0$ is always a corner of the feasible set.	
Each corner is uniquely determined by its zero entries.	
In the simplex method, the number of zero entries of the corners passed through strictly decreases.	

Application 7.1

(i) A product is stored at the locations $A_1, .., A_m$ in the respective quantities $a_1, \ldots, a_m$ and it is needed at the locations $B_1, \ldots, B_n$ in the quantities $b_1, \ldots, b_n$. Let c_{ij} denote the costs for transporting a unit quantity of the product from A_i to B_j. Formulate a linear program in standard form to minimise the total cost of transporting the product.

(ii) A producer of road salt receives an order to deliver 50 tonnes of road salt to Rome, 20 to Paris and 30 to Berlin. There are 40 and 60 tonnes available in warehouses in Prague and Amsterdam respectively. What are the optimal transport quantities, if the costs per 10 tonnes of transport quantity in Euros are given according to Table 7.1? Use the MATLAB routine linprog for the solution.

Table 7.1 Transport costs per tonne of road salt

	Rome	Paris	Berlin
Prague	700	600	200
Amsterdam	800	300	400

Chapter 8
Eigenvalue Problems

8.1 Localisation

The calculation of individual or all eigenvalues of a matrix and, if applicable, associated eigenvectors is referred to as *eigenvalue problems*. In general, it is difficult and inefficient to determine the roots of a characteristic polynomial, as even the evaluation of the polynomial is associated with high computational effort.

Proposition 8.1 *Let $A \in \mathbb{R}^{n \times n}$ and $\lambda \in \mathbb{C}$ be an eigenvalue of A. Then we have*

$$\lambda \in \bigcup_{i=1}^{n} K_i, \quad K_i = \{z \in \mathbb{C} : |z - a_{ii}| \leq \sum_{j=1,\dots,n,\ j \neq i} |a_{ij}|\}.$$

The sets K_i are called Gershgorin circles.

Proof Let $Ax = \lambda x$ hold for an $x \in \mathbb{C}^n \setminus \{0\}$. Then there exists an i with $|x_j| \leq |x_i|$ for all $j = 1, 2, \dots, n$, and $x_i \neq 0$. We have

$$\lambda x_i = (Ax)_i = \sum_{j=1}^{n} a_{ij} x_j$$

and after division by $x_i \neq 0$ it follows

$$\lambda - a_{ii} = \sum_{j=1,\dots,n,\ j \neq i} a_{ij} \frac{x_j}{x_i}.$$

The triangle inequality and $|x_j|/|x_i| \leq 1$ imply $\lambda \in K_i$ and thus the assertion. □

Fig. 8.1 Gershgorin circles in Example 8.1

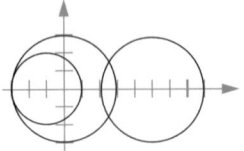

Example 8.1 For the following matrix $A \in \mathbb{R}^{3 \times 3}$, the Gershgorin circles K_1, K_2, K_3 result, see Fig. 8.1.

$$A = \begin{bmatrix} 5 & 1 & 2 \\ 1 & -1 & 1 \\ 2 & 1 & 0 \end{bmatrix}, \qquad \begin{aligned} K_1 &= \{z \in \mathbb{C} : |z - 5| \leq 3\}, \\ K_2 &= \{z \in \mathbb{C} : |z + 1| \leq 2\}, \\ K_3 &= \{z \in \mathbb{C} : |z| \leq 3\}. \end{aligned}$$

In the case of symmetric matrices, the eigenvalues can be characterised by the extreme values of the so-called *Rayleigh quotient*.

Proposition 8.2 *Let $A \in \mathbb{R}^{n \times n}$ be symmetric. For the maximum and minimum eigenvalue of A we have*

$$\lambda_{min} = \min_{x \in \mathbb{R}^n \setminus \{0\}} \frac{x^T A x}{\|x\|_2^2}, \qquad \lambda_{max} = \max_{x \in \mathbb{R}^n \setminus \{0\}} \frac{x^T A x}{\|x\|_2^2}.$$

Proof Let $(v_1, v_2, \ldots, v_n) \subset \mathbb{R}^n$ be an orthonormal basis of $\mathbb{R}^n$ consisting of eigenvectors corresponding to the eigenvalues $\lambda_1 \geq \lambda_2 \geq \cdots \geq \lambda_n \in \mathbb{R}$ of the matrix A. For $x \in \mathbb{R}^n$ there exist $\alpha_1, \alpha_2, \ldots, \alpha_n \in \mathbb{R}$, such that $x = \alpha_1 v_1 + \alpha_2 v_2 + \cdots + \alpha_n v_n$ and we have

$$Ax = \alpha_1 \lambda_1 v_1 + \alpha_2 \lambda_2 v_2 + \cdots + \alpha_n \lambda_n v_n.$$

The orthonormality $v_i \cdot v_j = \delta_{ij}$, $1 \leq i, j \leq n$, of the vectors $v_1, v_2, \ldots, v_n$ implies

$$x^T x = \left(\sum_{i=1}^n \alpha_i v_i\right) \cdot \left(\sum_{j=1}^n \alpha_j v_j\right) = \sum_{i,j=1}^n \alpha_i \alpha_j v_i \cdot v_j = \sum_{i=1}^n \alpha_i^2,$$

$$x^T A x = \left(\sum_{i=1}^n \alpha_i v_i\right) \cdot \left(\sum_{j=1}^n \alpha_j A v_j\right) = \sum_{i,j=1}^n \alpha_i \lambda_j \alpha_j v_i \cdot v_j = \sum_{i=1}^n \lambda_i \alpha_i^2.$$

It follows that

$$x^T A x \geq \lambda_n \sum_{i=1}^n \alpha_i^2 = \lambda_n \|x\|_2^2 = \lambda_{min} \|x\|_2^2,$$

where equality holds for $x = v_n$. The statement for $\lambda_1 = \lambda_{max}$ follows analogously. □

8.2 Conditioning

A matrix $A \in \mathbb{R}^{n \times n}$ is complex diagonalisable if there is a regular matrix $V \in \mathbb{C}^{n \times n}$ and a diagonal matrix $D \in \mathbb{C}^{n \times n}$ such that $A = VDV^{-1}$ holds. In this situation, the following result from Bauer and Fike can be proven.

Proposition 8.3 *Let $A \in \mathbb{R}^{n \times n}$ be complex diagonalisable with $A = VDV^{-1}$, let $E \in \mathbb{R}^{n \times n}$ and let $\tilde{\lambda} \in \mathbb{C}$ be an eigenvalue of $A + E$. Then there exists an eigenvalue $\lambda \in \mathbb{C}$ of A, such that*

$$|\tilde{\lambda} - \lambda| \leq \mathrm{cond}_2(V) \|E\|_2,$$

where the operator norm and condition number are generalised in an obvious way for complex matrices.

Proof If $\tilde{\lambda}$ is also an eigenvalue of A, then the statement is trivial. In the following, let $\tilde{\lambda}$ not be an eigenvalue of A, so that $\tilde{\lambda} I_n - A$ is invertible. If $x \in \mathbb{C}^n$ is an eigenvector of $A + E$ corresponding to the eigenvalue $\tilde{\lambda}$, then

$$Ex = (A + E)x - Ax = \tilde{\lambda} x - Ax = (\tilde{\lambda} I_n - A)x,$$

so $x = (\tilde{\lambda} I_n - A)^{-1} E x$, that is, 1 is an eigenvalue of $(\tilde{\lambda} I_n - A)^{-1} E$. From this follows

$$1 \leq \|(\tilde{\lambda} I_n - A)^{-1} E\|_2 = \|(\tilde{\lambda} V V^{-1} - VDV^{-1})^{-1} E\|_2$$
$$= \|V(\tilde{\lambda} I_n - D)^{-1} V^{-1} E\|_2 \leq \|V\|_2 \|(\tilde{\lambda} I_n - D)^{-1}\|_2 \|V^{-1}\|_2 \|E\|_2$$
$$= \mathrm{cond}_2(V) \max_{\lambda \in \sigma(A)} |\tilde{\lambda} - \lambda|^{-1} \|E\|_2,$$

where the maximum is formed over all complex eigenvalues λ of A. With the identity $\max_{x \in X} |x|^{-1} = (\min_{x \in X} |x|)^{-1}$ the claim follows. □

Remarks 8.1

(i) Not every matrix is complex diagonalisable, however every matrix is complex triangularisable.
(ii) Normal matrices, that is, matrices with the property $AA^\top = A^\top A$, are complex diagonalisable with unitary transformation matrix V, that is, V fulfils $\overline{V}^\top V = I_n$.

Corollary 8.1 *Let $A \in \mathbb{R}^{n \times n}$ be normal, $E \in \mathbb{R}^{n \times n}$ and $\tilde{\lambda}$ be an eigenvalue of $A + E$. Then there exists an eigenvalue of A with*

$$|\lambda - \tilde{\lambda}| \leq \|E\|_2$$

Fig. 8.2 The eigenvalues of the matrix A_ε from Example 8.2 lie on a circle with centre a and radius $\varepsilon^{1/n}$

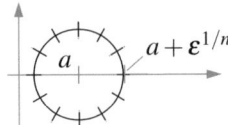

Proof Since A is diagonalisable with a unitary matrix $V \in \mathbb{C}^{n \times n}$ and since $\text{cond}_2(V) = 1$ holds, the estimate follows from the previous result. □

For normal, in particular symmetric matrices, the determination of the eigenvalues is thus a well conditioned problem with respect to absolute errors. Generally, this is not the case.

Example 8.2 If $p(t) = t^n + a_{n-1} t^{n-1} + \cdots + a_1 t + a_0$ is any polynomial, then $p(t) = (-1)^n \det(A - t I_n)$ with the Frobenius companion matrix

$$A = \begin{bmatrix} 0 & & & & -a_0 \\ 1 & 0 & & & -a_1 \\ & \ddots & \ddots & & \vdots \\ & & 1 & 0 & -a_{n-2} \\ & & & 1 & -a_{n-1} \end{bmatrix}.$$

In particular, the complex roots of p correspond to the complex eigenvalues of A. For $a \in \mathbb{R} \setminus \{0\}$ and $\varepsilon > 0$, the polynomial $p_0(t) = (t-a)^n$ has the n-fold root $\lambda = a$, while the polynomial $p_\varepsilon(t) = (t-a)^n - \varepsilon$ has the roots $\lambda_k = a + \varepsilon^{1/n} e^{i 2\pi k / n}$, $k = 1, 2, \ldots, n$, see Fig. 8.2. The polynomials p_0 and p_ε differ only in the constant coefficient and for the difference $A - A_\varepsilon$ of the associated companion matrices we have $\|A - A_\varepsilon\|_\ell = \varepsilon$ for $\ell \in \{1, 2, \infty\}$. We have $|\lambda - \lambda_k| = \varepsilon^{1/n}$ and for the relative errors it follows that

$$\frac{|\lambda - \lambda_k|}{|\lambda|} = \frac{\varepsilon^{1/n}}{|a|} \frac{\|A\|_\ell}{\|A\|_\ell} \frac{\|A - A_\varepsilon\|_\ell}{\varepsilon}$$

$$= \frac{\varepsilon^{1/n}}{\varepsilon} \frac{\|A\|_\ell}{|a|} \frac{\|A - A_\varepsilon\|_\ell}{\|A\|_\ell}.$$

The factor $\varepsilon^{(1-n)/n}$ is unbounded for $\varepsilon \to 0$, provided $n > 1$. Multiplying the equations with $|\lambda| = |a|$ shows that the ill conditioning also applies to absolute errors in the case $n > 1$ for small numbers $\varepsilon > 0$.

8.3 Power Method

Let $A \in \mathbb{R}^{n \times n}$ be real diagonalisable with eigenvalues $\lambda_1, \lambda_2, \ldots, \lambda_n \in \mathbb{R}$ and associated linearly independent eigenvectors $v_1, v_2, \ldots, v_n \in \mathbb{R}^n$, for which $\|v_i\|_2 = 1$, $i = 1, 2, \ldots, n$, holds. For every $x \in \mathbb{R}^n$ with

$$x = \sum_{i=1}^{n} \alpha_i v_i$$

the k-fold application of A to x results in

$$A^k x = A^{k-1}\left(\sum_{i=1}^{n} \lambda_i \alpha_i v_i\right) = \cdots = \sum_{i=1}^{n} \lambda_i^k \alpha_i v_i.$$

If λ_1 is the absolute largest eigenvalue, then for sufficiently large k, it follows that

$$A^k x \approx \alpha_1 \lambda_1^k v_1.$$

We consider the norms $\|A^k x\|_2$ and $\|A^{k+1} x\|_2$ and form their quotient, so due to $\|v_1\|_2 = 1$ it follows

$$\frac{\|A^{k+1} x\|_2}{\|A^k x\|_2} \approx |\lambda_1|.$$

With this observation, one can determine the dominant eigenvalue of a matrix.

Algorithm 8.1 (Von Mises Power Method) Let $A \in \mathbb{R}^{n \times n}$, $x \in \mathbb{R}^n \setminus \{0\}$ and $\varepsilon_{stop} > 0$. Set $x_0 = x/\|x\|_2$, $\mu_0 = 0$ and $k = 0$.

(1) Calculate $\tilde{x}_{k+1} = A x_k$, $\mu_{k+1} = \|\tilde{x}_{k+1}\|_2$ and $x_{k+1} = \tilde{x}_{k+1}/\mu_{k+1}$.
(2) Stop if $|\mu_{k+1} - \mu_k| \le \varepsilon_{stop}$; otherwise increase $k \to k+1$ and repeat step (1).

Remark 8.2 Inductively, it is shown that $x_k = A^k x / \|A^k x\|_2$ holds. To avoid leaving the range of representable numbers, a normalisation must be carried out in each iteration step.

We show that the sequence $(x_k)_{k \in \mathbb{N}}$ approximates a normalised eigenvector to the absolute maximum eigenvalue of A, whose magnitude is approximated by the sequence $(\mu_k)_{k \in \mathbb{N}}$.

Proposition 8.4 Let $|\lambda_1| > |\lambda_2| \ge \cdots \ge |\lambda_n| \ge 0$ and let $x = \sum_{i=1}^{n} \alpha_i v_i$ with the normalised eigenvectors $v_1, v_2, \ldots, v_n$ of A. If $\alpha_1 \ne 0$ then it follows with $q = |\lambda_2|/|\lambda_1| < 1$ and $k \ge K$, that

$$\left|\|A x_k\|_2 - |\lambda_1|\right| \le 4 \|A\|_2 c q^k$$

with a constant $c \ge 0$ independent of k.

Proof For each $k \geq 0$ we have

$$A^k x = \lambda_1^k \alpha_1 \left(v_1 + \sum_{i=2}^{n} \frac{\lambda_i^k}{\lambda_1^k} \frac{\alpha_i}{\alpha_1} v_i \right) = \lambda_1^k \alpha_1 (v_1 + w_k),$$

where $w_k \in \mathbb{R}^n$ is defined by the sum. It follows

$$\|w_k\|_2 \leq q^k \sum_{i=2}^{n} \frac{|\alpha_i|}{|\alpha_1|} = cq^k.$$

Furthermore,

$$x_k = \frac{A^k x}{\|A^k x\|_2} = \frac{\lambda_1^k \alpha_1 (v_1 + w_k)}{|\lambda_1^k \alpha_1| \|v_1 + w_k\|_2}$$

$$= \operatorname{sign}(\lambda_1^k \alpha_1) v_1 + \operatorname{sign}(\lambda_1^k \alpha_1) \left(\frac{v_1 + w_k}{\|v_1 + w_k\|_2} - v_1 \right)$$

$$= \operatorname{sign}(\lambda_1^k \alpha_1) v_1 + r_k.$$

With the reverse triangle inequality $\|a\|_2 - \|b\|_2 \leq \|a + b\|_2$ as well as the ordinary triangle inequality $\|a + b\|_2 \leq \|a\|_2 + \|b\|_2$ it follows

$$1 - cq^k \leq \|v_1\|_2 - \|w_k\|_2 \leq \|v_1 + w_k\|_2 \leq \|v_1\|_2 + \|w_k\|_2 \leq 1 + cq^k$$

and it follows for k sufficiently large, so that $cq^k \leq 1/2$ holds,

$$\|r_k\|_2 = \left\| \frac{v_1(1 - \|v_1 + w_k\|_2) + w_k}{\|v_1 + w_k\|_2} \right\|_2$$

$$\leq \frac{|1 - \|v_1 + w_k\|_2| + \|w_k\|_2}{\|v_1 + w_k\|_2} \leq \frac{2cq^k}{1 - cq^k} \leq 4cq^k.$$

For $\widetilde{x}_{k+1} = A x_k$ the above representation of x_k leads to

$$\widetilde{x}_{k+1} = \lambda_1 \operatorname{sign}(\lambda_1^k \alpha_1) v_1 + A r_k$$

and thus

$$\|\widetilde{x}_{k+1} - \lambda_1 \operatorname{sign}(\lambda_1^k \alpha_1) v_1\|_2 \leq \|A r_k\|_2 \leq 4 \|A\|_2 cq^k.$$

Another application of the reverse triangle inequality combined with $\widetilde{x}_{k+1} = A x_k$ and $\|v_1\|_2 = 1$ shows the assertion. □

8.3 Power Method

Remarks 8.3

(i) In each step of the iteration, the approximation error is reduced by the factor $q < 1$.
(ii) The last inequality in the proof shows that $\lambda_1 < 0$ holds exactly when the signs of $(x_k)_{k=1,2,\ldots}$ alternate, and that the sequence $(x_k)_{k=1,2,\ldots}$ converges to an eigenvector up to the factor $\operatorname{sign} \lambda_1^k$.
(iii) The condition $\alpha_1 \neq 0$ must be ensured in the specific case. Due to rounding errors, this can be assumed, but the constant $c \sim 1/|\alpha_1|$ may then be large.

In the case of symmetric matrices, a better convergence statement can be proven, which shows that the error is reduced by the factor q^2 in each step.

Proposition 8.5 *If $A \in \mathbb{R}^{n \times n}$ is symmetric, then under the conditions of the previous proposition*

$$|\lambda_1 - x_k^T A x_k| \leq 2 \|A\|_2 c^2 q^{2k}.$$

Proof If A is symmetric, the eigenvectors $(v_1, v_2, \ldots, v_n)$ in the proof of the previous result can be assumed to be orthonormal and we have

$$A^k x = \lambda_1^k \alpha_1 (v_1 + w_k)$$

with $v_1 \cdot w_k = 0$. Let

$$\gamma_k = \operatorname{sign}(\lambda_1^k \alpha_1) \|v_1 + w_k\|_2^{-1}$$

and since $\|v_1 + w_k\|_2^2 = \|v_1\|_2^2 + \|w_k\|_2^2 \geq 1$, we have $|\gamma_k| \leq 1$. It follows

$$x_k = \frac{A^k x}{\|A^k x\|_2} = \frac{\lambda_1^k \alpha_1 (v_1 + w_k)}{|\lambda_1^k \alpha_1| \|v_1 + w_k\|_2} = \gamma_k v_1 + \gamma_k w_k$$

and thus

$$(\lambda_1 I_n - A) x_k = \gamma_k \lambda_1 v_1 + \gamma_k \lambda_1 w_k - \gamma_k A v_1 - \gamma_k A w_k$$
$$= \gamma_k (\lambda_1 I_n - A) w_k.$$

Since $A w_k \in \operatorname{span}\{v_2, v_3, \ldots, v_n\}$, the vector on the right side is orthogonal to v_1. With the Cauchy-Schwarz inequality $a^T b \leq \|a\|_2 \|b\|_2$ and $\gamma_k \leq 1$, it follows that

$$|x_k^T (\lambda_1 I_n - A) x_k| = \gamma_k^2 |(v_1 + w_k)^T (\lambda_1 I_n - A) w_k|$$
$$\leq \|w_k\|_2 \|\lambda_1 (I_n - A)\|_2 \|w_k\|_2$$
$$\leq (|\lambda_1| + \|A\|_2) \|w_k\|_2^2$$
$$= 2\|A\|_2 \|w_k\|_2^2,$$

where in the last equality $|\lambda_1| = \|A\|_2$ was used. From $\|w_k\|_2 \leq cq^k$ and $x_k^T x_k = 1$ the assertion follows. □

Remarks 8.4

(i) If $0 < |\lambda_n| < |\lambda_{n-1}| \leq \cdots \leq |\lambda_1|$, then the power method with A^{-1} instead of A provides an approximation of $|\lambda_n|^{-1}$. This is referred to as *inverse iteration*.

(ii) If one applies the power method to the matrix $(A - \mu I_n)^{-1}$, it converges under suitable conditions towards the eigenvalue that is closest to μ.

(iii) The dominant eigenvalue can occur multiple times, meaning the condition of the propositions can be weakened to $|\lambda_1| = \cdots = |\lambda_p| > |\lambda_{p+1}| \geq \cdots \geq |\lambda_n| \geq 0$.

8.4 QR Method

The QR method calculates approximations of all eigenvalues of a matrix simultaneously under suitable conditions. To this end, the power method is applied to several vectors, which are orthogonalised at each step. We follow the argument from [30] and calculate two sequences of vectors $(x_k)_{k=0,1,\ldots}$ and $(y_k)_{k=0,1,\ldots}$ using the power method, and make a correction at each step, which ensures that the vectors x_k and y_k are orthogonal to each other for all $k \geq 0$. Let $x_0, y_0 \in \mathbb{R}^n \setminus \{0\}$ be orthogonal and determine for $k = 1, 2, \ldots$ the iterates x_k and y_k through

$$x_k = A^k x_0, \qquad y_k = A^k y_0 - \gamma_k x_k,$$

with $\gamma_k = (A^k y_0) \cdot x_k / \|x_k\|_2^2$, so that $x_k \cdot y_k = 0$ holds. If $(v_1, v_2, \ldots, v_n)$ is a basis of $\mathbb{R}^n$ consisting of eigenvectors corresponding to the eigenvalues $|\lambda_1| > |\lambda_2| > |\lambda_3| \geq \cdots \geq |\lambda_n|$ of A, it follows

$$x_k = \sum_{i=1}^n a_i \lambda_i^k v_i, \qquad y_k = \sum_{j=1}^n (b_j - \gamma_k a_j) \lambda_j^k v_j,$$

with the coefficients $a_1, a_2, \ldots, a_n$ and $b_1, b_2, \ldots, b_n$ of the vectors x_0 and y_0 with respect to the basis $(v_1, v_2, \ldots, v_n)$. We assume that a_1 and $r = b_2 - b_1 a_2/a_1$ are different from zero. The orthogonality $x_k \cdot y_k = 0$ implies that

$$0 = \sum_{i=1}^n \sum_{j=1}^n a_i (b_j - \gamma_k a_j) \lambda_i^k \lambda_j^k v_i \cdot v_j.$$

Assuming that the sequence γ_k, $k = 1, 2, \ldots$, is bounded, a division of this identity by λ_1^{2k} as well as the condition $|\lambda_\ell / \lambda_1| < 1$ for $\ell = 2, 3, \ldots, n$, implies that

8.4 QR Method

$$b_1 - \gamma_k a_1 \to 0$$

as $k \to \infty$. This yields for large numbers k, that

$$y_k \approx \sum_{j=2}^{n} (b_j - \gamma_k a_j) \lambda_j^k v_j.$$

Because $b_2 - \gamma_k a_2 \to r \neq 0$ and $|\lambda_2| > |\lambda_j|$, $j \geq 3$, the first term dominates in this sum, so that y_k converges to a multiple of the eigenvector v_2 as $k \to \infty$. For the practical implementation we use normalisations, meaning we set

$$\tilde{x}_k = \frac{x_k}{\|x_k\|}, \qquad \tilde{y}_k = \frac{y_k}{\|y_k\|}$$

and obtain the relations

$$c_k \tilde{x}_k = A \tilde{x}_{k-1}, \qquad d_k \tilde{y}_k = A \tilde{y}_{k-1} - e_k \tilde{x}_k$$

with suitable numbers $c_k, d_k, e_k \in \mathbb{R}$. With the definitions

$$U_k = [\tilde{x}_k \ \tilde{y}_k] \in \mathbb{R}^{n \times 2}, \qquad R_k = \begin{bmatrix} c_k & e_k \\ 0 & d_k \end{bmatrix} \in \mathbb{R}^{2 \times 2}$$

we deduce

$$A U_{k-1} = U_k R_k.$$

Due to the orthogonality and normalisation of the vectors $\tilde{x}_k$ and $\tilde{y}_k$, U_k is an orthogonal matrix, meaning we have that $U_k^T U_k = I_2$. If we define

$$A_{k+1} = U_k^T A U_k, \qquad Q_k = U_{k-1}^T U_k$$

then in the case $n = 2$, it follows that $Q_k \in O(2)$, and

$$A_k = U_{k-1}^T A U_{k-1} = (U_{k-1}^T U_k) R_k = Q_k R_k,$$
$$A_{k+1} = U_k^T A U_k = U_k^T A U_{k-1} U_{k-1}^T U_k = R_k Q_k,$$

where in the last step $R_k = U_k^T U_{k-1} U_{k-1}^T A U_{k-1} = U_k^T A U_{k-1}$ was used. Thus, a QR factorisation of A_k is determined and subsequently A_{k+1} is defined by swapping the factors. This procedure can be generalised to the case $n > 2$ and leads to the QR method, which under suitable conditions provides approximations of all eigenvalues of a matrix. The above-derived (approximate) similarities of A_k and R_k to A indicate, that the iterates A_k converge to an upper triangular matrix and their diagonal entries thus define approximations of the eigenvalues. As starting vectors for the QR method the canonical basis vectors are chosen.

Algorithm 8.2 (QR Method) *Let $A \in \mathbb{R}^{n \times n}$ be regular. Set $A_0 = A$ and $k = 0$.*

(1) Determine the QR decomposition $A_k = Q_k R_k$ and set $A_{k+1} = R_k Q_k$.
(2) Stop if $\|\mathrm{diag}(A_{k+1}) - \mathrm{diag}(A_k)\| \leq \varepsilon_{stop}$; otherwise increase $k \to k+1$ and repeat step (1).

The iterates of the QR method are similar to each other.

Lemma 8.1 *We have*

$$A_{k+1} = Q_k^T A_k Q_k = (Q_0 \ldots Q_k)^T A (Q_0 \ldots Q_k).$$

Proof From $A_{k+1} = R_k Q_k$ and $A_k = Q_k R_k$ respectively $R_k = Q_k^T A_k$ it follows $A_{k+1} = Q_k^T A_k Q_k$ and the repeated application of this argument proves the second equation. □

With the help of this lemma and a stability property of the QR decomposition the following result can be proven, see for example [7].

Proposition 8.6 *Let $A \in \mathbb{R}^{n \times n}$ be diagonalisable with $A = VDV^{-1}$ such that, for the eigenvalues $\lambda_1, \lambda_2, \ldots, \lambda_n \in \mathbb{R}$, we have*

$$|\lambda_1| > |\lambda_2| > \cdots > |\lambda_n| > 0$$

and the inverse of the matrix V has an LU decomposition. Then we have

$$\|\mathrm{diag}(A_k) - \mathrm{diag}(D)\|_2 \leq c q^k$$

with $q = \max_{1 \leq i < j \leq n} |\lambda_j|/|\lambda_i|$ and a constant $c \geq 0$.

Remarks 8.5

(i) In practice, convergence is observed under significantly weaker conditions on A.
(ii) In general, a step in the QR method leads to an effort of order $\mathcal{O}(n^3)$. If A is first transformed by Householder transformations into a so-called Hessenberg matrix

$$\widehat{A} = H^T A H = \begin{bmatrix} \widehat{a}_{11} & \cdots & & \widehat{a}_{1n} \\ \widehat{a}_{21} & \widehat{a}_{22} & \cdots & \widehat{a}_{2n} \\ & \widehat{a}_{32} & \cdots & \widehat{a}_{3n} \\ & & \ddots & \vdots \\ & & \widehat{a}_{n,n-1} & \widehat{a}_{nn} \end{bmatrix}$$

8.5 Jacobi Method

The result about Gershgorin circles shows that the diagonal entries of a matrix define approximations of the eigenvalues, which are particularly accurate when the off-diagonal elements are small. In the Jacobi method, these entries of a symmetric matrix are successively reduced with similarity transformations. We follow the presentation in [9].

Definition 8.1 For $A \in \mathbb{R}^{n \times n}$, let

$$\mathcal{N}(A) = \|A\|_{\mathcal{F}}^2 - \sum_{i=1}^n a_{ii}^2 = \sum_{1 \leq i,j \leq n, i \neq j} a_{ij}^2.$$

Obviously, A is a diagonal matrix exactly when $\mathcal{N}(A) = 0$ holds. More generally, it can be shown that for each diagonal entry a_{jj}, $1 \leq j \leq n$, there exists an eigenvalue λ with the property $|\lambda - a_{jj}| \leq \sqrt{\mathcal{N}(A)}$. From the Gershgorin circle result, one obtains the weaker statement that for each eigenvalue λ of A there exists a diagonal entry a_{jj} such that $|\lambda - a_{jj}| \leq (n-1)^{1/2}\sqrt{\mathcal{N}(A)}$.

Definition 8.2 For $c, s \in \mathbb{R}$ with $c^2 + s^2 = 1$ and $1 \leq p, q \leq n$ with $p \neq q$, a Givens rotation $G_{pq} \in O(n)$ is defined by

$$(G_{pq})_{ij} = \begin{cases} 1, & i = j, i \neq p, \\ 1, & i = j, i \neq q, \\ c, & i = p, j = p, \\ c, & i = q, j = q, \\ s, & i = q, j = p, \\ -s & i = p, j = q, \\ 0 & \text{otherwise.} \end{cases} \qquad G_{pq} = \begin{bmatrix} 1 & & & & & \\ & \ddots & & & & \\ & & c & & -s & \\ & & & \ddots & & \\ & & s & & c & \\ & & & & & \ddots \\ & & & & & & 1 \end{bmatrix}$$

The following lemma uses the fact that the Frobenius norm is invariant under orthogonal transformations, i.e. that $\|Q^T M\|_{\mathcal{F}} = \|MQ\|_{\mathcal{F}} = \|M\|_{\mathcal{F}}$ for all $M \in \mathbb{R}^{n \times n}$ and $Q \in O(n)$. This follows, for example, from $\|M\|_{\mathcal{F}}^2 = \operatorname{tr}(M^T M)$ and $\|M\|_{\mathcal{F}} = \|M^T\|_{\mathcal{F}}$.

---

that is, $\widehat{a}_{ij} = 0$ for $i > j + 1$, then the QR decomposition can be determined in $\mathcal{O}(n^2)$ steps using Givens rotations.

Lemma 8.2 *If $A \in \mathbb{R}^{n \times n}$ is symmetric and G_{pq} is any Givens rotation, then for $B = G_{pq}^T A G_{pq}$, we have that*

$$\mathcal{N}(B) = \mathcal{N}(A) - 2(a_{pq}^2 - b_{pq}^2),$$

where $b_{pq} = cs(a_{qq} - a_{pp}) + (c^2 - s^2)a_{pq}$.

Proof One directly verifies that the entries of the symmetric matrix B are given by $b_{ij} = a_{ij}$, provided $i, j \notin \{p, q\}$, as well as

$$b_{pp} = c^2 a_{pp} + 2cs a_{pq} + s^2 a_{qq},$$
$$b_{qq} = s^2 a_{pp} - 2cs a_{pq} + c^2 a_{qq},$$
$$b_{pq} = b_{qp} = cs(a_{qq} - a_{pp}) + (c^2 - s^2)a_{pq},$$
$$b_{ip} = c a_{ip} + s a_{iq}, \quad i \in \{1, 2, \ldots, n\} \setminus \{p, q\},$$
$$b_{iq} = -s a_{ip} + c a_{iq}, \quad i \in \{1, 2, \ldots, n\} \setminus \{p, q\}.$$

With $\|B\|_{\mathscr{F}} = \|A\|_{\mathscr{F}}$ it follows

$$\mathcal{N}(B) = \|B\|_{\mathscr{F}}^2 - \sum_{i=1}^n a_{ii}^2 + \sum_{i=1}^n (a_{ii}^2 - b_{ii}^2)$$

$$= \|A\|_{\mathscr{F}}^2 - \sum_{i=1}^n a_{ii}^2 + (a_{pp}^2 - b_{pp}^2 + a_{qq}^2 - b_{qq}^2)$$

$$= \mathcal{N}(A) + a_{pp}^2 + a_{qq}^2 - b_{pp}^2 - b_{qq}^2.$$

The formulas for the entries of B show that

$$\begin{bmatrix} b_{pp} & b_{pq} \\ b_{pq} & b_{qq} \end{bmatrix} = \begin{bmatrix} c & s \\ -s & c \end{bmatrix} \begin{bmatrix} a_{pp} & a_{pq} \\ a_{pq} & a_{qq} \end{bmatrix} \begin{bmatrix} c & -s \\ s & c \end{bmatrix}.$$

Identifying this identity with $\widehat{b} = g^T \widehat{a} g$, it follows $\|\widehat{b}\|_{\mathscr{F}}^2 = \|\widehat{a}\|_{\mathscr{F}}^2$ or

$$b_{pp}^2 + b_{qq}^2 + 2b_{pq}^2 = a_{pp}^2 + a_{qq}^2 + 2a_{pq}^2,$$

thus

$$a_{pp}^2 + a_{qq}^2 - b_{pp}^2 - b_{qq}^2 = 2(b_{pq}^2 - a_{pq}^2)$$

and this implies the assertion. $\square$

8.5 Jacobi Method

If the Givens rotation G_{pq} can be chosen so that $b_{pq} = 0$ applies, this results in a reduction of the non-diagonal entries. By considering $c = \cos(\alpha)$, $s = \pm\sin(\alpha)$ and $D = \cos(2\alpha)$ one obtains the following result.

Lemma 8.3 *If $a_{pq} \neq 0$ and the matrix G_{pq} is defined by $c = \sqrt{(1+D)/2}$ and $s = \text{sign}(a_{pq})\sqrt{(1-D)/2}$ with*

$$D = \frac{a_{pp} - a_{qq}}{\left((a_{pp} - a_{qq})^2 + 4a_{pq}^2\right)^{1/2}} \in [-1, 1]$$

then $b_{pq} = 0$ applies.

Proof Exercise. □

To achieve the greatest possible reduction of $\mathcal{N}(A)$, it is obvious to choose the absolute largest non-diagonal element of A.

Proposition 8.7 *If a_{pq} is the absolute largest non-diagonal element of A, then with the Givens rotation G_{pq} defined in the previous lemma for the matrix $B = G_{pq}^\mathsf{T} A G_{pq}$ and with $\varepsilon_n = 2/(n(n-1))$, we have that*

$$\mathcal{N}(B) \leq (1 - \varepsilon_n)\mathcal{N}(A),$$

Proof After choosing a_{pq}, we have $\mathcal{N}(A) \leq n(n-1)a_{pq}^2$. This implies

$$\mathcal{N}(B) = \mathcal{N}(A) - 2a_{pq}^2 \leq \left(1 - \frac{2}{n(n-1)}\right)\mathcal{N}(A),$$

thus the claimed estimate. □

From the proposition follows the convergence of the following method.

Algorithm 8.3 (Jacobi Method) *Let $A \in \mathbb{R}^{n \times n}$ be symmetric. Set $A_0 = A$ and $k = 0$.*

(1) Let p, q be the indices of the largest absolute non-diagonal element of A_k and choose the Givens rotation G_{pq}, so that for $A_{k+1} = G_{pq}^\mathsf{T} A_k G_{pq}$ the entry $(A_{k+1})_{pq}$ vanishes.
(2) Stop if $\mathcal{N}(A_{k+1}) \leq \varepsilon_{stop}$; otherwise increase $k \to k+1$ and repeat step (1).

Remarks 8.6

(i) In general, $\mathcal{O}(n^2 \log(1/\varepsilon_{stop}))$ many iterations are needed to guarantee $\mathcal{N}(A_{k+1}) \leq \varepsilon_{stop}$.
(ii) An entry already transformed to zero can deviate from zero again during the method.
(iii) The method constructs a factorisation $A = GDG^\mathsf{T}$ with an orthogonal matrix G and an approximate diagonal matrix D. In particular, the column vectors of G are approximate eigenvectors of A.

(iv) Since the search for the maximum non-diagonal element is time-consuming, in practice, all non-diagonal elements are processed successively and this is repeated until $\mathcal{N}(A_k)$ is sufficiently small. This approach is referred to as *cyclic Jacobi method*.

8.6 Learning Objectives, Quiz and Application

You should be familiar with various eigenvalue problems and their conditioning. You should be able to derive various methods for the numerical solution of eigenvalue problems and be able to specify their convergence and complexity properties.

Quiz 8.1 Decide for each of the following statements whether it is true or false. You should be able to justify your answer.

If $A \in \mathbb{R}^{n \times n}$ is regular, then the calculation of the eigenvalues of A is a well conditioned problem.	
The matrices A and A^T have the same eigenvalues and eigenvectors.	
The convergence speed of the power method depends on the ratio of the largest absolute to the smallest absolute eigenvalue.	
The execution of a step of the QR method requires a complexity of order $\mathcal{O}(n^3)$.	
The Jacobi method is feasible and convergent for every diagonalisable matrix.	

Application 8.1 The numbers 1, 2, 3 are indicators for the comprehensibility of a mathematics lecture, where 1 stands for good, 2 for medium and 3 for low comprehensibility. Assume that the probability that a lecture of value j is followed by a lecture of value i is denoted by p_{ij} and we have

$$P = \begin{bmatrix} 0.1 & 0.3 & 0.6 \\ 0.5 & 0.2 & 0.1 \\ 0.4 & 0.5 & 0.3 \end{bmatrix}.$$

A very comprehensible lecture is followed by a little comprehensible lecture with 40% probability. Given the vector $x_0 \in [0, 1]^3$ for the current lecture, the probabilities of the comprehensibility indicators k lectures later is given by $x_k = P^k x_0$.

(i) Experimentally test the convergence of the sequence $(x_k)_{k \geq 0}$, where x_0 is defined by canonical basis vectors in $\mathbb{R}^3$, that is, after how many steps does $\|x_k - x_{k+1}\|_1 \leq 10^{-5}$ hold? What does this mean for the comprehensibility of the lectures?

8.6 Learning Objectives, Quiz and Application

(ii) Suppose the sequence $(x_k)_{k\geq 0}$ becomes stationary, that is we have $x_k \approx x^*$ for all $k \geq K$ with a sufficiently large number $K \geq 0$. How can x^* be characterised?

(iii) Test five starting vectors $x_0 \in [0, 1]^3$ generated with `rand(3,1)`, scaled with $\|x_0\|_1 = 1$ and characterise the stationary points. Consider the eigenvalues and eigenvectors of P, which you can determine in MATLAB with `[V,D] = eig(P)`.

Chapter 9
Iterative Solution Methods

9.1 Inexact Solution

Due to model and data errors as well as numerical rounding, it is generally neither necessary nor sensible to solve a system of linear equations exactly in the sense of computer arithmetic. We will approximate the solution of a system of equations through a sequence of approximate solutions and stop the iteration when the equation is sufficiently well fulfilled. This approach leads to a significant reduction in effort in many cases. In this chapter, we follow the presentation in [10].

9.2 Banach's Fixed Point Theorem

Banach's fixed point theorem defines a method that approximates the solution of a fixed point equation under suitable conditions.

Definition 9.1 A mapping $\Phi : \mathbb{R}^n \to \mathbb{R}^n$ is called a *contraction* with respect to a norm $\|\cdot\|$ on $\mathbb{R}^n$, if there is a number $q < 1$ such that for all $x, y \in \mathbb{R}^n$ we have

$$\|\Phi(x) - \Phi(y)\| \leq q\|x - y\|.$$

For contractions, the following fixed point iteration leads to convergent approximations.

Algorithm 9.1 (Fixed Point Iteration) Let $\Phi : \mathbb{R}^n \to \mathbb{R}^n$ be a contraction and $x^0 \in \mathbb{R}^n$. Set $k = 0$.

(1) Define $x^{k+1} = \Phi(x^k)$.
(2) Stop if $\|x^{k+1} - x^k\| \leq \varepsilon_{stop}$; otherwise increase $k \to k+1$ and repeat step (1).

The iteration converges for every choice of the initial x^0.

Proposition 9.1 *If $\Phi : \mathbb{R}^n \to \mathbb{R}^n$ is a contraction, then Φ has a unique fixed point $x^* \in \mathbb{R}^n$, that is, we have $\Phi(x^*) = x^*$. For any initial value $x^0 \in \mathbb{R}^n$, the fixed point iteration $x^{k+1} = \Phi(x^k)$ for $k = 0, 1, 2, \ldots$, defines a sequence of approximations of x^* with the property*

$$\|x^k - x^*\| \leq \frac{q^k}{1-q} \|x^1 - x^0\|,$$

in particular, the sequence $(x^k)_{k \in \mathbb{N}}$ converges to x^*.

Proof The mapping Φ has at most one fixed point, because if $x^*, y^* \in \mathbb{R}^n$ are fixed points, then

$$\|x^* - y^*\| = \|\Phi(x^*) - \Phi(y^*)\| \leq q\|x^* - y^*\|$$

and since $q < 1$, it follows that $x^* = y^*$. The sequence defined by the procedure $x^{k+1} = \Phi(x^k)$ is a Cauchy sequence, because from

$$\|x^k - x^{k+1}\| = \|\Phi(x^{k-1}) - \Phi(x^k)\| \leq q\|x^{k-1} - x^k\|$$

it follows inductively that

$$\|x^k - x^{k+1}\| \leq q^k \|x^0 - x^1\|$$

and with the triangle inequality for $n \geq m$

$$\|x^m - x^n\| = \|x^m - x^{m+1} + x^{m+1} - x^{m+2} + x^{m+2} - \cdots - x^{n-1} + x^{n-1} - x^n\|$$

$$\leq \sum_{k=m}^{n-1} \|x^k - x^{k+1}\| \leq \sum_{k=m}^{n-1} q^k \|x^0 - x^1\| = \|x^0 - x^1\| q^m \sum_{k=0}^{n-m-1} q^k$$

$$= \|x^0 - x^1\| q^m \frac{1 - q^{n-m}}{1-q} \leq \|x^0 - x^1\| \frac{q^m}{1-q}.$$

As a Cauchy sequence, $(x^k)_{k \in \mathbb{N}}$ has a limit $x^* \in \mathbb{R}^n$ and for this it follows with the Lipschitz continuity of Φ, that

$$x^* = \lim_{k \to \infty} x^{k+1} = \lim_{k \to \infty} \Phi(x^k) = \Phi(x^*).$$

Thus, x^* is a fixed point of Φ. The error bound follows from the above estimates by considering the limit transition $n \to \infty$. □

Remarks 9.1

(i) From the error estimation, it can be determined how many iteration steps are necessary to achieve a given error tolerance.
(ii) The fact that the method converges to the solution for any choice of the initial value x^0 is referred to as *global convergence*.

9.3 Linear Iteration Methods

We want to investigate the contraction property for affine-linear mappings $\Phi(x) = Mx + c$. Obviously, the mapping Φ is a contraction, if there exists an operator norm $\|\cdot\|_{op}$ on $\mathbb{R}^{n \times n}$ with $\|M\|_{op} < 1$. The *spectral radius* of a matrix $M \in \mathbb{R}^{n \times n}$ is defined by

$$\rho(M) = \max\{|\lambda| : \lambda \text{ is a complex eigenvalue of } M\}.$$

The following proposition shows that it is sufficient to show $\rho(M) < 1$ to guarantee a contraction property. Note that $\rho(M)$ for $n \geq 2$ does not define a norm on $\mathbb{R}^{n \times n}$.

Proposition 9.2 *For $M \in \mathbb{R}^{n \times n}$ we have*

$$\rho(M) = \inf\left\{\|M\|_{op} : \|\cdot\|_{op} \text{ is an induced operator norm on } \mathbb{C}^{n \times n}\right\}.$$

Proof

(i) Let $\lambda \in \mathbb{C}$ be an eigenvalue of M with $\rho(M) = |\lambda|$ and $x \in \mathbb{C}^n \setminus \{0\}$ a corresponding eigenvector. Then for every norm on $\mathbb{C}^n$, we have that

$$\rho(M)\|x\| = \|\lambda x\| = \|Mx\| \leq \|M\|_{op}\|x\|,$$

thus $\rho(M) \leq \|M\|_{op}$.

(ii) The matrix M is complex triangularizable, i.e. there exist an invertible matrix $T \in \mathbb{C}^{n \times n}$ and an upper triangular matrix $R \in \mathbb{C}^{n \times n}$ with

$$R = T^{-1}MT = \begin{bmatrix} \lambda_1 & r_{12} & \cdots & r_{1n} \\ & \ddots & & \vdots \\ & & \lambda_{n-1} & r_{n-1,n} \\ & & & \lambda_n \end{bmatrix}$$

and the complex eigenvalues $\lambda_1, \lambda_2, \ldots, \lambda_n$ of M. For $\varepsilon > 0$ let $D_\varepsilon \in \mathbb{R}^{n \times n}$ be the diagonal matrix with diagonal elements $1, \varepsilon, \varepsilon^2, \ldots, \varepsilon^{n-1}$. Then by

$$\|x\|_\varepsilon = \|D_\varepsilon^{-1}T^{-1}x\|_\infty$$

a norm on $\mathbb{C}^n$ is defined. For the corresponding operator norm we have

$$\|M\|_\varepsilon = \sup_{x\neq 0} \frac{\|D_\varepsilon^{-1} T^{-1} Mx\|_\infty}{\|D_\varepsilon^{-1} T^{-1} x\|_\infty} \stackrel{x=TD_\varepsilon y}{=} \sup_{y\neq 0} \frac{\|D_\varepsilon^{-1} T^{-1} MTD_\varepsilon y\|_\infty}{\|y\|_\infty}$$

$$= \sup_{y\neq 0} \frac{\|D_\varepsilon^{-1} R D_\varepsilon y\|_\infty}{\|y\|_\infty} = \|D_\varepsilon^{-1} R D_\varepsilon\|_\infty$$

with the row sum norm $\|\cdot\|_\infty$. Direct calculation shows

$$D_\varepsilon^{-1} R D_\varepsilon = \left[\varepsilon^{-(i-1)} r_{ij} \varepsilon^{j-1}\right]_{i,j=1,\ldots,n} = \begin{bmatrix} \lambda_1 & \varepsilon r_{1,2} & \cdots & \varepsilon^{n-1} r_{1,n} \\ & \ddots & & \vdots \\ & & \lambda_{n-1} & \varepsilon r_{n-1,n} \\ & & & \lambda_n \end{bmatrix}$$

and thus, provided $\varepsilon \leq 1$,

$$\|M\|_\varepsilon = \|D_\varepsilon^{-1} R D_\varepsilon\|_\infty = \max_{i=1,2,\ldots,n}\left(|\lambda_i| + \sum_{j=i+1}^n \varepsilon^{j-i}|r_{ij}|\right)$$

$$\leq \max_{i=1,2,\ldots,n} |\lambda_i| + \varepsilon \|R\|_\infty = \rho(M) + \varepsilon \|R\|_\infty.$$

Since $\varepsilon > 0$ can be chosen arbitrarily small, the assertion follows. $\square$

Corollary 9.1 *If $\rho(M) < 1$, then the mapping $\Phi : x \mapsto Mx + c$ is a contraction.*

Example 9.1 The *Richardson method* for the approximate solution of the linear system $Ax = b$ is for $\omega > 0$ defined by $M = I_n - \omega A$ and $c = \omega b$, that is

$$x^{k+1} = Mx^k + c = x^k - \omega(Ax^k - b).$$

If A is symmetric and positive definite, then all eigenvalues of A are positive, and for ω sufficiently small, $\rho(I_n - \omega A) < 1$. If $x^{k+1} = x^k$, then x^k is a solution of $Ax = b$.

9.4 Jacobi and Gauss-Seidel Methods

Based on simple decompositions of matrices, iterative methods can be defined.

9.5 Diagonal Dominance and Irreducibility

Definition 9.2 For $A \in \mathbb{R}^{n \times n}$, the lower, diagonal and upper part $L, U, D \in \mathbb{R}^{n \times n}$ of A are defined by

$$d_{ij} = \begin{cases} a_{ii}, & i = j, \\ 0, & i \neq j, \end{cases} \quad \ell_{ij} = \begin{cases} a_{ij}, & i > j, \\ 0, & i \leq j, \end{cases} \quad u_{ij} = \begin{cases} a_{ij}, & i < j, \\ 0, & i \geq j. \end{cases}$$

Since $A = L + D + U$, the linear system $Ax = b$ is equivalent to

$$Lx + Dx + Ux = b$$

and iteration methods can be defined by replacing x in the different terms by x^k or x^{k+1}, for example

$$Lx^k + Dx^{k+1} + Ux^k = b \iff x^{k+1} = -D^{-1}(A - D)x^k + D^{-1}b.$$

For a stationary point or in the case $x^{k+1} = x^k$, the linear system is obviously fulfilled. An alternative to this approach is

$$Lx^{k+1} + Dx^{k+1} + Ux^k = b \iff x^{k+1} = -(L + D)^{-1}Ux^k + (L + D)^{-1}b.$$

Definition 9.3 The *Jacobi* and *Gauss-Seidel methods* are defined by

$$M^J = -D^{-1}(A - D), \quad c^J = D^{-1}b,$$
$$M^{GS} = -(L + D)^{-1}U, \quad c^{GS} = (L + D)^{-1}b,$$

provided D is regular.

Remarks 9.2

(i) In the Jacobi method, a linear system with a diagonal matrix is to be solved in each iteration step, and in the Gauss-Seidel method with a lower triangular matrix.
(ii) It is expected that the Gauss-Seidel method has better convergence properties than the Jacobi method, as the matrix $L + D$ is generally a better approximation of A than the matrix D.

9.5 Diagonal Dominance and Irreducibility

We want to formulate sufficient conditions for a matrix that imply the contraction property of an iteration procedure.

Definition 9.4 The matrix $A \in \mathbb{R}^{n \times n}$ is called *diagonally dominant*, if for $i = 1, 2, \ldots, n$ we have

$$\sum_{j=1,\ldots,n,\ j \neq i} |a_{ij}| \leq |a_{ii}|$$

and this inequality is strict for an $i_0 \in \{1, 2, \ldots, n\}$. If it is strict for all $i = 1, 2, \ldots, n$, then A is called *strictly diagonally dominant*.

Example 9.2 For the matrices

$$A_1 = \begin{bmatrix} 2 & -1 & & \\ -1 & \ddots & \ddots & \\ & \ddots & \ddots & -1 \\ & & -1 & 2 \end{bmatrix}, \quad A_2 = \begin{bmatrix} 4 & -1 & & \\ -1 & \ddots & \ddots & \\ & \ddots & \ddots & -1 \\ & & -1 & 4 \end{bmatrix}$$

we have that A_1 is diagonally dominant but not strictly diagonally dominant and A_2 is strictly diagonally dominant.

Remarks 9.3

(i) If A is strictly diagonally dominant, then $a_{ii} \neq 0$ for $i = 1, 2, \ldots, n$ and D is regular. For the iteration matrix $M^J = -D^{-1}(A - D)$ of the associated Jacobi method, $m_{ii}^J = 0$ holds and thus due to the diagonal dominance for $i = 1, 2, \ldots, n$

$$\sum_{j=1}^{n} |m_{ij}^J| = \sum_{j=1,\ldots,n,\ j \neq i} \frac{|a_{ij}|}{|a_{ii}|} = \frac{1}{|a_{ii}|} \sum_{j=1,\ldots,n,\ j \neq i} |a_{ij}| < 1.$$

This means $\|M^J\|_\infty < 1$ and thus $\rho(M^J) < 1$.

(ii) For strictly diagonally dominant matrices, $\rho(M^{GS}) \leq \rho(M^J)$ holds.

Strict diagonal dominance is generally too restrictive a condition.

Definition 9.5 The matrix $A \in \mathbb{R}^{n \times n}$ is called *reducible*, if disjoint, non-empty index sets $I, J \subset \{1, 2, \ldots, n\}$ exist, such that $I \cup J = \{1, 2, \ldots, n\}$ and $a_{ij} = 0$ for all pairs $(i, j) \in I \times J$. Otherwise, A is called *irreducible*.

Example 9.3

(i) The matrix

$$A = \begin{bmatrix} 1 & 0 & 2 \\ 3 & 4 & 5 \\ 6 & 0 & 7 \end{bmatrix}$$

is reducible with $I = \{1, 3\}$ and $J = \{2\}$.

(ii) The band matrices from Example 9.2 are irreducible, because from $i \in I$ it also follows $i + 1 \in I$ for $1 \le i \le n - 1$ as well as $i - 1 \in I$ for $2 \le i \le n$ and thus $I = \{1, 2, \ldots, n\}$ or $I = \emptyset$.

Remark 9.4 For reducible matrices, the solution of the linear system $Ax = b$ can be decomposed into smaller systems. If for $X, Y \subset \{1, 2, \ldots, n\}$ the submatrix A_{XY} is defined by $A_{XY} = (a_{ij})_{i \in X, j \in Y}$ and the subvector x_Y by $x_Y = (x_k)_{k \in Y}$, then we have $A_{IJ} = 0$ and thus $A_{II} x_I = b_I$ and $A_{JJ} x_J = b_J - A_{JI} x_I$.

Lemma 9.1 *If M is irreducible and diagonally dominant, then M is regular with $m_{ii} \ne 0$ for $i = 1, 2, \ldots, n$.*

Proof If M is not regular, then there exists $x \in \mathbb{R}^n \setminus \{0\}$ with $Mx = 0$ and from the i-th row of the identity it follows

$$|m_{ii} x_i| \le \sum_{j=1,\ldots,n,\, j \ne i} |m_{ij}||x_j|.$$

Define $I = \{i : |x_i| = \|x\|_\infty\}$ and $J = \{j : |x_j| < \|x\|_\infty\}$. Then $I \ne \emptyset$ and $I \cup J = \{1, 2, \ldots, n\}$. Also, $J \ne \emptyset$, because otherwise $|x_j| = \|x\|_\infty$ for $j = 1, 2, \ldots, n$ and thus

$$|m_{ii}| \le \sum_{j=1,\ldots,n,\, j \ne i} |m_{ij}|$$

would hold, which contradicts the diagonal dominance, which guarantees strict inequality in the opposite direction for an i_0. Therefore, $J \ne \emptyset$ and due to the irreducibility there exist $i \in I$ and $j \in J$ with $m_{ij} \ne 0$ and thus

$$|m_{ii}| \le \sum_{j=1,\ldots,n,\, j \ne i} |m_{ij}| \frac{|x_j|}{\|x\|_\infty} < \sum_{j=1,\ldots,n,\, j \ne i} |m_{ij}|$$

contradicting the diagonal dominance of M. Consequently, M is regular. The regularity and the diagonal dominance of M imply $m_{ii} \ne 0$ for $i = 1, 2, \ldots, n$, because otherwise a row of M would be identically zero, which would contradict the regularity of M. □

9.6 Convergence

The preceding lemma allows us to prove the convergence of the Jacobi and Gauss-Seidel methods.

Proposition 9.3 *If A is irreducible and diagonally dominant, then the Jacobi and Gauss-Seidel methods are feasible and convergent, that is M^J and M^{GS} are well-defined and satisfy $\rho(M^J) < 1$ and $\rho(M^{GS}) < 1$.*

Proof

(i) According to the previous lemma, $a_{ii} \neq 0$ for $i = 1, 2, \ldots, n$ and thus $M^J = -D^{-1}(A - D)$ is well-defined. We show, that $M^J - \mu I_n$ for all $\mu \in \mathbb{C}$ with $|\mu| \geq 1$ is regular, so that $\rho(M^J) < 1$ follows. Since irreducibility is independent of the diagonal elements of a matrix, A as well as $A - D$ and $M^J = -D^{-1}(A - D)$ are irreducible. Likewise, $M = M^J - \mu I_n$ is irreducible. With the diagonal dominance of A, it follows for $i = 1, 2, \ldots, n$, that

$$\sum_{j=1,\ldots,n,\ j \neq i} |m_{ij}| = \sum_{j=1,\ldots,n,\ j \neq i} |m_{ij}^J| = \sum_{j=1,\ldots,n,\ j \neq i} \frac{|a_{ij}|}{|a_{ii}|} \leq 1 \leq |\mu| = |m_{ii}|,$$

where the inequality is strict for an $i_0 \in \{1, 2, \ldots, n\}$. Consequently, M is diagonally dominant for every $\mu \in \mathbb{C}$ with $|\mu| \geq 1$ and together with the irreducibility, the regularity of M follows.

(ii) Again, $a_{ii} \neq 0$ for $i = 1, 2, \ldots, n$ implies that $M^{GS} = -(L + D)^{-1}U$ is well-defined. For $\mu \in \mathbb{C}$ with $|\mu| \geq 1$ let $M = M^{GS} - \mu I_n$. Since $L + D$ is regular, M is regular if and only if

$$\widetilde{M} = -(L + D)M = -(L + D)\bigl(-(L + D)^{-1}U - \mu I_n\bigr) = U + \mu L + \mu D$$

is regular. With $A = U + L + D$, the matrix $\widetilde{M}$ is also irreducible. Furthermore, $\widetilde{M}$ is diagonally dominant, because for $i = 1, 2, \ldots, n$ due to the diagonal dominance of A, we have that

$$\sum_{j=1,\ldots,n,\ j \neq i} |\widetilde{m}_{ij}| = |\mu| \sum_{j=1}^{i-1} |a_{ij}| + \sum_{j=i+1}^{n} |a_{ij}| \leq |\mu| \sum_{j=1,\ldots,n,\ j \neq i} |a_{ij}|$$

$$\leq |\mu||a_{ii}| = |\widetilde{m}_{ii}|,$$

where strict inequality holds for an $i_0 \in \{1, 2, \ldots, n\}$. So $\widetilde{M}$ is diagonally dominant and the preceding lemma implies the regularity of $\widetilde{M}$. □

Remark 9.5

(i) In the case of convergence, often a few iteration steps lead to a good approximate solution. Since each iteration step in the Jacobi and Gauss-Seidel method requires $\mathcal{O}(n^2)$ many operations, this can reduce the typical effort of $\mathcal{O}(n^3)$ of direct solution methods like Gauss elimination.

(ii) The conditions are sufficient but not necessary, as for example regular diagonal matrices are diagonally dominant and reducible, but the Jacobi and Gauss-

9.6 Convergence

Seidel method determine the exact solution in one step. In general, both conditions are needed, because for the matrices

$$A_2 = \begin{bmatrix} 1 & 1 \\ -1 & 1 \end{bmatrix}, \quad A_3 = \begin{bmatrix} 1 & 1 & 0 \\ -1 & 1 & 0 \\ 0 & 0 & 1 \end{bmatrix},$$

we have that A_2 is irreducible but not diagonally dominant, while A_3 is diagonally dominant but not irreducible. In both cases, the iteration matrix of the Jacobi method realises a rotation by $\pi/2$ in the (x, y)-plane, so in general, convergence does not have to occur.

An alternative proof for the convergence of the Gauss-Seidel method can be given for symmetric, positive definite matrices.

Proposition 9.4 *If $A \in \mathbb{R}^{n \times n}$ is symmetric and positive definite, then $\rho(M^{GS}) < 1$.*

Proof First, we obtain with $U = A - (D + L)$ and $Q = 2A^{-1}(D + L) - I_n$, that for $M^{GS} = -(D + L)^{-1} U$ we have

$$M^{GS} = I_n - (D + L)^{-1} A = I_n - 2(2A^{-1}(D + L))^{-1}$$
$$= I_n - 2(Q + I_n)^{-1} = (Q - I_n)(Q + I_n)^{-1}.$$

From this identity, it follows that $\mu = 1$ is not an eigenvalue of M^{GS}, and if $\mu \in \mathbb{C} \setminus \{1\}$ is an eigenvalue of M^{GS}, then there exists an eigenvalue $\lambda \in \mathbb{C} \setminus \{-1\}$ of Q, such that $\mu = (\lambda - 1)/(\lambda + 1)$ holds. If $z \in \mathbb{C}^n$ is an eigenvector associated to the eigenvalue $\lambda \in \mathbb{C}$ of Q, then

$$Qz = \lambda z \quad \Longleftrightarrow \quad \lambda A z = 2(D + L)z - Az.$$

We multiply the second equation from the left with $\bar{z}^\mathsf{T}$ and use the fact that the symmetry and positive definiteness of A and D imply that

$$\bar{z}^\mathsf{T} A z > 0, \quad \bar{z}^\mathsf{T} D z > 0$$

hold, in particular both expressions are real-valued. With $a^\mathsf{T} b = b^\mathsf{T} a$ it follows

$$2 \operatorname{Re}(\bar{z}^\mathsf{T} L z) = \bar{z}^\mathsf{T} L z + z^\mathsf{T} L \bar{z} = \bar{z}^\mathsf{T} L z + (L^\mathsf{T} z)^\mathsf{T} \bar{z} = \bar{z}^\mathsf{T} (L z + L^\mathsf{T} z).$$

Because $L^\mathsf{T} = U$ and $A = D + L + U$ it follows that

$$\bar{z}^\mathsf{T} A z \operatorname{Re}(\lambda) = \operatorname{Re}(\lambda \bar{z}^\mathsf{T} A z) = 2\bar{z}^\mathsf{T} D z + 2 \operatorname{Re}(\bar{z}^\mathsf{T} L z) - \bar{z}^\mathsf{T} A z$$
$$= \bar{z}^\mathsf{T} (2D + L + U)z - \bar{z}^\mathsf{T} A z = \bar{z}^\mathsf{T} D z > 0,$$

and thus $\mathrm{Re}(\lambda) > 0$. Any such number $\lambda \in \mathbb{C}$ has a strictly smaller distance to the point 1 than to the point -1, that means we have $|\lambda - 1| < |\lambda - (-1)| = |\lambda + 1|$, and thus for every eigenvalue $\mu = (\lambda - 1)/(\lambda + 1)$ of M^{GS}, that $|\mu| < 1$ or $\rho(M^{GS}) < 1$ holds. □

Remark 9.6 The argumentation of the proof goes back to Ostrowski and Reich and can be generalised to a family of so-called *relaxation methods*.

9.7 Learning Objectives, Quiz and Application

You should be able to derive iterative methods for solving linear systems and be able to demonstrate their advantages compared to other methods. You should be able to name sufficient conditions for the convergence of linear iteration methods and to explain structural properties of matrices that ensure the convergence of the methods and illustrate their significance.

Quiz 9.1 Decide for each of the following statements whether it is true or false. You should be able to justify your answer.

If $A \in \mathbb{R}^{n \times n}$ is irreducible and $D \in \mathbb{R}^{n \times n}$ is a diagonal matrix, then $A - D$ is also irreducible.
If $A \in \mathbb{R}^{n \times n}$ is diagonally dominant, then A is regular.
For symmetric matrices, the Jacobi and Gauss-Seidel methods agree.
The property $a_{ii} \neq 0$ of a matrix $A \in \mathbb{R}^{n \times n}$ is necessary for the well-definedness of the Jacobi and Gauss-Seidel methods.
If $A - \mu I_n$ is regular for all $\mu \in \mathbb{C}$ with $|\mu| \geq 1/4$, then $|\lambda| < 1/4$ holds for all eigenvalues of A.

Application 9.1 In applications such as the description of the elastic behaviour of truss structures, regular matrices $A \in \mathbb{R}^{n \times n}$ occur, in which many entries vanish. In these cases, it is often sensible to implement an iterative method without completely storing the matrix A. Show that the Jacobi and the Gauss-Seidel method can be written in the form

$$x_i^{k+1} = a_{ii}^{-1}\left(b_i - \sum_{j \neq i} a_{ij} x_j^k\right),$$

respectively

$$x_i^{k+1} = a_{ii}^{-1}\left(b_i - \sum_{j<i} a_{ij} x_j^{k+1} - \sum_{j>i} a_{ij} x_j^k\right)$$

for $i = 1, 2, \ldots, n$. Simplify these formulas for the case of matrices with finite bandwidth $w > 0$, that is, in the case that $a_{ij} = 0$ for $|i - j| > w$.

Part II
Numerical Analysis

Chapter 10
General Condition Number and Floating Point Numbers

10.1 Conditioning

We consider the effects of perturbations in the evaluation of a mathematical operation $\phi(x)$, which is defined by a mapping $\phi : X \to Y$ between normed vector spaces. Here, perturbations $\widetilde{x}$ of x are additively represented as the sum $\widetilde{x} = x + \Delta x$ with $\Delta x = \widetilde{x} - x$. The following definition from [2] generalises the concept of the condition number for general mathematical operations.

Definition 10.1 The *(relative) condition number* $\kappa_\phi(x)$ of the function $\phi : X \to Y$ at $x \neq 0$ with $\phi(x) \neq 0$ is the infimum of all $\kappa \geq 0$, for which a $\delta > 0$ exists, such that

$$\varepsilon_\phi = \frac{\|\phi(x + \Delta x) - \phi(x)\|}{\|\phi(x)\|} \leq \kappa \frac{\|\Delta x\|}{\|x\|} = \kappa \varepsilon_x$$

for all $\Delta x \in X$ with $\varepsilon_x = \|\Delta x\|/\|x\| \leq \delta$. The operation $\phi(x)$ is called *ill conditioned*, if $\kappa_\phi(x) \gg 1$, and *ill-posed*, if $\kappa_\phi(x)$ is not defined. It is called *well conditioned* otherwise.

Examples 10.1

(i) The operation defined by $\phi(x) = |x - 1|^s + 1$ is well conditioned at $x = 1$ exactly for $s \geq 1$, that is, when left- and right-hand derivatives at $x = 1$ exist.
(ii) The operation defined by $\phi(x) = a|x - 1| + 1$ is ill conditioned at $x = 1$ when $a \gg 1$.
(iii) The operation defined by the discontinuous function $\phi(x) = \text{sign}(x - 1) + 2$ is ill-posed at $x = 1$.

Fig. 10.1 Well conditioned, ill conditioned and ill-posed operations $\phi(x)$

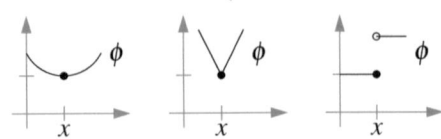

The examples are sketched in Fig. 10.1.

Proposition 10.1 *If ϕ is differentiable at x, then*

$$\kappa_\phi(x) = \frac{\|D\phi(x)\|\|x\|}{\|\phi(x)\|}.$$

Proof It holds

$$\phi(x + \Delta x) - \phi(x) = D\phi(x)[\Delta x] + \psi(\Delta x),$$

with a function ψ, which fulfils $\psi(\Delta x)/\|\Delta x\| \to 0$ for $\Delta x \to 0$. Thus, for every $\varepsilon > 0$ there exists a $\delta > 0$, such that for all Δx with $\|\Delta x\|/\|x\| \le \delta$ it holds

$$\left\| \frac{\phi(x + \Delta x) - \phi(x)}{\|\Delta x\|} - \frac{D\phi(x)[\Delta x]}{\|\Delta x\|} \right\| \le \varepsilon.$$

From this follows

$$\frac{\|\phi(x + \Delta x) - \phi(x)\|}{\|\phi(x)\|} \le \left(\varepsilon + \frac{\|D\phi(x)[\Delta x]\|}{\|\Delta x\|} \right) \frac{\|\Delta x\|}{\|\phi(x)\|}.$$

By definition of the operator norm, it holds $\|D\phi(x)[\Delta x]\| \le \|D\phi(x)\|\|\Delta x\|$, where equality occurs for suitable Δx. Since $\varepsilon > 0$ can be chosen arbitrarily small, the assertion follows. $\square$

In the case of systems of linear equations, the condition number is bounded by the special condition number of the matrix.

Remarks 10.1

(i) For $\phi(b) = A^{-1}b$ it holds $D\phi(b) = A^{-1}$ and with the identity $\|b\| = \|A(A^{-1}b)\| \le \|A\|\|A^{-1}b\|$ implies that

$$\kappa_\phi(b) = \frac{\|A^{-1}\|}{\|A^{-1}b\|}\|b\| \le \|A^{-1}\|\|A\| = \text{cond}(A).$$

Furthermore, there exists a $b \in \mathbb{R}^n$ such that equality holds.

(ii) To investigate the influences of perturbations of the matrix A, we consider the mapping $\phi(A) = A^{-1}b$. From the constancy of $A \mapsto A\phi(A) = b$ it follows that $D\phi(A)[E] = -A^{-1}EA^{-1}b$ and with the estimate $\|D\phi(A)\| \le$

$\|A^{-1}\| \|A^{-1}b\|$ we get

$$\kappa_\phi(A) \leq \frac{\|A^{-1}\| \|A^{-1}b\| \|A\|}{\|A^{-1}b\|} = \text{cond}(A),$$

that is, errors in A are also amplified by the factor $\text{cond}(A)$.

Cancellation effects are also captured by the condition number.

Examples 10.2

(i) For $\phi(x_1, x_2) = x_1 + x_2$ we have $D\phi(x_1, x_2) = [1, 1]$ and thus

$$\kappa_\phi(x_1, x_2) = \frac{\|[1, 1]\|_1 \|(x_1, x_2)\|_1}{|x_1 + x_2|} = \frac{|x_1| + |x_2|}{|x_1 + x_2|},$$

so the operation is ill-conditioned if $x_1 \approx -x_2$, that is, when cancellation effects can occur.

(ii) Intuitively, standing a pen upright is an ill conditioned problem, while standing a can is generally well conditioned.

10.2 Floating Point Numbers

On digital computers, only a finite number of numbers are available, which are determined according to certain rules. We follow the presentation in [9].

Definition 10.2 For a *base* $b \geq 2$, a *precision* $p \geq 1$ and *exponent limits* $e_{min} \leq e_{max}$ with $b, p, e_{min}, e_{max} \in \mathbb{Z}$ the set of *floating point* or *machine numbers* is defined by

$$G = \left\{ \pm m b^{e-p} : m, e \in \mathbb{Z},\ 0 \leq m \leq b^p - 1,\ e_{min} \leq e \leq e_{max} \right\}.$$

A floating point number $g \in G$ is called *normalised*, if $m \geq b^{p-1}$, and we let G_{nor} denote the set of normalised floating point numbers. In the cases $b = 2$ and $b = 10$ we speak of the *binary* and *decimal system*.

Examples 10.3

(i) For $b = 2$, $e_{min} = 0$, $e_{max} = 3$, $p = 2$ the normalised floating point numbers are given by $\pm \{2/4, 3/4, 2/2, 3/2, 2/1, 3/1, 2 \cdot 2, 3 \cdot 2\}$; the set of all floating point numbers additionally contains the numbers $\pm \{0, 1/4\}$.

(ii) For $b = 10$, $p = 3$, $e_{min} = -2$ and $e_{max} = 2$ G consists of all numbers of the form $\pm m \cdot 10^{-r}$ with $0 \leq m \leq 999$ and $1 \leq r \leq 5$, for example

$$-783 \cdot 10^{-5},\quad 400 \cdot 10^{-3},\quad 40 \cdot 10^{-2},$$

where only the first two numbers are normalised.

Remarks 10.2

(i) Each floating point number $g \in G$ can be represented as a b-adic sum, that is, it holds

$$g = \pm b^e \left(d_1 b^{-1} + d_2 b^{-2} + \cdots + d_p b^{-p} \right)$$

with digits $d_1, d_2, \ldots, d_p \in \{0, 1, \ldots, b-1\}$ and $e_{min} \leq e \leq e_{max}$. For normalised floating point numbers, this representation is uniquely defined with $d_1 \neq 0$.

(ii) For normalised floating point numbers g, $g_{min} \leq |g| \leq g_{max}$ with $g_{min} = b^{e_{min}-1}$ and $g_{max} = b^{e_{max}}(1 - b^{-p})$.

(iii) For $b = 10$, $g = \pm 10^e \cdot 0.d_1 d_2 \ldots d_p$ and the decimal point is floating depending on e.

Example 10.4 In the 754R standard of the *Institute of Electrical and Electronics Engineers* (IEEE), the formats *single* and *double precision* are defined by

$$b = 2, \ e_{min} = -125, \ e_{max} = 128, \ p = 24,$$
$$b = 2, \ e_{min} = -1021, \ e_{max} = 1024, \ p = 53.$$

The relative error in the approximation of real numbers by machine numbers is limited by the so-called machine precision.

Definition 10.3 The *machine precision* is defined by the smallest number $g_{eps} \in G_{nor}$, for which $g_{eps} > 1$, as $eps = g_{eps} - 1 = \min_{g>1} g - 1$

Remark 10.3 It holds that $g_{eps} = b^1(b^{-1} + 0 \cdot b^{-2} + \cdots + 0 \cdot b^{-p+1} + b^{-p}) = 1 + b^{1-p}$ and thus $eps = b^{1-p}$.

Examples 10.5

(i) The normalised floating point numbers between b^e and b^{e+1} for $e_{min} - 1 \leq e \leq e_{max} - 1$ are uniformly arranged at a distance of $b^e eps$, see Fig. 10.2.
(ii) For the IEEE-754R formats *single* and *double*, $eps = 2^{-23} \approx 1.2 \cdot 10^{-7}$ and $eps = 2^{-52} \approx 2.2 \cdot 10^{-16}$ respectively.

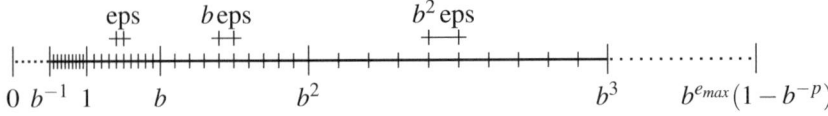

Fig. 10.2 Schematic representation of the arrangement of machine numbers

10.3 Rounding

Rounding functions approximate real numbers by machine numbers.

Definition 10.4 For a set of machine numbers G, a mapping rd : $[-g_{max}, g_{max}] \to G_{nor}$ is called a *rounding function*, if for every real number $x \in [-g_{max}, g_{max}]$, $|x - \text{rd}(x)| = \min_{g \in G_{nor}} |x - g|$.

Remarks 10.4

(i) If x is exactly between two machine numbers, the IEEE standards select the machine number whose last digit is even.
(ii) We speak of *overflow* and *underflow*, when $|x| > g_{max}$ or $|x| < g_{min}$ respectively. In the second case, it is usually rounded to zero, but a large error occurs. In the denormalised IEEE standard, additional machine numbers are used in a neighbourhood of zero.
(iii) In addition to the numbers in G, there is usually also the value NaN, which is used for undefined expressions such as $1/0$ and stands for *not-a-number*.

Lemma 10.1 *For every $x \in \mathbb{R}$ with $|x| \in [g_{min}, g_{max}]$,*

$$\frac{|x - \text{rd}(x)|}{|x|} \leq \frac{1}{2}\text{eps},$$

that is, there exists a $\delta \in \mathbb{R}$ with $|\delta| \leq \text{eps}/2$ and $\text{rd}(x) = (1+\delta)x$.

Proof Since the normalised floating point numbers are uniformly spaced in every interval $[b^e, b^{e+1}]$ at a distance $b^e \text{eps}$, there exists a $\ell \geq 0$ with $b^e + \ell b^e \text{eps} \leq x \leq b^e + (\ell+1)b^e \text{eps}$ and let g be the upper or lower bound with $|x - g| \leq (1/2)b^e \text{eps}$. Since $|x| \geq b^e$ the assertion follows. □

Definition 10.5 The *standard model of floating point arithmetics* requires that for all $x, y \in \mathbb{R}$ with $|x|, |y| \in [g_{min}, g_{max}]$ and every standard arithmetic operation op $\in \{+, -, *, :\}$ with $|x \text{ op } y| \in [g_{min}, g_{max}]$ as well as their numerical realisation $\text{op}_G : G \times G \to G$ there exists a $\delta \in \mathbb{R}$ with $|\delta| \leq \text{eps}/2$ such that

$$\text{rd}(x) \text{ op}_G \text{ rd}(y) = (x \text{ op } y)(1 + \delta).$$

Remarks 10.5

(i) It is often further simplified to assume that $\text{rd}(x) \text{ op}_G \text{ rd}(y) = \text{rd}(x \text{ op } y)$.
(ii) The standard model is fulfilled by the IEEE standards, which are implemented on common computers.
(iii) In many operations, rounding errors can accumulate and become relevant. This is also referred to as error propagation.

10.4 Stability

Let $\tilde{\phi} : X \to Y$ denote a numerical method, that is a finite sequence of operations subject to rounding errors. When rounding the argument x it holds

$$\phi(x) - \tilde{\phi}(x + \Delta x) = \big(\phi(x) - \phi(x + \Delta x)\big) + \big(\phi(x + \Delta x) - \tilde{\phi}(x + \Delta x)\big),$$

where the first term on the right-hand side is controlled by the conditioning of ϕ and the second describes the stability of the method. The latter depends on the theoretically freely selectable rounding accuracy ε and in the following we also write $\tilde{\phi}_\varepsilon$ instead of $\tilde{\phi}$, to indicate this. If instead of the impact of erroneous operations the approximation of an operation ϕ by an approximation $\tilde{\phi}$ is considered, for example when approximating a series by a finite sum, the difference $\phi - \tilde{\phi}$ is also referred to as the *consistency term*. The following definition from [2] specifies the concept of stability in the present context. It is measured relative to the condition number.

Definition 10.6 The *stability indicator* $\sigma_{\tilde{\phi}}(x)$ of the numerical method $\tilde{\phi}$ is the infimum of all $\sigma \geq 0$ for which a $\delta > 0$ exists, such that

$$\frac{\|\phi(x) - \tilde{\phi}_\varepsilon(x)\|}{\|\phi(x)\|} \leq \sigma \kappa_\phi(x)\varepsilon$$

for every $0 \leq \varepsilon \leq \delta$. The method $\tilde{\phi}$ is called *unstable*, if $\sigma_{\tilde{\phi}}(x) \gg 1$. Otherwise, the method is called *(forward-) stable*.

The number $\sigma_{\tilde{\phi}}(x)$ describes the error amplification caused by the method. Precise stability analyses are generally extremely complex. The following concepts are usually applied in practice.

Remark 10.6 In the *linear forward analysis*, each intermediate result z_i is considered as subject to rounding and replaced by $(1 + \varepsilon_i)z_i$ with $|\varepsilon_i| \leq \varepsilon$. Products of the form $\varepsilon_i \varepsilon_j$ are neglected in the calculation. The division is linearised with respect to ε, that is for example

$$(x(1+\varepsilon))^{-1} \approx (1 - \varepsilon)x^{-1}.$$

A simple to check, but very restrictive stability criterion is the so-called backward stability, where the method error is represented by a perturbation of the argument x.

Definition 10.7 The *backward stability indicator* $\rho_{\tilde{\phi}}(x)$ of an operation $\tilde{\phi}_\varepsilon : X \to Y$ at x is the infimum of all $\rho \geq 0$, for which a $\delta > 0$ exists, such that for all $0 \leq \varepsilon \leq \delta$ there is a $\Delta x \in X$ with $\phi(x + \Delta x) = \tilde{\phi}_\varepsilon(x)$ and

$$\frac{\|\phi^{-1}(\phi(x)) - \phi^{-1}(\tilde{\phi}_\varepsilon(x))\|}{\|\phi^{-1}(\phi(x))\|} = \frac{\|\Delta x\|}{\|x\|} \leq \rho\varepsilon.$$

The method is called *backward stable*, provided that $\rho_{\tilde{\phi}}(x) \gg 1$ does not hold.

Remark 10.7 If $\widetilde{\phi}_\varepsilon$ is backward stable, then $\widetilde{\phi}_\varepsilon$ is stable, because it holds

$$\frac{1}{\kappa_\phi(x)\varepsilon} \frac{\|\widetilde{\phi}_\varepsilon(x) - \phi(x)\|}{\|\phi(x)\|} = \frac{1}{\kappa_\phi(x)\varepsilon} \frac{\|\phi(x+\Delta x) - \phi(x)\|}{\|\phi(x)\|} \leq \frac{1}{\varepsilon} \frac{\|\Delta x\|}{\|x\|} \leq \rho_{\widetilde{\phi}}(x)$$

and thus by definition of the stability indicator $\sigma_{\widetilde{\phi}}(x) \leq \rho_{\widetilde{\phi}}(x)$.

Examples 10.6

(i) The floating-point realisation of the operation $\phi(x) = 1 + x$ is not backward stable for small numbers x, because it holds

$$|\phi^{-1}(1 + x + \Delta x) - \phi^{-1}(1 + x)|/|\phi^{-1}(1 + x)| = |\Delta x|/|x| \gg |\Delta x|.$$

Obviously, $\widetilde{\phi} = \phi$ is however stable for small numbers x.

(ii) Cramer's rule is not backward stable but forward stable for systems of linear equations of dimension 2.

10.5 Learning Objectives, Quiz and Application

You should be able to understand the general condition number of a mathematical operation and illustrate it with examples. You should moreover be able to clarify the importance of floating point numbers and their accuracy for numerical calculations and motivate the concept of the stability indicator of a numerical method.

Quiz 10.1 Decide for each of the following statements whether it is true or false. You should be able to justify your answer.

If $\phi : \mathbb{R}^n \to \mathbb{R}^m$ is Lipschitz continuous and $\|\phi(x)\| \geq c\|x\|$ for all $x \in \mathbb{R}^n$ with $c > 0$, then ϕ is well conditioned.	
For $b = 10$, $p = 4$, $e_{min} = -3$, $e_{max} = 3$, $-13 \cdot 10^{-2}$ is a normalised floating point number.	
If $\text{rd}(x) = 0$ and $e_{min} \leq 2 - p$, then $|x| < \text{eps}$.	
The machine precision eps limits the absolute error in the approximation of real numbers by floating point numbers.	
If ϕ is ill conditioned, then every numerical method $\widetilde{\phi}$ is stable.	

Fig. 10.3 Conditioning examination of a billiard shot

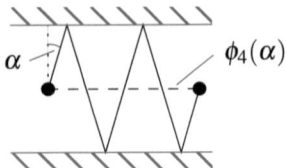

Application 10.1 On a billiard table of width 1, a ball is placed on the centre line and played at an angle $\alpha \in (0, \pi/2)$ to it. For a fixed number $n \in \mathbb{N}$, let $\phi_n(\alpha)$ be the distance to the starting position on the centre line, with which the ball crosses this after n wall contacts, see Fig. 10.3. Derive a formula for $\phi_n(\alpha)$, determine the condition number $\kappa_{\phi_n}(\alpha)$ and interpret this.

Chapter 11
Polynomial Interpolation

11.1 Lagrange Interpolation

Interpolation refers to the approximation of a given function in a finite-dimensional space of functions, such as polynomials of limited degree. Since only the coefficients with respect to a basis need to be stored, this is advantageous for the numerical processing or tabular recording of a function. In the following, let

$$\mathscr{P}_n = \Big\{ \sum_{i=0}^{n} a_i x^i : a_0, a_1, \ldots, a_n \in \mathbb{R} \Big\}$$

be the vector space of polynomials of maximum degree $n \in \mathbb{N}_0$.

Remark 11.1 We have that $\dim \mathscr{P}_n = n+1$ and the monomials $(x^0, x^1, \ldots, x^n)$ form a basis of $\mathscr{P}_n$. Here, x^0 denotes the constant function with value 1.

Definition 11.1 The *Lagrange interpolation task* seeks for given, pairwise different *nodes* $a \le x_0 < x_1 < \cdots < x_n \le b$ and associated *values* $y_0, y_1, \ldots, y_n$ a polynomial $p \in \mathscr{P}_n$ with $p(x_i) = y_i$ for $i = 0, 1, \ldots, n$, see Fig. 11.1.

The interpolation task can be solved directly with a special basis of $\mathscr{P}_n$.

Definition 11.2 The *Lagrange polynomials* $L_0, L_1, \ldots, L_n \in \mathscr{P}_n$ associated with the nodes $x_0 < x_1 < \cdots < x_n$ are defined by

$$L_i(x) = \prod_{\substack{j=0 \\ j \ne i}}^{n} \frac{x - x_j}{x_i - x_j} = \frac{(x - x_1)}{(x_i - x_1)} \cdots \frac{(x - x_{i-1})}{(x_i - x_{i-1})} \frac{(x - x_{i+1})}{(x_i - x_{i+1})} \cdots \frac{(x - x_n)}{(x_i - x_n)}.$$

Remark 11.2 We have that $L_i(x_j) = \delta_{ij}$ for all $0 \le i, j \le n$, see Fig. 11.2.

Fig. 11.1 Lagrange interpolation task

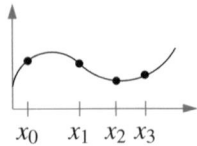

Fig. 11.2 The interpolation task can be solved with the Lagrange polynomials L_i

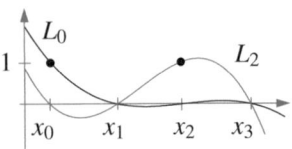

Proposition 11.1 *The Lagrange interpolation task is uniquely solved by*

$$p = \sum_{i=0}^{n} y_i L_i$$

This polynomial is referred to as the (Lagrange) interpolation polynomial.

Proof From $L_i(x_j) = \delta_{ij}$ it follows that $p(x_j) = y_j$ for $j = 0, 1, \ldots, n$, i.e. p is a solution. If $q \in \mathscr{P}_n$ is another solution, then for $r = p - q \in \mathscr{P}_n$, $r(x_j) = 0$ for $j = 0, 1, \ldots, n$, i.e. r has $n+1$ zeros from which $r = 0$ and thus $p = q$ follows. □

Remark 11.3 If $(q_0, q_1, \ldots, q_n)$ is a basis of $\mathscr{P}_n$, then the solution of the Lagrange interpolation task can be represented as a linear combination $p = \sum_{i=0}^{n} c_i q_i$, where the coefficient vector $c = [c_0, \ldots, c_n]^T$ solves the regular system of linear equations $Vc = y$ with $y = [y_0, y_1, \ldots, y_n]^T$ and the *Vandermonde matrix* $V \in \mathbb{R}^{(n+1) \times (n+1)}$ with entries $v_{ij} = q_i(x_j)$. For the choice of Lagrange polynomials, it follows that $V = I_n$. If, on the other hand, one chooses the monomial basis $(x^0, x^1, \ldots, x^n)$, then V is generally ill conditioned.

11.2 Interpolation Error

Often the values $y_0, y_1, \ldots, y_n$ represent function values of a function f and one is interested in the size of the error $f - p$.

Proposition 11.2 *Let $f \in C^{n+1}([a, b])$ and let $f(x_i) = y_i$ for $i = 0, 1, \ldots, n$. For the solution $p \in \mathscr{P}_n$ of the Lagrange interpolation problem and every $x \in [a, b]$ there exists a $\xi \in [a, b]$, such that*

$$f(x) - p(x) = \frac{f^{(n+1)}(\xi)}{(n+1)!} \prod_{j=0}^{n}(x - x_j).$$

Proof Let $x \in [a, b]$. If $x \in \{x_0, x_1, \ldots, x_n\}$, then the statement is clear, so assume $x \neq x_i$ for $i = 0, 1, \ldots, n$ in the following. With the node polynomial

$$w(y) = \prod_{j=0}^{n}(y - x_j) = y^{n+1} + a_n y^n + \cdots + a_0 \in \mathscr{P}_{n+1}$$

for $y \in [a, b]$ let

$$F(y) = \big(f(x) - p(x)\big)w(y) - \big(f(y) - p(y)\big)w(x).$$

Then $F(x_i) = 0$ for $i = 0, 1, \ldots, n$ and $F(x) = 0$, that is F has at least $n + 2$ different zeros. According to Rolle's theorem, F' has a zero between two zeros of F, thus F' has at least $n + 1$ different zeros. The repeated application of this argument shows that the derivative $F^{(n+1)}$ has at least one zero $\xi \in [a, b]$. This gives

$$0 = F^{(n+1)}(\xi) = \big(f(x) - p(x)\big)(n + 1)! - f^{(n+1)}(\xi)w(x)$$

and this is the claimed identity. □

Corollary 11.1 *For the interpolation error we have*

$$\|f - p\|_{C^0([a,b])} \leq \frac{\|f^{(n+1)}\|_{C^0([a,b])}}{(n+1)!}(b-a)^{n+1}.$$

The corollary implies that the Lagrange interpolation polynomials converge uniformly to f as $n \to \infty$, provided the distance $b - a$ is reduced or the number of nodes is increased and the derivatives of f do not grow too rapidly. However, the latter is generally not the case.

Example 11.1 Let $f : [-1, 1] \to \mathbb{R}$ be defined by $f(x) = (1 + 25x^2)^{-1}$ and let the nodes be chosen equidistantly, i.e. $x_i = -1 + 2i/n$ for $i = 0, 1, \ldots, n$. Then the sequence of Lagrange interpolation polynomials $(p_n)_{n \in \mathbb{N}}$ of f does not converge pointwise to f as $n \to \infty$, since the expression $\|f^{(n+1)}\|_{C^0([-1,1])}$ grows too rapidly. The interpolation polynomial is shown in Fig. 11.6.

11.3 Neville's Algorithm

The direct evaluation of the interpolation polynomial at a point $x \in [a, b]$ is expensive and potentially unstable. Neville's algorithm allows a calculation of $p(x)$ with $\mathcal{O}(n^2)$ computational operations. We follow the presentations in [9, 10].

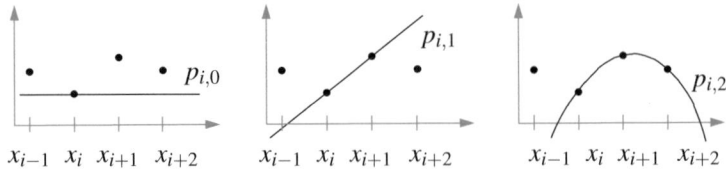

Fig. 11.3 The interpolation polynomial is constructed step by step in the Neville scheme

Definition 11.3 For $n + 1$ nodes and values $(x_0, y_0), (x_1, y_1), \ldots, (x_n, y_n)$ as well as $0 \leq j \leq n$ and $0 \leq i \leq n - j$, let $p_{i,j} \in \mathscr{P}_j$ be the uniquely determined Lagrange interpolation polynomial with $p_{i,j}(x_k) = y_k$, $k = i, i+1, \ldots, i+j$, see Fig. 11.3.

Remark 11.4 We have that $p_{i,0}(x) = y_i$ for $i = 0, 1, \ldots, n$ and $p_{0,n}(x) = p(x)$ for $x \in [a, b]$.

Proposition 11.3 *The polynomials $p_{i,j}$ are given by $p_{i,0}(x) = y_i$ for $i = 0, 1, \ldots, n$ as well as*

$$p_{i,j}(x) = \frac{(x - x_i) p_{i+1,j-1}(x) - (x - x_{i+j}) p_{i,j-1}(x)}{x_{i+j} - x_i}$$

for $i = 0, 1, \ldots, n - j$ and $j = 1, 2, \ldots, n$.

Proof For $j = 0$ the statement holds by definition, so assume that it is true for $j - 1$ with $j \geq 1$. Let $0 \leq i \leq n - j$ and let $q(x)$ denote the right-hand side of the claimed identity for $p_{i,j}$. Since $p_{i+1,j-1}, p_{i,j-1} \in \mathscr{P}_{j-1}$, it follows that $q \in \mathscr{P}_j$. Moreover, $q(x_i) = p_{i,j-1}(x_i) = y_i$ and $q(x_{i+j}) = p_{i+1,j-1}(x_{i+j}) = y_{i+j}$. For $k = i+1, i+2, \ldots, i+j-1$, since $p_{i+1,j-1}(x_k) = p_{i,j-1}(x_k) = y_k$, it follows that

$$q(x_k) = \frac{(x_k - x_i) p_{i+1,j-1}(x_k) - (x_k - x_{i+j}) p_{i,j-1}(x_k)}{x_{i+j} - x_i}$$

$$= \frac{(x_k - x_i) y_k - (x_k - x_{i+j}) y_k}{x_{i+j} - x_i} = y_k.$$

The uniqueness of the interpolation polynomial implies $q = p_{i,j}$. □

Remark 11.5 The Neville scheme should not be implemented backwards in recursive form, as many quantities would be calculated multiple times. Instead, the values $p_{i,j}(x)$ should be evaluated successively forwards, which leads to a computational cost of $\mathcal{O}(n^2)$ and is illustrated in Fig. 11.4.

Remark 11.6 Closely connected with the Neville scheme is the method of *divided differences*, which determines the coefficients λ_j, $j = 0, 1, \ldots, n$, of the Lagrange

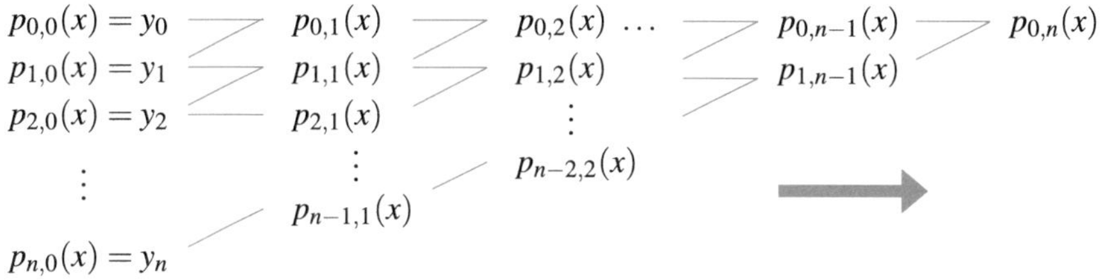

11.3 Neville's Algorithm

$$
\begin{array}{llll}
p_{0,0}(x) = y_0 & p_{0,1}(x) & p_{0,2}(x) \ldots & p_{0,n-1}(x) \quad p_{0,n}(x) \\
p_{1,0}(x) = y_1 & p_{1,1}(x) & p_{1,2}(x) & p_{1,n-1}(x) \\
p_{2,0}(x) = y_2 & p_{2,1}(x) & \vdots \\
\vdots & \vdots & p_{n-2,2}(x) \\
& p_{n-1,1}(x) \\
p_{n,0}(x) = y_n
\end{array}
$$

Fig. 11.4 Schematic representation of the Neville scheme; the evaluation is carried out from left to right

interpolation polynomial with respect to the *Newton basis* $(q_0, q_1, \ldots, q_n)$ defined by $q_0 = 1$ and

$$q_j(x) = \prod_{k=0}^{j-1} (x - x_k),$$

$j = 1, 2, \ldots, n$, that is $p(x) = \sum_{j=0}^{n} \lambda_j q_j(x)$. With the initialisation $y_{i,0} = y_i$, $i = 0, 1, \ldots, n$, and the iteration rule

$$y_{i,j} = \frac{y_{i+1,j-1} - y_{i,j-1}}{x_{i+j} - x_i}$$

for $1 \leq j \leq n$ and $0 \leq i \leq n - j$, we have that $\lambda_j = y_{0,j}$, $j = 0, 1, \ldots, n$. The evaluation of the interpolation polynomial is then efficiently done with effort $\mathcal{O}(n)$ via *Horner's scheme*, that is by means of the representation

$$p(x) = \lambda_0 + (x - x_0)\bigl[\lambda_1 + (x - x_1)\bigl[\lambda_2 + \ldots \bigl[\lambda_{n-1} + (x - x_{n-1})\lambda_n\bigr]\ldots\bigr]\bigr].$$

This type of evaluation of the interpolation polynomial has the useful property that additional interpolation points can be easily added. The scheme is also well suited when the value of the polynomial p is needed at several points.

Remark 11.7 An alternative approach to evaluating an interpolation polynomial p is obtained using *barycentric representations* of the basis functions. Letting

$$\gamma_i = \prod_{j=0,\ldots,n,\ j\neq i} (x_i - x_j)^{-1}, \quad w(x) = \prod_{j=0}^{n} (x - x_j),$$

one has $L_i(x) = \gamma_i w(x)(x - x_i)^{-1}$ for $x \neq x_i$, and correspondingly

$$p(x) = w(x) \sum_{i=0}^{n} (x - x_i)^{-1} \gamma_i y_i$$

for $x \notin \{x_0, x_1, \ldots, x_n\}$ and $p(x_i) = y_i$ otherwise. If $y_0 = y_1 = \cdots = y_n = 1$, then we have $1 = w(x) \sum_{i=0}^{n} \gamma_i (x - x_i)^{-1}$ which leads to

$$p(x) = \frac{\sum_{i=0}^{n} (x - x_i)^{-1} \gamma_i y_i}{\sum_{i=0}^{n} (x - x_i)^{-1} \gamma_i}.$$

These representations of p can be evaluated with $\mathcal{O}(n)$ operations provided that the weights $(\gamma_i)_{i=0,\ldots,n}$ have been pre-computed.

11.4 Chebyshev Nodes

One way to reduce the interpolation error in Lagrange interpolation is to optimise the position of the nodes, so that the node polynomial

$$w(x) = \prod_{j=0}^{n} (x - x_j)$$

takes as uniformly small values as possible in the interval $[a, b]$. Without loss of generality, we consider the case $[a, b] = [-1, 1]$.

Definition 11.4 For $n \in \mathbb{N}_0$, the n-th *Chebyshev polynomial* for $t \in [-1, 1]$ is defined by

$$T_n(t) = \cos\left(n \arccos t\right),$$

see Fig. 11.5. The roots of a Chebyshev polynomial are called *Chebyshev nodes*.

The Chebyshev polynomials have remarkable properties.

Lemma 11.1

(i) *We have $|T_n(t)| \leq 1$ for all $t \in [-1, 1]$.*
(ii) *With $T_0(t) = 1$ and $T_1(t) = t$ we have*

$$T_{n+1}(t) = 2t\, T_n(t) - T_{n-1}(t)$$

Fig. 11.5 The roots of the Chebyshev polynomials define the Chebyshev nodes

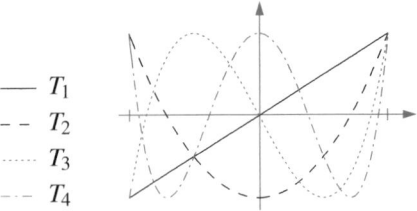

11.4 Chebyshev Nodes

for all $t \in [-1, 1]$. In particular, $T_n \in \mathscr{P}_n|_{[-1,1]}$ and for $n \geq 1$ it follows $T_n(t) = 2^{n-1}t^n + q_{n-1}(t)$ with $q_{n-1} \in \mathscr{P}_{n-1}|_{[-1,1]}$.

(iii) For $n \geq 1$, T_n has the roots $t_j = \cos((j + 1/2)\pi/n)$, $j = 0, 1, \ldots, n-1$, and the $n+1$ extreme points $s_j = \cos(j\pi/n)$, $j = 0, 1, \ldots, n$.

Proof Exercise. □

The roots of the Chebyshev polynomials define an optimal choice of nodes in the sense of the following proposition, which shows that for them the supremum norm of the node polynomial is minimal.

Proposition 11.4 *Let $t_0, t_1, \ldots, t_n \in [-1, 1]$ be the roots of the Chebyshev polynomial T_{n+1}. Then we have*

$$\min_{x_0,\ldots,x_n \in [-1,1]} \max_{x \in [-1,1]} \prod_{j=0}^{n} |x - x_j| = \max_{x \in [-1,1]} \prod_{j=0}^{n} |x - t_j| = 2^{-n}.$$

Proof From the preceding lemma it follows $T_{n+1}(x) = 2^n \prod_{j=0}^{n}(x - t_j)$ as well as $\max_{x \in [-1,1]} |T_{n+1}(x)| = 1$, which proves the second claimed identity. Suppose the nodes $t_0, t_1, \ldots, t_n$ are not optimal, that is there exist $x_0, x_1, \ldots, x_n$ such that for $w(x) = \prod_{j=0}^{n}(x - x_j)$, we have that $\max_{x \in [-1,1]} |w(x)| < 2^{-n}$. Since $w(x) = x^{n+1} + r_n(x)$ and $T_{n+1}(x) = 2^n x^{n+1} + q_n(x)$ with $q_n, r_n \in \mathscr{P}_n$, it follows that $p = 2^{-n}T_{n+1} - w = 2^{-n}q_n - r$ is a polynomial of degree n, thus $p \in \mathscr{P}_n$. Since T_{n+1} takes the values ± 1 with alternating signs at its $n+2$ extreme points $s_0, s_1, \ldots, s_{n+1}$ and $|w(x)| < 2^{-n}$ holds, it follows that $s_0, s_1, \ldots, s_{n+1}$ are not roots of p and their function values $p(s_i)$ have alternating signs. This implies that p has at least $n+1$ roots in $[-1, 1]$, which results in $p = 0$ and thus contradicts $p(s_i) \neq 0$. □

Remarks 11.8

(i) For the interpolation error with Chebyshev nodes in the interval $[-1, 1]$ we have the estimate

$$\|f - p\|_{C^0([-1,1])} \leq 2^{-n} \frac{\|f^{(n+1)}\|_{C^0([-1,1])}}{(n+1)!}.$$

(ii) For general intervals $[a, b]$ the optimal nodes are constructed using an affine-linear transformation $\psi : [-1, 1] \to [a, b]$.
(iii) The Chebyshev nodes correspond to the vertical projection of uniformly distributed $n+1$ points on a semicircle, see Fig. 11.6.
(iv) For interpolation with Chebyshev nodes, it can be shown that uniform convergence applies for Lipschitz-continuous functions. In particular, pointwise convergence applies for the function $f(x) = 1/(1 + 25x^2)$. The interpolation polynomial with 9 Chebyshev nodes is shown in Fig. 11.6.

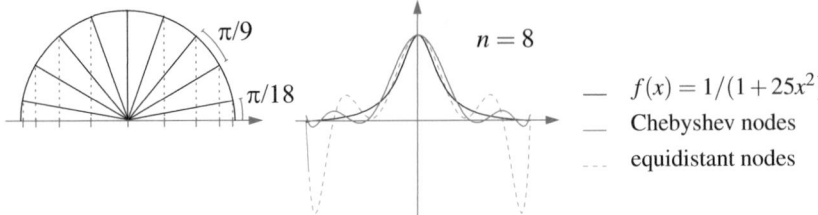

Fig. 11.6 Interpolation of the function $f(x) = 1/(1 + 25x^2)$ with equidistantly distributed as well as Chebyshev nodes $t_0, t_1, \ldots, t_8$

11.5 Hermite Interpolation

In the interpolation of smooth, i.e. very often continuously differentiable, functions, it makes sense to prescribe derivatives at nodes in order to reduce the approximation error while keeping the number of nodes constant, see Fig. 11.7.

Definition 11.5 The *Hermite interpolation task* looks for a polynomial $p \in \mathscr{P}_N$ such that for given nodes $a \leq x_0 < x_1 < \cdots < x_n \leq b$ and values $y_i^{(0)}, y_i^{(1)}, \ldots, y_i^{(\ell_i)}$ for $i = 0, 1, \ldots, n$ with numbers $\ell_i \in \mathbb{N}_0$ we have

$$p(x_i) = y_i^{(0)}, \quad p'(x_i) = y_i^{(1)}, \quad \ldots, \quad p^{(\ell_i)}(x_i) = y_i^{(\ell_i)}$$

for $i = 0, 1, \ldots, n$ with $N = \sum_{i=0}^{n}(\ell_i + 1) - 1$ holds.

In the Hermite interpolation task, $N + 1 = \sum_{i=0}^{n}(\ell_i + 1)$ conditions are to be met, so it is canonical to use the polynomial space $\mathscr{P}_N$.

Proposition 11.5 *The Hermite interpolation task is uniquely solvable.*

Proof The linear mapping $T : \mathscr{P}_N \to \mathbb{R}^{N+1}$ is defined by

$$Tp = \left[p(x_0), p'(x_0), \ldots, p^{(\ell_0)}(x_0), \ldots, p(x_n), p'(x_n), \ldots, p^{(\ell_n)}(x_n)\right]^{\mathsf{T}}.$$

If $Tp = 0$, then p has the roots x_i, $i = 0, 1, \ldots, n$, with multiplicities $\ell_i + 1$. Considering the multiplicities, $p \in \mathscr{P}_N$ thus has a total of $N + 1$ roots and the fundamental theorem of algebra implies $p = 0$. Thus, T is injective and as a linear mapping between spaces of the same dimension also bijective. This implies the unique solvability of the Hermite interpolation problem. □

Fig. 11.7 In Hermite interpolation, derivatives are also prescribed at the nodes in addition to function values

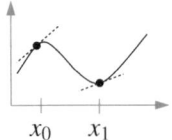

11.5 Hermite Interpolation

Remark 11.9 In the case of a single node x_0 and values $y_0^{(j)} = f^{(j)}(x_0)$, $j = 0, 1, \ldots, \ell_0$, the Hermite interpolation problem yields the ℓ_0-th Taylor polynomial of f at the point x_0.

To derive an error estimate, we restrict ourselves to the case $\ell_0 = \ell_1 = \cdots = \ell_n = \ell$ for an $\ell \geq 0$, so that $N + 1 = (\ell + 1)(n + 1)$ holds.

Proposition 11.6 For $f \in C^{N+1}([a, b])$, let $p \in \mathscr{P}_N$ be the Hermite polynomial with $p^{(k)}(x_i) = f^{(k)}(x_i)$, $0 \leq i \leq n$, $0 \leq k \leq \ell$, for given nodes $a \leq x_0 < x_1 < \cdots < x_n \leq b$. For each $x \in [a, b]$ there exists a $\xi \in [a, b]$ with

$$f(x) - p(x) = \frac{f^{(N+1)}(\xi)}{(N+1)!} \prod_{i=0}^{n} (x - x_i)^{\ell+1}.$$

In particular, it holds that

$$\|f - p\|_{C^0([a,b])} \leq \frac{\|f^{(N+1)}\|_{C^0([a,b])}}{(N+1)!} (b - a)^{N+1}.$$

Proof If $x \in \{x_0, x_1, \ldots, x_n\}$, the statement is clear, so let $x \in [a, b] \setminus \{x_0, x_1, \ldots, x_n\}$ in the following. For $y \in [a, b]$ define

$$w(y) = \prod_{i=0}^{n} (y - x_i)^{\ell+1} \in \mathscr{P}_{N+1}$$

and

$$F(y) = \big(f(x) - p(x)\big) w(y) - \big(f(y) - p(y)\big) w(x).$$

The function F has the $(\ell + 1)$-fold zeros $x_0, x_1, \ldots, x_n$ as well as the simple zero x and between two adjacent zeros F' has a zero different from them. Thus, F' according to Rolle's theorem has besides the ℓ-fold zeros at $x_0, x_1, \ldots, x_n$ further $n + 1$ zeros, thus in total at least $2n + 2$ zeros. Between all these zeros F'' has further zeros, provided $\ell \geq 2$ holds, that is F'' has $(n + 1) + (2n + 1) = 3n + 2$ zeros. Inductively it follows, that $F^{(\ell)}$ has at least $(n + 1) + (\ell n + 1)$ many zeros. When differentiating $F^{(\ell)}$, the number of zeros decreases by one and thus the derivative $F^{(\ell+n+1+\ell n)} = F^{(N+1)}$ still has one zero $\xi \in [a, b]$. Thus,

$$0 = F^{(N+1)}(\xi) = \big(f(x) - p(x)\big)(N + 1)! - \big(f^{(N+1)}(\xi) - 0\big) w(x)$$

and this implies the claimed statements. □

Remark 11.10 In the case of Hermite interpolation with 3 nodes and specification of the function values as well as two derivatives at each node, one obtains a comparable accuracy to that of Lagrange interpolation with 9 nodes.

11.6 Learning Objectives, Quiz and Application

You should be familiar with various interpolation tasks and be able to prove corresponding error estimates. You should be able to explain and illustrate with examples the possibilities of improving interpolation results through different choices of nodes.

Quiz 11.1 Decide for each of the following statements whether it is true or false. You should be able to justify your answer.

The Lagrange polynomials satisfy the identity $\sum_{i=0}^{n} L_i(x) = 1$ for all $x \in [a, b]$.	
To capture the function $f(x) = \sin(x)$ in the interval $[0, 1]$ in tabular form with an error of at most 0.01, the specification of four function values is sufficient.	
Chebyshev nodes are the extreme points of the Chebyshev polynomials.	
The Neville scheme calculates the coefficients of the Lagrange interpolation polynomial with respect to the monomial basis.	
The Hermite interpolation task with four nodes and specification of the first and second derivatives at the nodes leads to 8 conditions.	

Application 11.1 A press for the production of mechanical components is driven by a spindle. In this case, the spindle travel $0 \leq s \leq \ell$ leads to a diameter $d(s)$ of the component. To produce components of a specified diameter, a suitable spindle travel must therefore be specified. Tests with the machine lead to the measured values in millimetres

$$(s, d(s)) = (0.10, 0.098),\ (0.20, 0.043),\ (0.35, 0.122),\ (0.40, 0.157).$$

Construct a function that provides a sensible spindle travel based on these data for a desired radius.

Chapter 12
Interpolation with Splines

12.1 Splines

Interpolation with polynomials requires high regularity properties of functions to guarantee small errors. To also approximate functions that only satisfy $f \in C^2([a,b])$ with high accuracy, the interval $[a,b]$ is divided into subintervals and a polynomial interpolation is performed on each subinterval. At the transitions between the subintervals, suitable continuity and differentiability conditions must be imposed. In this chapter, we follow the presentations in [7–9].

Definition 12.1 For a partitioning $\mathcal{T}_n$ of $[a,b]$ defined by $a = x_0 < x_1 < \cdots < x_n = b$, a function $s : [a,b] \to \mathbb{R}$ is called a *spline of (polynomial) degree* $m \in \mathbb{N}_0$ *and of (differentiability) order* $k \in \mathbb{N}_0$, if $s \in C^k([a,b])$ and $s|_{[x_{i-1},x_i]} \in \mathcal{P}_m|_{[x_{i-1},x_i]}$, $i = 1, 2, \ldots, n$, holds. Let $\mathscr{S}^{m,k}(\mathcal{T}_n)$ denote the space of all splines of degree m with respect to $\mathcal{T}_n$. Splines of degree $m = 1, 2, 3$ and of order $m - 1$ are called *linear, quadratic* or *cubic* splines, respectively.

Remark 12.1 Often, with splines, only the differentiability order $k = m - 1$ is considered and then $\mathscr{S}^m(\mathcal{T}_n)$ is written instead of $\mathscr{S}^{m,m-1}(\mathcal{T}_n)$. This is the maximum order for which the polynomial space $\mathcal{P}_m$ is a proper subspace of $\mathscr{S}^{m,k}(\mathcal{T}_n)$. For $k \geq m$, however, $\mathcal{P}_m|_{[a,b]} = \mathscr{S}^{m,k}(\mathcal{T}_n)$.

Proposition 12.1 *For given values $y_0, y_1, \ldots, y_n$ there exists exactly one linear spline $s \in \mathscr{S}^{1,0}(\mathcal{T}_n)$ with $s(x_i) = y_i$ for $i = 0, 1, \ldots, n$. This is given by $s = \sum_{i=0}^n y_i \varphi_i$ with the hat functions $(\varphi_0, \varphi_1, \ldots, \varphi_n) \in \mathscr{S}^{1,0}(\mathcal{T}_n)$, which are defined by $\varphi_i(x_j) = \delta_{ij}$ for $0 \leq i, j \leq n$, see Fig. 12.1.*

Fig. 12.1 Hat functions φ_0, φ_2 associated with the nodes x_0 and x_2 and linear spline function s

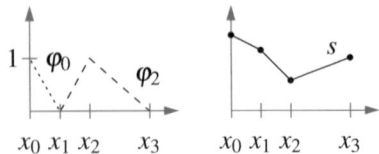

Proof The function $\varphi_i \in \mathscr{S}^{1,0}(\mathscr{T}_n)$ is given for $x \in [a,b]$ by

$$\varphi_i(x) = \begin{cases} (x - x_{i-1})/(x_i - x_{i-1}), & x \in [x_{i-1}, x_i], \\ (x_{i+1} - x)/(x_{i+1} - x_i), & x \in [x_i, x_{i+1}], \\ 0, & otherwise. \end{cases}$$

From the representation $s = \sum_{i=0}^{n} y_i \varphi_i$ the unique solvability of the interpolation task follows. □

Proposition 12.2 *The dimension of the space* $\mathscr{S}^{m,m-1}(\mathscr{T}_n)$ *is* $n + m$.

Proof For $m = 1$ the statement follows from the previous result. Let $(\varphi_0, \varphi_1, \ldots, \varphi_n)$ be the basis of $\mathscr{S}^{1,0}(\mathscr{T}_n)$ consisting of the hat functions. For $i = 0, 1, \ldots, n$ let r_i be an $(m-1)$-th antiderivative of φ_i, that is $r_i^{(m-1)} = \varphi_i$. Then, $(r_0, r_1, \ldots, r_n) \subset \mathscr{S}^{m,m-1}(\mathscr{T}_n)$ holds. Moreover, the monomial basis $(x^0, x^1, \ldots, x^{m-2})$ is contained in $\mathscr{S}^{m,m-1}(\mathscr{T}_n)$ and we show that

$$(r_0, r_1, \ldots, r_n, x^0, x^1, \ldots, x^{m-2})$$

is a basis of $\mathscr{S}^{m,m-1}(\mathscr{T}_n)$. Let $s \in \mathscr{S}^{m,m-1}(\mathscr{T}_n)$. Since $s^{(m-1)} \in \mathscr{S}^{1,0}(\mathscr{T}_n)$, there exist $c_0, c_1, \ldots, c_n$ such that

$$s^{(m-1)} = \sum_{i=0}^{n} c_i \varphi_i$$

and integrating $(m-1)$ times leads to

$$s(x) = \sum_{i=0}^{n} c_i r_i(x) + \sum_{j=0}^{m-2} d_j x^j$$

with integration constants $d_0, d_1, \ldots, d_{m-2}$. To establish linear independence, let $c_0, c_1, \ldots, c_n$ and $d_0, d_1, \ldots, d_{m-2}$ be such that

$$\sum_{i=0}^{n} c_i r_i(x) + \sum_{j=0}^{m-2} d_j x^j = 0$$

holds for all $x \in [a, b]$. By differentiating $(m-1)$ times it follows that

$$\sum_{i=0}^{n} c_i r_i^{(m-1)} = \sum_{i=0}^{n} c_i \varphi_i = 0$$

and thus $c_0 = c_1 = \cdots = c_n = 0$. This implies $\sum_{j=0}^{m-2} d_j x^j = 0$ in $[a, b]$, which in turn implies $d_0 = d_1 = \cdots = d_{m-2} = 0$. □

Remarks 12.2

(i) With $n+1$ nodes, in addition to $n+1$ interpolation conditions $s(x_i) = y_i$, $i = 0, 1, \ldots, n$, a further $m-1$ conditions must be imposed, to uniquely determine $s \in \mathscr{S}^{m,m-1}(\mathscr{T}_n)$.

(ii) If m is odd and $f \in C^{m+1}([a, b])$, an interpolating spline $s \in \mathscr{S}^{m,(m-1)/2}(\mathscr{T}_n)$ can be defined by piecewise Lagrange or Hermite interpolation with the property

$$\|f - s\|_{C^0([a,b])} \leq \frac{h^{m+1}}{(m+1)!} \|f^{(m+1)}\|_{C^0([a,b])},$$

where $h = \max_{i=1,\ldots,n}(x_i - x_{i-1})$ is the maximum grid width of the partition $\mathscr{T}_n$.

(iii) More generally, by integrating piecewise polynomial (typically discontinuous) functions of degree $m - (k+1)$ repeatedly $(k+1)$ times, it can be shown that $\dim \mathscr{S}^{m,k}(\mathscr{T}_n) = n(m-k) + (k+1)$ holds for $k \leq m$.

12.2 Cubic Splines

While linear splines have kinks and quadratic splines have discontinuous second derivatives, which can be well perceived at practically relevant resolutions, cubic splines appear very smooth.

Definition 12.2 For a partition $\mathscr{T}_n = \{x_0, x_1, \ldots, x_n\}$ of the interval $[a, b]$ and support values $y_0, y_1, \ldots, y_n$, the *interpolation task with cubic splines* consists in determining a function $s \in \mathscr{S}^{3,2}(\mathscr{T}_n)$ with $s(x_i) = y_i$ for $i = 0, 1, \ldots, n$ taking into account one of the following boundary conditions:

- *natural boundary conditions*, i.e. $s''(a) = 0$ and $s''(b) = 0$;
- *complete* or *Hermite boundary conditions*, i.e. $s'(a) = y_0^{(1)}$ and $s'(b) = y_n^{(1)}$ with given numbers $y_0^{(1)}, y_n^{(1)} \in \mathbb{R}$;
- *periodic boundary conditions*, that is $s'(a) = s'(b)$ and $s''(a) = s''(b)$, where additionally $y_0 = y_n$ applies.

Fig. 12.2 Cubic splines are piecewise smooth, running through given points

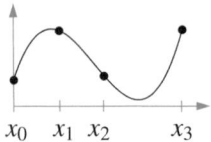

Remark 12.3 The cubic spline interpolation can be interpreted as the fixing of a thin wooden strip through given points, see Fig. 12.2. The word *spline* refers to a long, very flexible ruler used in shipbuilding.

Interpolating cubic splines are minimal for a linearised bending energy, as the following statement shows.

Proposition 12.3 *Let* $s \in \mathscr{S}^{3,2}(\mathscr{T}_n)$ *be a solution of a cubic spline interpolation task and let* $g \in C^2([a, b])$ *be any function that satisfies the interpolation conditions* $g(x_i) = y_i$, $i = 0, 1, \ldots, n$. *Assume that (i)* $s''(x) = 0$ *for* $x \in \{a, b\}$, *(ii)* $(g - s)'(x) = 0$ *for* $x \in \{a, b\}$ *or (iii)* $s''(a) = s''(b)$ *and* $(g - s)'(a) = (g - s)'(b)$. *Then it holds*

$$\int_a^b |s''(x)|^2 \, dx + \int_a^b |(s - g)''(x)|^2 \, dx = \int_a^b |g''(x)|^2 \, dx.$$

Proof It holds

$$\int_a^b |g''|^2 \, dx = \int_a^b |s'' + (g - s)''|^2 \, dx$$

$$= \int_a^b |s''|^2 \, dx + \int_a^b |(g - s)''|^2 \, dx + 2 \int_a^b s''(g - s)'' \, dx$$

and it suffices to show that the last integral on the right side vanishes. From the boundary conditions it follows

$$s''(a)\bigl(g'(a) - s'(a)\bigr) = s''(b)\bigl(g'(b) - s'(b)\bigr).$$

Partial integration on each subinterval $[x_{i-1}, x_i]$, $i = 1, 2, \ldots, n$, shows

$$\int_a^b s''(g - s)'' \, dx = \sum_{i=1}^n \int_{x_{i-1}}^{x_i} s''(g - s)'' \, dx$$

$$= \sum_{i=1}^n \left(-\int_{x_{i-1}}^{x_i} s'''(g - s)' \, dx + \bigl(s''(g - s)'\bigr)\bigl|_{x_{i-1}}^{x_i} \right).$$

For the sum of the boundary terms, using $s \in C^2([a, b])$ and the boundary conditions at $x_0 = a$ and $x_n = b$, it follows that

12.2 Cubic Splines

$$\sum_{i=1}^{n} \left(s''(g-s)'\right)\Big|_{x_{i-1}}^{x_i} = \sum_{i=1}^{n} \left(s''(x_i)(g-s)'(x_i) - s''(x_{i-1})(g-s)'(x_{i-1})\right)$$

$$= s''(x_n)(g-s)'(x_n) - s''(x_0)(g-s)'(x_0) = 0.$$

Since s''' is constant on each interval (x_{i-1}, x_i), for example with value c_i, it follows using the main theorem of differential and integral calculus and the interpolation conditions $s(x_i) = g(x_i)$ for $i = 0, 1, \ldots, n$, that

$$\sum_{i=1}^{n} \int_{x_{i-1}}^{x_i} s'''(g-s)' \, dx = \sum_{i=1}^{n} c_i \left((g-s)(x_i) - (g-s)(x_{i-1})\right) = 0.$$

In total, it is thus shown that

$$\int_a^b s''(g-s)'' \, dx = 0$$

and the statement of the proposition is proven. □

The preceding result implies the well-posedness of the interpolation task, which is stated here only for complete and natural boundary conditions.

Proposition 12.4 *There exists a unique solution to the interpolation task with cubic splines and natural or complete boundary conditions.*

Proof If $s, g \in \mathscr{S}^{3,2}(\mathscr{T}_n)$ are two solutions to the interpolation task, it follows from the repeated application of the previous result with swapped roles of s and g and addition of the two resulting equations, that

$$\int_a^b |(s-g)''|^2 \, dx = 0$$

and thus $(s-g)'' = 0$ or $s(x) - g(x) = p + qx$ in $[a, b]$. From $s(x_i) - g(x_i) = 0$ for $i = 0$ and $i = n$ it follows $p = q = 0$ and thus $s = g$. In the case of complete boundary conditions, with $\dim \mathscr{S}^{3,2}(\mathscr{T}_n) = n + 3$, the linear mapping

$$T_H : \mathscr{S}^{3,2}(\mathscr{T}_n) \to \mathbb{R}^{n+3}, \ s \mapsto \left(s(x_0), \ldots, s(x_n), s'(x_0), s'(x_n)\right)$$

is injective and thus also bijective. The case of natural boundary conditions follows by replacing $s'(x_0)$ and $s'(x_n)$ by $s''(x_0)$ and $s''(x_n)$ in the mapping T_H. □

12.3 Calculation of Cubic Splines

Due to the regularity condition $s \in C^2([a,b])$, interpolating cubic splines cannot be determined locally and a system of linear equations must be solved to obtain a representation in the monomial basis on each subinterval.

Proposition 12.5 *For a partitioning $x_0 < x_1 < \cdots < x_n$ and given interpolation values $y_0, y_1, \ldots, y_n$, let $s \in \mathscr{S}^{3,2}(\mathscr{T}_n)$ with $s(x_i) = y_i$, $i = 0, 1, \ldots, n$. Then the quantities $\gamma_i = s''(x_i)$, $i = 0, 1, \ldots, n$, satisfy the system of linear equations*

$$h_i \frac{\gamma_i}{6} + \frac{(h_{i+1} + h_i) 4\gamma_{i+1}}{2} + h_{i+1} \frac{\gamma_{i+2}}{6} = \frac{y_{i+2} - y_{i+1}}{h_{i+1}} - \frac{y_{i+1} - y_i}{h_i},$$

for $i = 0, 1, \ldots, n-2$, where $h_i = x_{i+1} - x_i$. With the quantities

$$b_i = \frac{y_{i+1} - y_i}{h_i} - \frac{\gamma_i}{2} h_i - \frac{d_i}{6} h_i^2, \quad d_i = \frac{\gamma_{i+1} - \gamma_i}{h_i}$$

we have on each subinterval $[x_i, x_{i+1}]$, $i = 0, 1, \ldots, n-1$, the representation

$$s|_{[x_i, x_{i+1}]}(x) = y_i + b_i(x - x_i) + \frac{\gamma_i}{2}(x - x_i)^2 + \frac{d_i}{6}(x - x_i)^3.$$

Proof If $s \in \mathscr{S}^{3,2}(\mathscr{T}_n)$ with $s(x_i) = y_i$ and $s''(x_i) = \gamma_i$ then there exist $b_i, d_i \in \mathbb{R}$, $i = 0, 1, \ldots, n-1$, with

$$s|_{[x_i, x_{i+1}]}(x) = p_i(x) = y_i + b_i(x - x_i) + \frac{\gamma_i}{2}(x - x_i)^2 + \frac{d_i}{6}(x - x_i)^3.$$

(i) The continuity of s at x_{i+1}, that is the identity $p_i(x_{i+1}) = p_{i+1}(x_{i+1})$ leads to the equation

$$y_i + b_i h_i + \frac{\gamma_i}{2} h_i^2 + \frac{d_i}{6} h_i^3 = y_{i+1} \iff b_i = \frac{y_{i+1} - y_i}{h_i} - \frac{\gamma_i}{2} h_i - \frac{d_i}{6} h_i^2$$

for $i = 0, 1, \ldots, n-1$, with b_i being determined by d_i, y_i, y_{i+1} and γ_i.

(ii) The continuity of s'' at x_{i+1} or the identity $p_i''(x_{i+1}) = p_{i+1}''(x_{i+1})$ as well as $s''(x_n) = \gamma_n$ leads to the equation

$$\gamma_i + d_i h_i = \gamma_{i+1} \iff d_i = \frac{\gamma_{i+1} - \gamma_i}{h_i}$$

for $i = 0, 1, \ldots, n-1$, whereby d_i is determined by γ_i and γ_{i+1}.

(iii) The continuity of s' at x_{i+1}, that is the identity $p_i'(x_{i+1}) = p_{i+1}'(x_{i+1})$, is equivalent to

$$b_i + h_i \gamma_i + \frac{d_i}{2} h_i^2 = b_{i+1} \iff b_{i+1} - b_i = h_i \gamma_i + \frac{d_i}{2} h_i^2$$

for $i = 0, 1, \ldots, n - 2$. If one uses in this identity the above representations of b_i and b_{i+1} as well as d_i and d_{i+1}, the asserted $n - 1$ equations for the coefficients γ_i, $i = 0, 1, 2, \ldots, n$ are obtained. □

In the preceding result, $n - 1$ equations were derived, which must be satisfied by the $n + 1$ derivatives $\gamma_i = s''(x_i)$, $i = 0, 1, \ldots, n$. The addition of two boundary conditions completes the system of equations.

Example 12.1 For an equidistant grid, that is $h_i = h$ for $i = 0, 1, \ldots, n - 1$, and the natural boundary conditions $s''(x_0) = s''(x_n) = 0$ or $\gamma_0 = \gamma_n = 0$, the quantities $\gamma_1, \gamma_2, \ldots, \gamma_{n-1}$ are given as the solution of the tridiagonal system of linear equations

$$\frac{1}{6} \begin{bmatrix} 4 & 1 & & \\ 1 & 4 & \ddots & \\ & \ddots & \ddots & 1 \\ & & 1 & 4 \end{bmatrix} \begin{bmatrix} \gamma_1 \\ \gamma_2 \\ \vdots \\ \gamma_{n-1} \end{bmatrix} = \begin{bmatrix} r_1 \\ r_2 \\ \vdots \\ r_{n-1} \end{bmatrix}$$

with $r_i = (y_{i+1} - 2y_i + y_{i-1})/h^2$, $i = 1, 2, \ldots, n - 1$. The strictly diagonally dominant system matrix is regular and thus there exists a unique solution.

12.4 Interpolation Error

Error estimates for spline interpolation are usually provided in norms, which are given by integrals. We consider the L^2 norm induced by the L^2 scalar product, that is

$$(f, g)_{L^2(I)} = \int_I f(x) g(x) \, dx, \quad \|f\|_{L^2(I)} = \left(\int_I f^2(x) \, dx \right)^{1/2}.$$

An important tool for working with this norm is the Cauchy–Schwarz or Hölder inequality

$$(f, g)_{L^2(I)} \leq \|f\|_{L^2(I)} \|g\|_{L^2(I)},$$

which holds for functions $f, g \in C^0(I)$.

Proposition 12.6 Let $s_1 \in \mathscr{S}^{1,0}(\mathscr{T}_h)$ be the continuous, piecewise linear spline interpolant of the function $f \in C^2([a, b])$, that is $s_1(x_i) = f(x_i)$, $i = 0, 1, \ldots, N$. Then

$$\|f - s_1\|_{L^2([a,b])} \leq \frac{h^2}{2} \|f''\|_{L^2([a,b])}$$

with $h = \max_{i=1,\ldots,N} x_i - x_{i-1}$.

Proof The error $e = f - s_1$ satisfies $e(x_i) = 0$ for $i = 0, 1, \ldots, N$, as well as $e|_{I_i} \in C^2(I_i)$ on all subintervals $I_i = [x_{i-1}, x_i]$. Under these conditions, the Poincaré and the seminorm interpolation estimates apply

$$\|e\|_{L^2(I_i)} \leq \frac{h}{\sqrt{2}} \|e'\|_{L^2(I_i)}, \quad \|e'\|_{L^2(I_i)} \leq \|e\|_{L^2(I_i)}^{1/2} \|e''\|_{L^2(I_i)}^{1/2}$$

for $i = 1, 2, \ldots, N$. From these two estimates it follows that

$$\|e\|_{L^2(I_i)}^2 \leq \frac{h^2}{2} \|e'\|_{L^2(I_i)}^2 \leq \frac{h^2}{2} \|e\|_{L^2(I_i)} \|e''\|_{L^2(I_i)}$$

and after division by $\|e\|_{L^2(I_i)}$ and using $e''|_{I_i} = f''|_{I_i}$, it follows that

$$\|e\|_{L^2(I_i)} \leq \frac{h^2}{2} \|e''\|_{L^2(I_i)} = \frac{h^2}{2} \|f''\|_{L^2(I_i)}.$$

Squaring and summing this inequality leads to

$$\|e\|_{L^2([a,b])}^2 = \sum_{i=1}^{N} \|e\|_{L^2(I_i)}^2 \leq \frac{h^4}{4} \sum_{i=1}^{N} \|f''\|_{L^2(I_i)}^2 = \frac{h^4}{4} \|f''\|_{L^2([a,b])}^2,$$

thus the claimed error estimate. The Poincaré inequality used here results from the representation

$$e(x) = \int_{x_{i-1}}^{x} e'(y)\,dy,$$

the application of the Hölder inequality to this integral, that is

$$|e(x)| \leq \left(\int_{x_{i-1}}^{x} 1\,dy\right)^{1/2} \left(\int_{x_{i-1}}^{x} (e'(y))^2\,dy\right)^{1/2} \leq (x - x_{i-1})^{1/2} \|e'\|_{L^2(I_i)},$$

and subsequent squaring and integrating

$$\|e\|_{L^2(I_i)}^2 \leq \|e'\|_{L^2(I_i)}^2 \int_{x_{i-1}}^{x_i} (x - x_{i-1})\,dx \leq \frac{h^2}{2} \|e'\|_{L^2(I_i)}^2.$$

The seminorm interpolation estimate is obtained with partial integration, the identities $e(x_{i-1}) = e(x_i) = 0$ and the Hölder inequality,

$$\|e'\|_{L^2(I_i)}^2 = -\int_{I_i} e(x)e''(x)\,dx + \bigl(e(x_i)e'(x_i) - e(x_{i-1})e'(x_{i-1})\bigr)$$

$$\leq \|e\|_{L^2(I_i)} \|e''\|_{L^2(I_i)}.$$

This completes the proof. □

The proposition implies an error estimate for cubic spline interpolation.

Corollary 12.1 *Let $s_3 \in \mathscr{S}^{3,2}(\mathscr{T}_n)$ be an interpolating cubic spline function of the function $f \in C^4([a,b])$. Assume that (i) $f''(x) = 0$ for $x \in \{a,b\}$, (ii) $s_3'(x) = f'(x)$ for $x \in \{a,b\}$ or (iii) $f^{(i)}(a) = f^{(i)}(b)$ for $i = 0, 1, 2$, in the case of natural, complete, and periodic boundary conditions, respectively. Then it holds that*

$$\|f - s_3\|_{L^2([a,b])} \leq \frac{h^4}{4} \|f^{(4)}\|_{L^2([a,b])}.$$

Proof Let $I_1(g) \in \mathscr{S}^{1,0}(\mathscr{T}_n)$ denote the interpolating linear spline function of a function $g \in C^2([a,b])$. Then $I_1(f - s_3) = 0$ and according to the previous result

$$\|f - s_3\|_{L^2([a,b])} = \|(f - s_3) - I_1(f - s_3)\|_{L^2([a,b])} \leq \frac{h^2}{2} \|(f - s_3)''\|_{L^2([a,b])}.$$

Let $r \in \mathscr{S}^{3,2}(\mathscr{T}_n)$ be such that $r'' = I_1(f'')$ holds and r has the same boundary values as f and s_3. By construction we then have $r''(x) = f''(x)$ for $x \in \{a,b\}$. For $s = s_3 - r \in \mathscr{S}^{3,2}(\mathscr{T}_n)$ and $g = f - r \in C^2([a,b])$ the conditions of Proposition 12.3 apply, so that

$$\|(s_3 - r)''\|_{L^2([a,b])}^2 + \|(f - r)'' - (s_3 - r)''\|_{L^2([a,b])}^2 = \|(f - r)''\|_{L^2([a,b])}^2.$$

This implies that

$$\|(f - s_3)''\|_{L^2([a,b])} \leq \|(f - r)''\|_{L^2([a,b])} = \|f'' - I_1(f'')\|_{L^2([a,b])}.$$

With another application of the previous proposition, we obtain in total

$$\|f - s_3\|_{L^2([a,b])} \leq \frac{h^2}{2} \|f'' - I_1(f'')\|_{L^2([a,b])} \leq \frac{h^4}{4} \|f^{(4)}\|_{L^2([a,b])},$$

which proves the claimed inequality. □

Remark 12.4 A key aspect of interpolation with cubic spline functions is that, unlike Lagrange interpolation with piecewise cubic polynomials, we obtain an interpolating function in the space $C^2([a,b])$.

12.5 Learning Objectives, Quiz and Application

You should be able to define spline spaces and determine their dimensions. For cubic splines, you should be able to specify a minimality property and its derivation.

Quiz 12.1 Decide for each of the following statements whether it is true or false. You should be able to justify your answer.

Every spline function is once continuously differentiable.
It holds that $\mathscr{S}^{1,0}(\mathscr{T}_n) \cap \mathscr{S}^{3,2}(\mathscr{T}_n) = \{0\}$, where 0 denotes the constant function with value 0.
If $q \in \mathscr{P}_m$ and $x_0 < x_1 < \cdots < x_n$ is a partition $\mathscr{T}_n$ of $[a, b]$, then $q\|_{[a,b]} \in \mathscr{S}^{m,m-1}(\mathscr{T}_n)$.
Interpolating cubic spline functions minimise a linearised bending energy among interpolating C^2 functions.
The calculation of a cubic spline leads to a system of linear equations with a diagonally dominant, irreducible system matrix.

Application 12.1 Smooth curves such as cubic spline functions find diverse applications in computer graphics for the calculation and representation of curves and surfaces. With few pieces of information, complex graphic objects such as CAD models or *postscript* fonts can be described. In addition to the memory requirement, the efficient further processing such as scaling or rotation of the objects is an important aspect. Closely related to spline functions are so-called *Bézier curves*, which are defined for given points $P_0, P_1, \ldots, P_n \in \mathbb{R}^2$ and $t \in [0, 1]$ by

$$z(t) = \sum_{i=0}^{n} \binom{n}{i} t^i (1-t)^{n-i} P_i.$$

(i) Show that for $n = 2$ the representation

$$z(t) = (1-t)\left[(1-t)P_0 + tP_1\right] + t\left[(1-t)P_1 + tP_2\right]$$

holds and interpret this formula geometrically.

(ii) Show that with the initialisation $z_i^0(t) = P_i$, $i = 0, 1, \ldots, n$, and the recursion rule

$$z_i^j(t) = (1-t)z_i^{j-1}(t) + tz_{i+1}^{j-1}(t)$$

for $t \in [0, 1]$, $j = 1, 2, \ldots, n$, and $i = 0, 1, \ldots, n-j$ the identity $z = z_0^n$ follows.

(iii) Implement a recursive function y = de_casteljau(j,i,P) for evaluating the curve z for given points $P_0, P_1, \ldots, P_n$ at a point $t \in [0, 1]$ using the formula from (ii). Use your program to graphically represent the curve defined by the points $P_0 = (0, 0)$, $P_1 = (1, 1)$, $P_2 = (2, 0)$ and $P_3 = (3, 2)$.

Chapter 13
Discrete Fourier Transform

13.1 Trigonometric Interpolation

Many signals or functions that occur in applications are created by superpositions of fundamental oscillations of different frequencies, that is, after suitable transformation to the interval $[0, 2\pi]$

$$f(x) = \sum_{\ell=0}^{\infty} \bigl(c_\ell \cos(\ell x) + d_\ell \sin(\ell x)\bigr),$$

see Fig. 13.1. In fact, every Riemann-integrable function can be represented in this way and this motivates to interpolate functions with trigonometric functions. Compared to approximation for example with polynomials, many coefficients are small and in practice negligible.

Definition 13.1 For $m \in \mathbb{N}$, $n = 2m$ and nodes $x_j = 2\pi j/n$ and values $y_j \in \mathbb{R}$, $j = 0, 1, \ldots, n-1$, the *real trigonometric interpolation task* consists in determining $a_\ell, b_\ell \in \mathbb{R}$, $\ell = 1, \ldots, m-1$, and $a_0, a_m \in \mathbb{R}$, so that for

$$T(x) = \frac{a_0}{2} + \sum_{\ell=1}^{m-1} \bigl(a_\ell \cos(\ell x) + b_\ell \sin(\ell x)\bigr) + \frac{a_m}{2} \cos(mx)$$

the identity $T(x_j) = y_j$ holds for $j = 0, 1, \ldots, n-1$.

The real trigonometric interpolation task can be represented concisely in the complex plane. Let $\mathrm{i} = \sqrt{-1} \in \mathbb{C}$ denote the imaginary unit.

Definition 13.2 The *complex trigonometric interpolation task* consists in determining $\beta_k \in \mathbb{C}$, $k = 0, 1, \ldots, n-1$, so that for $x_j = 2\pi j/n$ and $y_j \in \mathbb{C}$,

Fig. 13.1 Functions can often be represented as a sum of sine oscillations

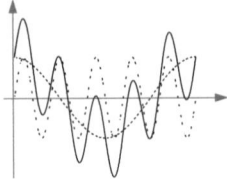

$j = 0, 1, \ldots, n - 1$, and

$$p(x) = \beta_0 + \beta_1 e^{ix} + \beta_2 e^{i2x} + \cdots + \beta_{n-1} e^{i(n-1)x} = \sum_{k=0}^{n-1} \beta_k e^{ikx}$$

the identity $p(x_j) = y_j$ holds for $j = 0, 1, \ldots, n - 1$.

The real and the complex trigonometric interpolation tasks are equivalent to each other in the following sense.

Proposition 13.1 *Let $n = 2m$ and $y_0, y_1, \ldots, y_{n-1} \in \mathbb{R}$. The coefficients β_k, $k = 0, \ldots, n - 1$, solve the complex trigonometric interpolation task exactly when the coefficients a_0, a_m and a_ℓ, b_ℓ, for $\ell = 1, 2, \ldots, m - 1$, defined by*

$$a_0 = 2\beta_0, \quad a_\ell = \beta_\ell + \beta_{2m-\ell}, \quad b_\ell = i(\beta_\ell - \beta_{2m-\ell}), \quad a_m = 2\beta_m,$$

solve the real trigonometric interpolation task defined by $y_0, y_1, \ldots, y_{n-1}$.

Proof It holds that $e^{-i\ell x_j} = e^{-i2\pi \ell j/n} = e^{i2\pi(n-\ell)j/n} = e^{i(n-\ell)x_j}$ and with $e^{ix} = \cos(x) + i\sin(x)$ it follows

$$\cos(\ell x_j) = \operatorname{Re}(e^{i\ell x_j}) = \frac{e^{i\ell x_j} + e^{-i\ell x_j}}{2} = \frac{e^{i\ell x_j} + e^{i(n-\ell)x_j}}{2},$$

$$\sin(\ell x_j) = \operatorname{Im}(e^{i\ell x_j}) = \frac{e^{i\ell x_j} - e^{-i\ell x_j}}{2i} = \frac{e^{i\ell x_j} - e^{i(n-\ell)x_j}}{2i}.$$

With $1/i = -i$ and $n = 2m$ and $\cos(mx_j) = e^{imx_j}$, this implies that

$$\frac{a_0}{2} + \sum_{\ell=1}^{m-1} \left(a_\ell \cos(\ell x_j) + b_\ell \sin(\ell x_j)\right) + \frac{a_m}{2} \cos(mx_j)$$

$$= \frac{a_0}{2} + \sum_{\ell=1}^{m-1} \frac{a_\ell - ib_\ell}{2} e^{i\ell x_j} + \sum_{\ell=1}^{m-1} \frac{a_\ell + ib_\ell}{2} e^{i(n-\ell)x_j} + \frac{a_m}{2} e^{imx_j}.$$

With the relations $\beta_0 = a_0/2$, $\beta_\ell = (a_\ell - ib_\ell)/2$ and $\beta_{n-\ell} = (a_\ell + ib_\ell)/2$ for $\ell = 1, 2, \ldots, m - 1$ and $\beta_m = a_m/2$ the assertion follows. □

Remarks 13.1

(i) In the situation of the previous result, $p(x_j) = T(x_j)$, $j = 0, 1, \ldots, n-1$, but in general $p \neq T$.
(ii) Due to the identity $e^{ikx} = (e^{ix})^k$, we also speak of trigonometric polynomials.
(iii) Better approximation properties are achieved with functions of the form $r(x) = \sum_{k=-m}^{m-1} \delta_k e^{ikx}$, which however can be obtained via the complex trigonometric interpolation task.

13.2 Fourier Bases

If we write the interpolation conditions $p(x_j) = y_j$ in vector form, we get

$$y = \begin{bmatrix} y_0 \\ y_1 \\ \ldots \\ y_{n-1} \end{bmatrix} = \sum_{k=0}^{n-1} \beta_k \begin{bmatrix} e^{ikx_0} \\ e^{ikx_1} \\ \ldots \\ e^{ikx_{n-1}} \end{bmatrix} = \sum_{k=0}^{n-1} \beta_k \omega^k.$$

This identity can be interpreted as a basis change from the representation of the vector y with respect to the canonical basis in $\mathbb{R}^n$ to a representation with the vectors

$$\omega^k = [e^{ikx_0}, e^{ikx_1}, \ldots, e^{ikx_{n-1}}]^\mathsf{T}$$

Necessary and sufficient for the solvability of the complex-valued trigonometric interpolation task is that the vectors $(\omega^k)_{k=0,\ldots,n-1}$ define a basis of the $\mathbb{C}$-vector space $\mathbb{C}^n$.

Definition 13.3 For $n \in \mathbb{N}$, let $\omega_n = e^{i2\pi/n}$ be the n-th unit root, see Fig. 13.2, and for $k = 0, 1, \ldots, n-1$, let $\omega^k \in \mathbb{C}^n$ be defined by

$$\omega^k = [\omega_n^{0k}, \omega_n^{1k}, \ldots, \omega_n^{(n-1)k}]^\mathsf{T}$$

The family $(\omega^0, \omega^1, \ldots, \omega^{n-1}) \subset \mathbb{C}^n$ is called *Fourier basis*.

The structure of the Fourier basis vectors motivates the numbering of vectors in $\mathbb{C}^n$ with the indices $j = 0, 1, \ldots, n-1$. Similarly, matrices will be numbered in the

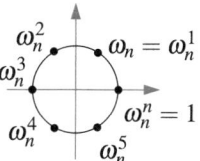

Fig. 13.2 The powers of the n-th unit root ω_n are evenly distributed complex numbers on the unit circle

following beginning with 0. The scalar product of two vectors $a, b \in \mathbb{C}^n$ is defined by $a \cdot b = a^\mathsf{T} \overline{b} = \sum_{j=0}^{n-1} a_j \overline{b}_j$.

Lemma 13.1 *The vectors $(\omega^k)_{k=0,\ldots,n-1}$ form an orthogonal basis of the $\mathbb{C}$-vector space $\mathbb{C}^n$, that is, $\omega^k \cdot \omega^\ell = n\delta_{k\ell}$ holds.*

Proof Exercise. □

To solve the complex trigonometric interpolation problem, the representing matrix of the basis change must therefore be determined.

Lemma 13.2 *The basis change from the Fourier basis to the basis $(e_0, e_1, \ldots, e_{n-1})$, consisting of the canonical basis vectors, is realised by the matrix*

$$T_n = [\omega^0, \omega^1, \ldots, \omega^{n-1}] \in \mathbb{C}^{n \times n}$$

with inverse $T_n^{-1} = (1/n)\overline{T}_n^\mathsf{T}$. For all $y = \sum_{j=0}^{n-1} y_j e_j \in \mathbb{C}^n$ we therefore have that $y = \sum_{k=0}^{n-1} \beta_k \omega^k$ with $\beta = (1/n)\overline{T}_n^\mathsf{T} y$.

Proof For $y \in \mathbb{C}^n$, let $\beta = [\beta_0, \beta_1, \ldots, \beta_{n-1}]^\mathsf{T}$ be the coefficient vector with respect to the Fourier basis $(\omega^k)_{k=0,\ldots,n-1}$, that is, $y = \sum_{k=0}^{n-1} \beta_k \omega^k$ holds. From this it follows that

$$y^\mathsf{T} \overline{\omega}^\ell = \left(\sum_{k=0}^{n-1} \beta_k \omega^k\right)^\mathsf{T} \overline{\omega}^\ell = \sum_{k=0}^{n-1} \beta_k (\omega^k)^\mathsf{T} \overline{\omega}^\ell = n\beta_\ell,$$

so $\beta_\ell = (1/n) y^\mathsf{T} \overline{\omega}^\ell = (1/n)(\overline{\omega}^\ell)^\mathsf{T} y$ or in vector notation

$$\beta = \begin{bmatrix} \beta_0 \\ \vdots \\ \beta_{n-1} \end{bmatrix} = \frac{1}{n} \begin{bmatrix} (\overline{\omega}^0)^\mathsf{T} \\ \vdots \\ (\overline{\omega}^{n-1})^\mathsf{T} \end{bmatrix} y = \frac{1}{n} \overline{T}_n^\mathsf{T} y.$$

The identity $\omega^k \cdot \omega^m = n\delta_{km}$ implies that $T_n \overline{T}_n^\mathsf{T} = nI_n$ or $T_n^{-1} = (1/n)\overline{T}_n^\mathsf{T}$ holds. □

Definition 13.4 The mapping $y \mapsto \beta = (1/n)\overline{T}_n^\mathsf{T} y$ is called *(discrete) Fourier transform* and the inverse mapping $\beta \mapsto y = T_n \beta$ is referred to as *Fourier synthesis*.

Remarks 13.2

(i) The Fourier transform can be represented using the Fourier synthesis and complex conjugations, i.e. since T_n is symmetric we have $\beta = \frac{1}{n}\overline{T}_n y = \frac{1}{n}\overline{(T_n \overline{y})}$.

(ii) The complex trigonometric interpolation problem is solved with the discrete Fourier transform. The Fourier synthesis realises the evaluation of a trigonometric polynomial at the nodes.

13.3 Fast Fourier Transform

The matrix T_n has only n different entries, which are arranged in a cyclic manner, so that the multiplication with T_n can be realised with a significantly lower effort than $\mathcal{O}(n^2)$.

Example 13.1 ([8]) Using $\omega_8^\ell = \omega_8^{\ell \bmod 8}$, the Fourier synthesis $y = T_8 \beta$ can be expressed as

$$\begin{bmatrix} y_0 \\ y_1 \\ y_2 \\ y_3 \\ y_4 \\ y_5 \\ y_6 \\ y_7 \end{bmatrix} = \begin{bmatrix} \omega_8^0 & \omega_8^0 & \omega_8^0 & \omega_8^0 & \omega_8^0 & \omega_8^0 & \omega_8^0 & \omega_8^0 \\ \omega_8^0 & \omega_8^1 & \omega_8^2 & \omega_8^3 & \omega_8^4 & \omega_8^5 & \omega_8^6 & \omega_8^7 \\ \omega_8^0 & \omega_8^2 & \omega_8^4 & \omega_8^6 & \omega_8^0 & \omega_8^2 & \omega_8^4 & \omega_8^6 \\ \omega_8^0 & \omega_8^3 & \omega_8^6 & \omega_8^1 & \omega_8^4 & \omega_8^7 & \omega_8^2 & \omega_8^5 \\ \omega_8^0 & \omega_8^4 & \omega_8^0 & \omega_8^4 & \omega_8^0 & \omega_8^4 & \omega_8^0 & \omega_8^4 \\ \omega_8^0 & \omega_8^5 & \omega_8^2 & \omega_8^7 & \omega_8^4 & \omega_8^1 & \omega_8^6 & \omega_8^3 \\ \omega_8^0 & \omega_8^6 & \omega_8^4 & \omega_8^2 & \omega_8^0 & \omega_8^6 & \omega_8^4 & \omega_8^2 \\ \omega_8^0 & \omega_8^7 & \omega_8^6 & \omega_8^5 & \omega_8^4 & \omega_8^3 & \omega_8^2 & \omega_8^1 \end{bmatrix} \begin{bmatrix} \beta_0 \\ \beta_1 \\ \beta_2 \\ \beta_3 \\ \beta_4 \\ \beta_5 \\ \beta_6 \\ \beta_7 \end{bmatrix}.$$

A rearrangement of the right side according to even and odd indices leads to

$$\begin{bmatrix} y_0 \\ y_1 \\ y_2 \\ y_3 \\ \hline y_4 \\ y_5 \\ y_6 \\ y_7 \end{bmatrix} = \left[\begin{array}{cccc|cccc} \omega_8^0 & \omega_8^0 & \omega_8^0 & \omega_8^0 & \omega_8^0 & \omega_8^0 & \omega_8^0 & \omega_8^0 \\ \omega_8^0 & \omega_8^2 & \omega_8^4 & \omega_8^6 & \omega_8^1 & \omega_8^3 & \omega_8^5 & \omega_8^7 \\ \omega_8^0 & \omega_8^4 & \omega_8^0 & \omega_8^4 & \omega_8^2 & \omega_8^6 & \omega_8^2 & \omega_8^6 \\ \omega_8^0 & \omega_8^6 & \omega_8^4 & \omega_8^2 & \omega_8^3 & \omega_8^1 & \omega_8^7 & \omega_8^5 \\ \hline \omega_8^0 & \omega_8^0 & \omega_8^0 & \omega_8^0 & \omega_8^4 & \omega_8^4 & \omega_8^4 & \omega_8^4 \\ \omega_8^0 & \omega_8^2 & \omega_8^4 & \omega_8^6 & \omega_8^5 & \omega_8^7 & \omega_8^1 & \omega_8^3 \\ \omega_8^0 & \omega_8^4 & \omega_8^0 & \omega_8^4 & \omega_8^6 & \omega_8^2 & \omega_8^6 & \omega_8^2 \\ \omega_8^0 & \omega_8^6 & \omega_8^4 & \omega_8^2 & \omega_8^7 & \omega_8^5 & \omega_8^3 & \omega_8^1 \end{array} \right] \begin{bmatrix} \beta_0 \\ \beta_2 \\ \beta_4 \\ \beta_6 \\ \hline \beta_1 \\ \beta_3 \\ \beta_5 \\ \beta_7 \end{bmatrix}.$$

With the identities $\omega_8^{2k} = e^{i 2\pi 2k/8} = e^{i 2\pi k/4} = \omega_4^k$ and $\omega_8^4 = e^{i\pi} = -1$ it follows

$$\begin{bmatrix} y_0 \\ \vdots \\ y_3 \\ \hline y_4 \\ \vdots \\ y_7 \end{bmatrix} = \left[\begin{array}{c|c} T_4 & D_4 T_4 \\ \hline T_4 & -D_4 T_4 \end{array} \right] \begin{bmatrix} \beta_0 \\ \vdots \\ \beta_6 \\ \hline \beta_1 \\ \vdots \\ \beta_7 \end{bmatrix},$$

where T_4 and D_4 are defined by

$$T_4 = \begin{bmatrix} \omega_4^0 & \omega_4^0 & \omega_4^0 & \omega_4^0 \\ \omega_4^0 & \omega_4^1 & \omega_4^2 & \omega_4^3 \\ \omega_4^0 & \omega_4^2 & \omega_4^0 & \omega_4^2 \\ \omega_4^0 & \omega_4^3 & \omega_4^2 & \omega_4^1 \end{bmatrix}, \quad D_4 = \begin{bmatrix} \omega_8^0 & & & \\ & \omega_8^1 & & \\ & & \omega_8^2 & \\ & & & \omega_8^3 \end{bmatrix}.$$

This implies

$$\begin{bmatrix} y_0 \\ y_1 \\ y_2 \\ y_3 \end{bmatrix} = T_4 \begin{bmatrix} \beta_0 \\ \beta_2 \\ \beta_4 \\ \beta_6 \end{bmatrix} + D_4 T_4 \begin{bmatrix} \beta_1 \\ \beta_3 \\ \beta_5 \\ \beta_7 \end{bmatrix}, \quad \begin{bmatrix} y_4 \\ y_5 \\ y_6 \\ y_7 \end{bmatrix} = T_4 \begin{bmatrix} \beta_0 \\ \beta_2 \\ \beta_4 \\ \beta_6 \end{bmatrix} - D_4 T_4 \begin{bmatrix} \beta_1 \\ \beta_3 \\ \beta_5 \\ \beta_7 \end{bmatrix}.$$

The Fourier synthesis $y = T_8 \beta$ can therefore be reduced to two Fourier syntheses of dimension $n = 4$.

The procedure of the example can be generalised.

Proposition 13.2 *For $\beta \in \mathbb{C}^{2m}$ let $D_m \in \mathbb{C}^{m \times m}$ be the diagonal matrix with entries $(D_m)_{\ell \ell} = \omega_{2m}^\ell$, $\ell = 0, 1, \ldots, m-1$. Then $y = T_{2m} \beta$ is given by $y = (y^1, y^2)$ with vectors $y^1, y^2 \in \mathbb{C}^m$ defined by*

$$y^1 = T_m \beta^{even} + D_m T_m \beta^{odd}, \quad y^2 = T_m \beta^{even} - D_m T_m \beta^{odd},$$

where $\beta^{even} = [\beta_0, \beta_2, \ldots, \beta_{2m-2}]^T$ and $\beta^{odd} = [\beta_1, \beta_3, \ldots, \beta_{2m-1}]^T$.

Proof For $0 \leq \ell \leq m-1$ we find, using $\omega_{2m}^{2k\ell} = \omega_m^{k\ell}$,

$$y_\ell = \sum_{j=0}^{2m-1} (T_{2m})_{\ell j} \beta_j = \sum_{j=0}^{2m-1} \omega_{2m}^{j\ell} \beta_j$$

$$= \sum_{k=0}^{m-1} \omega_{2m}^{2k\ell} \beta_{2k} + \sum_{k=0}^{m-1} \omega_{2m}^{(2k+1)\ell} \beta_{2k+1}$$

$$= \sum_{k=0}^{m-1} \omega_m^{k\ell} \beta_{2k} + \omega_{2m}^\ell \sum_{k=0}^{m-1} \omega_m^{k\ell} \beta_{2k+1}$$

$$= \sum_{k=0}^{m-1} (T_m)_{\ell k} \beta_{2k} + (D_m)_{\ell \ell} \sum_{k=0}^{m-1} (T_m)_{\ell k} \beta_{2k+1},$$

thus $y^1 = T_m \beta^{even} + D_m T_m \beta^{odd}$. For $\ell \geq m$ a similar calculation considering $\omega_{2m}^\ell = \omega_{2m}^{m+\ell \bmod m} = -\omega_{2m}^{\ell \bmod m}$ leads to the claimed identity. $\square$

The result reduces a problem of size n with effort $\mathscr{A}(n)$ to two problems of size $n/2$ with effort $\mathscr{A}(n/2)$. The assembly of the vectors $T_m \beta^{even}$ and $T_m \beta^{odd}$ to the subvectors y^1, y^2 of length $n/2$ according to the identities of the proposition requires the computational effort of $3n/2$. The procedure can be generalised and iterated for dimensions $n = 2^\ell$, $\ell = \log_2(n) \in \mathbb{N}$. For the computational effort we thus obtain

$$\mathscr{A}(n) \to 2\mathscr{A}(n/2) + \frac{3n}{2} \to 2\left(2\mathscr{A}(n/4) + \frac{3n}{4}\right) + \frac{3n}{2} \to \cdots \to 2^\ell \mathscr{A}(1) + \ell \frac{3n}{2}.$$

Since $\mathscr{A}(1) = 1$ applies, the effort of the resulting procedure is about $n(1 + (3/2)\log_2 n)$ (complex) arithmetic operations. This replaces the effort $\mathscr{O}(n^2)$ of a matrix-vector multiplication by the significantly lower effort $\mathscr{O}(n \log_2(n))$.

13.4 Learning Objectives, Quiz and Application

You should be able to explain the basic ideas of the discrete Fourier transform and describe the effort reduction of the fast Fourier transform.

Quiz 13.1 Decide for each of the following statements whether it is true or false. You should be able to justify your answer.

As a complex vector space, $\mathbb{C}^n$ has the dimension n and as a real vector space the dimension $2n$.
If ω_n is the n-th unit root, then $\omega_n^{n/2} = -1$ if and only if n is even.
The complex trigonometric interpolation problem is solved by $\beta = T_n y$ with the Fourier matrix T_n.
The matrix $S_n = (1/\sqrt{n}) T_n$ defines an isometry on $\mathbb{C}^n$, that is, we have $\|S_n y\|_2 = \|y\|_2$ for all $y \in \mathbb{C}^n$.
For real values $y_0, y_1, \ldots, y_{n-1}$, the solution of the complex trigonometric interpolation problem is real-valued.

Application 13.1 The discrete Fourier transform calculates a frequency decomposition of a given signal. In order to process only relevant information, calculated coefficients, which are small compared to others, can often be neglected. In addition, coefficients that belong to frequencies that are not perceptible in the respective application can be eliminated. The vector $y = [y_0, y_1, \ldots, y_{n-1}]^T \in \mathbb{R}^n$ is defined by $y_j = \sin(2\pi j/n) + (1/10)\xi_j$, $j = 0, 1, \ldots, n-1$, where ξ_j stands for a normally distributed random value that can be generated in MATLAB with randn. Use the

MATLAB routine `fft` to determine the Fourier transform $\beta \in \mathbb{C}^n$, and eliminate coefficients β_k, for which

$$|\beta_k| \leq \theta \max_{\ell=0,1,\ldots,n-1} |\beta_\ell|$$

applies, that is, replace such coefficients with zero. Use the inverse transform `ifft` to obtain a vector $\tilde{y} \in \mathbb{C}^n$. Interpret the vectors y and $\tilde{y}$ as values of a function and graphically represent them for $n = 256$ and various values of θ.

Chapter 14
Numerical Integration

14.1 Quadrature Formulas

The aim of numerical integration or quadrature is the approximation of proper integrals

$$I(f) = \int_a^b f(x)\,dx,$$

which cannot be explicitly calculated using an antiderivative.

Definition 14.1 A *quadrature formula* on the interval $[a, b]$ is a linear mapping $Q : C^0([a, b]) \to \mathbb{R}$ of the form

$$Q(f) = \sum_{i=0}^{n} w_i f(x_i)$$

with *(quadrature) points* $(x_i)_{i=0,\ldots,n}$ and *(quadrature) weights* $(w_i)_{i=0,\ldots,n}$. The number $\|Q\| = (b-a)^{-1} \sum_{i=0}^{n} |w_i|$ is its *stability indicator*.

Remarks 14.1

(i) If $a = x_0 < x_1 < \cdots < x_n = b$, the Riemann integral can be approximated by

$$\int_a^b f(x)\,dx \approx \sum_{i=0}^{n-1}(x_{i+1} - x_i) f(x_i)$$

and the right-hand side defines a quadrature formula with weights $w_i = x_{i+1} - x_i$ for $i = 0, 1, \ldots, n-1$ and $w_n = 0$, see Fig. 14.1.

Fig. 14.1 Riemann sums define simple quadrature formulas

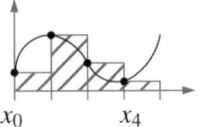

(ii) For every quadrature formula, we have that

$$|Q(f)| \leq \left(\sum_{i=0}^{n} |w_i|\right) \|f\|_{C^0([a,b])} = \|Q\|(b-a)\|f\|_{C^0([a,b])}.$$

Definition 14.2 The quadrature formula Q is called *exact of degree r*, if $Q(p) = I(p)$ for all $p \in \mathscr{P}_r$ holds.

If a function f can be well approximated by polynomials, then a quadrature formula with a high degree of exactness provides good approximations of the integral.

Proposition 14.1 Let Q be exact of degree $r \geq 0$. Then $\sum_{i=0}^{n} w_i = b - a$ holds and for all $f \in C^0([a, b])$

$$|I(f) - Q(f)| \leq (1 + \|Q\|)(b - a) \min_{p \in \mathscr{P}_r} \|f - p\|_{C^0([a,b])}.$$

In the case $w_i \geq 0$, $i = 0, 1, \ldots, n$, we have that $\|Q\| = 1$.

Proof According to the assumption, $\sum_{i=0}^{n} w_i = Q(1) = I(1) = b - a$ holds. Let $f \in C^0([a, b])$ and $p \in \mathscr{P}_r$ be arbitrary. With $I(p) = Q(p)$, the linearity of I and Q as well as the triangle inequality, it follows that

$$|I(f) - Q(f)| \leq (1 + \|Q\|)(b - a)\|f - p\|_{C^0([a,b])}.$$

Since $p \in \mathscr{P}_r$ is arbitrary, the assertion follows. □

Remarks 14.2

(i) With interpolation estimates, quantitative statements about the quadrature error are obtained, that is, for example, with Corollary 11.1

$$|I(f) - Q(f)| \leq (1 + \|Q\|)(b - a)\frac{\|f^{(r+1)}\|_{C^0([a,b])}}{(r+1)!}(b-a)^{r+1}.$$

By using Chebyshev nodes, this estimate can be further improved.

(ii) If Q is exact of degree $2q$ and the weights $(w_i)_{i=0,\ldots,n}$ and nodes $(x_i)_{i=0,\ldots,n}$ are symmetric with respect to the interval midpoint $(a+b)/2$, then Q is exact even of degree $2q + 1$.

(iii) If Q is a quadrature formula on $[a, b]$, then one obtains with the transformation $\varphi : [a, b] \to [c, d]$, $x \mapsto c + (x - a)(d - c)/(b - a)$, and

$$\int_c^d g(y)\,dy = \int_a^b g(\varphi(x))\varphi'(x)\,dx = \frac{d-c}{b-a}\int_a^b g(\varphi(x))\,dx$$

a quadrature formula on the interval $[c, d]$.

14.2 Newton-Cotes Formulas

A class of quadrature formulas is obtained by Lagrange interpolation of a function and subsequent exact integration of the interpolation polynomial. For given equidistant nodes $x_0 < x_1 < \cdots < x_n$ and the associated Lagrange basis polynomials

$$L_i(x) = \prod_{\substack{j=0 \\ j \neq i}}^{n} \frac{x - x_j}{x_i - x_j}$$

the Lagrange interpolation polynomial is given by $p = \sum_{i=0}^n f(x_i) L_i$. Hence, by

$$\int_a^b p(x)\,dx = \sum_{i=0}^n f(x_i) \int_a^b L_i(x)\,dx = \sum_{i=0}^n w_i f(x_i) = Q(f)$$

a quadrature formula Q with weights $w_i = \int_a^b L_i(x)\,dx$ is defined. Since $p = f$ for all $f \in \mathscr{P}_n$, this quadrature formula is exact of degree n. It is referred to as a *Newton-Cotes formula*.

Proposition 14.2 *The Newton–Cotes formula defined by nodes $x_0 < x_1 < \cdots < x_n$ and weights $w_i = \int_a^b L_i(x)\,dx$, $i = 0, 1, \ldots, n$, is exact of degree n.*

Proof The statement follows directly from the construction of the quadrature formula. □

For the cases $n = 0, 1, 2$, simple quadrature formulas are obtained, which are shown in Fig. 14.2.

Examples 14.1

(i) For $n = 0$ and $x_0 = (a + b)/2$, the *midpoint rule* $Q_{Mp}(f) = (b - a)f\big((a + b)/2\big)$ is obtained, which is exact of degree 1.
(ii) For $n = 1$ and $x_0 = a$, $x_1 = b$, the *trapezoidal rule*

$$\int_a^b f(x)\,dx \approx Q_{Trap}(f) = \frac{b-a}{2}\big[f(a) + f(b)\big].$$

is also exact of degree 1.

Fig. 14.2 Midpoint, trapezoidal and the Simpson rule as special cases of the Newton–Cotes formulas for $n = 0, 1, 2$

(iii) For $n = 2$ and $x_0 = a$, $x_1 = (a+b)/2$, $x_2 = b$, the *Simpson* or *Kepler's barrel rule*

$$\int_a^b f(x)\,dx \approx Q_{Sim}(f) = \frac{b-a}{6}\left[f(a) + 4f\left(\frac{a+b}{2}\right) + f(b)\right],$$

which due to its symmetry is exact of degree 3, is obtained.

(iv) For $n \geq 7$, negative weights occur, which can lead to stability problems as then $\|Q\| > 1$ holds.

14.3 Composite Quadrature Formulas

To achieve high accuracies without restrictive regularity assumptions, the interval $[a, b]$ can be divided into smaller subintervals, on which a quadrature formula of possibly low degree of exactness is applied.

Definition 14.3 Let $a = a_0 < a_1 < \cdots < a_N = b$ be the uniform partitioning of the interval $[a, b]$ with nodes $a_\ell = a + \ell(b-a)/N$, $\ell = 0, 1, \ldots, N$, and let $Q_\ell : C^0([a_{\ell-1}, a_\ell]) \to \mathbb{R}$ be a quadrature formula on the subinterval $[a_{\ell-1}, a_\ell]$ for $\ell = 1, 2, \ldots, N$. Then the mapping

$$Q^N(f) = \sum_{\ell=1}^N Q_\ell(f|_{[a_{\ell-1}, a_\ell]})$$

is a *composite quadrature formula*.

Example 14.2 With the trapezoidal rule on each subinterval $[a_{\ell-1}, a_\ell]$ we get

$$Q^N f = \sum_{\ell=1}^N \frac{a_\ell - a_{\ell-1}}{2}\left(f(a_{\ell-1}) + f(a_\ell)\right)$$

$$= \frac{b-a}{2N}\left(f(a_0) + 2f(a_1) + \cdots + 2f(a_{N-1}) + f(a_N)\right).$$

14.3 Composite Quadrature Formulas

The accuracy of composite quadrature formulas can be improved by reducing the length of the subintervals or by increasing the degree of exactness on each subinterval.

Proposition 14.3 *If the quadrature formulas on the subintervals have the degree of exactness $r \geq 0$, then*

$$\left| I(f) - Q^N(f) \right| \leq (b-a)^{r+2} (1 + \max_{\ell=1,\ldots,N} \|Q_\ell\|) \frac{N^{-(r+1)}}{(r+1)!} \|f^{(r+1)}\|_{C^0([a,b])}.$$

Proof On each subinterval $[a_{\ell-1}, a_\ell]$ we have

$$\left| \int_{a_{\ell-1}}^{a_\ell} f \, dx - Q_\ell(f) \right| \leq (1 + \|Q_\ell\|)(a_\ell - a_{\ell-1}) \min_{p \in \mathscr{P}_r} \|f - p\|_{C^0([a_{\ell-1}, a_\ell])}.$$

The error estimates for the Lagrange interpolation show

$$\min_{p \in \mathscr{P}_r} \|f - p\|_{C^0([a_{\ell-1}, a_\ell])} \leq \frac{\|f^{(r+1)}\|_{C^0([a_{\ell-1}, a_\ell])}}{(r+1)!} (a_\ell - a_{\ell-1})^{r+1}.$$

With $a_\ell - a_{\ell-1} = (b-a)/N$ we get

$$\left| I(f) - Q^N(f) \right| \leq \sum_{\ell=1}^{N} \left| \int_{a_{\ell-1}}^{a_\ell} f \, dx - Q_\ell(f) \right|$$

$$\leq \sum_{\ell=1}^{N} (1 + \|Q_\ell\|) \frac{(b-a)^{r+2}}{N^{r+2}} \frac{\|f^{(r+1)}\|_{C^0([a_{\ell-1}, a_\ell])}}{(r+1)!}$$

$$\leq (1 + \max_{\ell=1,\ldots,N} \|Q_\ell\|) N \frac{(b-a)^{r+2}}{N^{r+2}} \frac{\|f^{(r+1)}\|_{C^0([a,b])}}{(r+1)!}.$$

This implies the claimed estimate. □

Definition 14.4 A composite quadrature formula Q^N is called *convergent of order* $s \geq 0$, if

$$|Q^N(f) - I(f)| = \mathcal{O}(h^s)$$

for all $f \in C^s([a, b])$ and $h = (b-a)/N \to 0$ holds. In the cases $s = 1, 2, 3$ this is referred to as *linear*, *quadratic* and *cubic* convergence, respectively.

Examples 14.3

(i) The composite trapezoidal rule is quadratically convergent.
(ii) The composite Simpson rule has the order of convergence $s = 4$.

Remark 14.3 Often, faster convergence for the composite trapezoidal rule is observed. For periodic functions $f \in C^k([0, 2\pi])$, whose derivatives up to order k are also periodic, one can show for the quadrature error $\delta_N = |Q^N(f) - I(f)|$ that $\delta_N = \mathcal{O}(h^k)$. If the interval $[0, 2\pi]$ is identified with the unit circle in $\mathbb{C}$ and if f admits a holomorphic extension to an open neighbourhood of the circle then the exponential convergence property $\delta_N = \mathcal{O}(r^{-N})$ for some $r > 1$ can be proved.

14.4 Gaussian Quadrature

The choice of quadrature points and weights affects the accuracy of a quadrature formula. A certain degree of exactness cannot be exceeded with a given number of points.

Lemma 14.1 *A quadrature formula with $n+1$ weights and quadrature points possesses at most the degree of exactness $2n+1$.*

Proof Let $Q(f) = \sum_{i=0}^{n} w_i f(x_i)$ and define $p(x) = \prod_{i=0}^{n}(x - x_i)^2$. Then $p \in \mathscr{P}_{2n+2}$ and p is positive except at the quadrature points, where p vanishes. This implies $I(p) > 0$ as well as $Q(p) = 0$ and this implies the assertion. □

We will show in the following that there actually is a quadrature formula with the maximum degree of exactness $2n+1$. If a quadrature formula is exact of degree n, then the weights are already uniquely determined. If it is exact of degree $2n$, then these are positive.

Lemma 14.2 *A quadrature formula with $n+1$ weights and quadrature points $(x_i, w_i)_{i=0,\ldots,n}$ is exact of degree n if and only if we have*

$$w_i = \int_a^b L_i(x)\,\mathrm{d}x$$

for $i = 0, 1, \ldots, n$ with the Lagrange basis polynomials $(L_i)_{i=0,\ldots,n}$ defined by the points $(x_i)_{i=0,\ldots,n}$. If the quadrature formula is exact of degree $2n$, then $w_i > 0$ for $i = 0, 1, \ldots, n$.

Proof Exercise. □

In the Gauss quadrature, $n+1$ quadrature points are constructed so that the maximum degree of exactness $2n+1$ is achieved. More generally, weighted integrals of the form

$$I_\omega(f) = \int_a^b f(x)\omega(x)\,\mathrm{d}x$$

with a non-negative *weight function* $\omega \in C^0(a, b)$ are considered. This function is chosen so that through

14.4 Gaussian Quadrature

$$\langle f, g \rangle_\omega = \int_a^b f(x) g(x) \omega(x) \, dx$$

a scalar product on $C^0([a, b])$ is defined. This is exactly the case when ω is improperly integrable on (a, b) and from $\langle f, f \rangle_\omega = 0$ already $f = 0$ follows for every function $f \in C^0([a, b])$. With respect to this scalar product, an orthogonal basis of $\mathcal{P}_n$ is determined using the Gram–Schmidt process.

Proposition 14.4 *There exist orthogonal polynomials* $(\pi_j)_{j=0,\ldots,n}$ *such that* $\pi_j \in \mathcal{P}_j$ *and* $\langle \pi_j, \pi_k \rangle_\omega = \delta_{jk}$ *for all* $0 \leq j, k \leq n$. *In particular,* $\langle \pi_j, p \rangle_\omega = 0$ *holds for all* $p \in \mathcal{P}_{j-1}$ *and the polynomials form a basis of* $\mathcal{P}_n$.

Proof Exercise. □

The orthogonality implies the existence of roots.

Lemma 14.3 *Every orthogonal polynomial* π_j, $0 \leq j \leq n$, *has* j *simple roots in the interval* (a, b).

Proof Let us assume that the statement of the lemma is false for a $j \in \{0, 1, \ldots, n\}$. If π_j has a root $z \in \mathbb{R} \setminus (a, b)$, then $p(x) = \pi_j(x)/(x - z)$ is a polynomial in $\mathcal{P}_{j-1}$ and it follows

$$0 = \langle \pi_j, p \rangle_\omega = \int_a^b \frac{\pi_j^2(x)}{x - z} \omega(x) \, dx,$$

which is not possible, since $x - z$ has no root in (a, b) and π_j is not identically zero. If $z \in (a, b)$ is a multiple root or if $z \in \mathbb{C} \setminus \mathbb{R}$, then $\bar{z}$ is also a root of π_j and it follows $p(x) = \pi_j(x)/((x-z)(x-\bar{z})) = \pi_j(x)/|x - z|^2 \in \mathcal{P}_{j-2}$. Again, the identity $0 = \langle \pi_j, p \rangle_\omega$ leads to a contradiction. □

Examples 14.4

(i) For the weight function $\omega(x) = (1 - x^2)^{-1/2}$ in the interval $(-1, 1)$, the Chebyshev polynomials are obtained.
(ii) For $\omega(x) = 1$ in the interval $[-1, 1]$ the Legendre polynomials are obtained as derivatives of order n of the polynomial $(x^2 - 1)^n$, that is

$$P_n(x) = \frac{1}{2^n n!} \frac{d^n}{dx^n} (x^2 - 1)^n.$$

The zeros of the orthogonal polynomial π_n define a quadrature formula with degree of exactness $2n + 1$.

Proposition 14.5 *Let* $\pi_{n+1} \in \mathcal{P}_{n+1}$ *be the* $(n + 1)$-*th orthogonal polynomial with respect to the weight function* $\omega \in C^0(a, b)$. *The zeros* $(x_i)_{i=0,\ldots,n}$ *of* π_{n+1} *and the weights*

$$w_i = \int_a^b L_i(x) \omega(x) \, dx$$

for $i = 0, 1, \ldots, n$ define a quadrature formula $Q_\omega f = \sum_{i=0}^{n} w_i f(x_i)$ such that

$$Q_\omega(p) = I_\omega(p) = \int_a^b p(x)\omega(x)\,dx$$

for all $p \in \mathscr{P}_{2n+1}$.

Proof The quadrature formula defined in the proposition is well-defined and by choice of the weights, we have $I_\omega(r) = Q_\omega(r)$ for all $r \in \mathscr{P}_n$. If $p \in \mathscr{P}_{2n+1}$, one obtains by polynomial division polynomials $q, r \in \mathscr{P}_n$ with $p = q\pi_{n+1} + r$. Since $\langle q, \pi_{n+1} \rangle_\omega = 0$ holds, it follows

$$I_\omega(p) = \int_a^b q(x)\pi_{n+1}(x)\omega(x)\,dx + \int_a^b r(x)\omega(x)\,dx = \langle q, \pi_{n+1} \rangle_\omega + I_\omega(r) = I_\omega(r).$$

With $\pi_{n+1}(x_i) = 0$, $i = 0, 1, \ldots, n$, it follows

$$Q_\omega(p) = \sum_{i=0}^{n} w_i \big(q(x_i)\pi_{n+1}(x_i) + r(x_i)\big) = \sum_{i=0}^{n} w_i r(x_i) = Q_\omega(r).$$

In total, $I_\omega(p) = I_\omega(r) = Q_\omega(r) = Q_\omega(p)$. □

Example 14.5 For the weight function $\omega(x) = 1$ in the interval $[-1, 1]$ we have $P_0(x) = 1$, $P_1(x) = x$, $P_2(x) = (3x^2 - 1)/2$ and $P_3(x) = (5x^3 - 3x)/2$. Thus, for $n = 0, 1, 2$ we obtain quadrature formulas defined by

$$x_0 = 0, \quad w_0 = 2,$$
$$x_0 = -\sqrt{1/3}, \; x_1 = \sqrt{1/3}, \quad w_0 = 1, \; w_1 = 1,$$
$$x_0 = -\sqrt{3/5}, \; x_1 = 0, \; x_2 = \sqrt{3/5}, \quad w_0 = 5/9, \; w_1 = 8/9, \; w_2 = 5/9,$$

and which are Gaussian quadrature formulas.

14.5 Extrapolation

A composite quadrature formula defines a function $T(h)$, which for a given function $f \in C^k([a, b])$ and partitioning fineness $h = (b-a)/N$ provides an approximation of the integral, which is generally not directly accessible, and is denoted by $T(0) = \lim_{h \to 0} T(h)$. We assume that T is given as a function on $\mathbb{R}_{\geq 0}$. If the error of the quadrature formula is of order h^γ for a $\gamma \in \mathbb{N}$, it follows

$$T(h) = T(0) + \varphi(h)$$

14.5 Extrapolation

Fig. 14.3 Extrapolation of the calculated values $T(h_i)$, $i = 0, 1$, for a better approximation of the unknown value $T(0)$ through $p_1(0)$

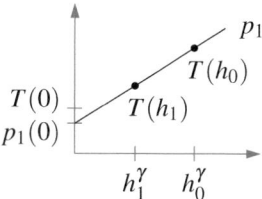

for a function φ with $\varphi(0) = 0$ and $|\varphi(h)| \leq ch^\gamma$. A Taylor expansion of φ around 0, that is

$$\varphi(z) = c_1 z^\gamma + c_2 z^{\gamma+1} + r(z)$$

with coefficients c_1, c_2 and a remainder term $r \in o(z^{\gamma+1})$ for $z \to 0$, leads to

$$T(h) = T(0) + c_1 h^\gamma + c_2 h^{\gamma+1} + r(h).$$

The evaluation of the quadrature formula for the fineness $h/2$ then yields

$$T(h/2) = T(0) + \frac{c_1}{2^\gamma} h^\gamma + \frac{c_2}{2^{\gamma+1}} h^{\gamma+1} + r(h/2).$$

With this equation, the term $c_1 h^\gamma$ in the identity for $T(h)$ can be eliminated and we obtain

$$T^*(h) = \frac{T(h) - 2^\gamma T(h/2)}{1 - 2^\gamma} = T(0) + c_2 \frac{1 - 2^{-1}}{1 - 2^\gamma} h^{\gamma+1} + o(h^{\gamma+1}).$$

The computable expression $T^*(h)$ thus defines an approximation of $T(0)$ with an error of the order $h^{\gamma+1}$, which is more accurate for small values of h than the approximations $T(h)$ and $T(h/2)$. The procedure is illustrated in Fig. 14.3.

Example 14.6 The extrapolation of the composite trapezoidal rule with convergence order $s = \gamma = 2$ leads to the composite Simpson rule, where due to symmetry effects $c_2 = 0$ can be assumed and thus instead of the expected improved convergence order $s = 3$ even the order $s = 4$ is obtained.

The described procedure can be generalised by performing a polynomial interpolation of the values $T(h_i)$, $i = 0, 1, \ldots, n$, for the nodes h_i^γ, $i = 0, 1, \ldots, n$, for example with $h_i = 2^{-i} h$ for a fixed $h > 0$. The interpolation polynomial $p_n \in \mathscr{P}_n$ is thus defined by the conditions

$$p_n(h_i^\gamma) = T(h_i)$$

for $i = 0, 1, \ldots, n$. The extrapolated value $p_n(0) \approx T(0)$ can be determined using the Neville scheme. Corresponding details can be found, for example, in [1, 9].

Remark 14.4 The *Aitkin delta-squared process* constructs from a given sequence $(x_k)_{k\geq 0}$ a sequence $(y_k)_{k\geq 2}$ with potentially improved convergence properties. If the sequence $(x_k)_{k\geq 0}$ converges linearly with factor $0 < q < 1$ to some x^*, then we have the approximations $(x^* - x_k) \approx q(x^* - x_{k-1})$ and $(x^* - x_{k+1}) \approx q(x^* - x_k)$, which can be combined to eliminate q and obtain an approximate formula for x^*, i.e.,

$$x^* \approx y_{k+1} = \frac{x_{k-1}x_{k+1} - x_k^2}{x_{k+1} - 2x_k + x_{k-1}} = x_{k+1} - \frac{(\delta x_k)^2}{\delta^2 x_k},$$

where $\delta x_k = x_{k+1} - x_k$ and $\delta^2 x_k = x_{k+1} - 2x_k + x_{k-1}$. Under certain conditions on the differences $x_k - x^*$ the sequence $(y_k)_{k\geq 2}$ converges quadratically to x^*.

14.6 Experimental Convergence Order

The convergence properties of a composite quadrature formula Q^N with step size $h = (b-a)/N$ can be experimentally analysed, by considering for a non-polynomial function $f \in C^k([a, b])$, for example $f(x) = \sin(x)$, whose exact integral $I(f)$ is explicitly known, the errors

$$e_h = |I(f) - Q^N(f)|$$

for some step sizes $h > 0$. From the approach $e_h \approx c_1 h^\gamma$ it follows by using two different step sizes $h, H > 0$, that

$$c_1 \approx \frac{e_h}{h^\gamma} \approx \frac{e_H}{H^\gamma}$$

and thus

$$\gamma \approx \frac{\log(e_h/e_H)}{\log(h/H)} = \frac{\log(e_h) - \log(e_H)}{\log(h) - \log(H)}.$$

In particular, if $H = h/2$, then $\gamma \approx \log(e_h/e_{h/2})/\log(2)$. If this expression is calculated for several step sizes h, then this defines an *(average) experimental convergence order* by means of a least squares fit or the arithmetic mean. It should be noted, that this can depend on the differentiability order of f. Additionally, one can graphically represent the convergence behaviour by using logarithmic scaling of the x- and y-axes and connecting experimentally determined pairs (h, e_h) via a polygonal chain. If there is indeed a relationship of the form $e_h \approx c_1 h^\gamma$, then the polygonal chain with respect to the logarithmic scaling will have the slope γ.

Example 14.7 We consider the pairs of values (h, e_h) given by $h = 2^{-\ell}$, $\ell = 1, \ldots, 5$, and $e_h = h^2/3$. Figure 14.4 shows that the slope of the polygonal chain

Fig. 14.4 The experimental convergence order results as the slope of a least squares line through measurement points with respect to logarithmic scaling

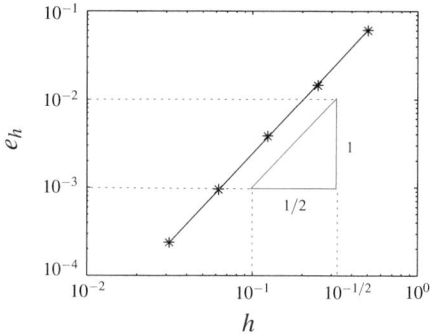

defined in this way in logarithmic scaling matches the slope of the line parallel to it through the points $(10^{-1}, 10^{-3})$ and $(10^{-1/2}, 10^{-2})$. The logarithmic slope of this line results from the difference quotient of the powers, that is

$$\gamma \approx \frac{\Delta_y^{\log}}{\Delta_x^{\log}} = \frac{(-2) - (-3)}{(-1/2) - (-1)} = 2.$$

Due to the logarithmic scaling, the value $10^{-1/2}$ on the x-axis is exactly in the middle of the values 10^0 and 10^{-1}, see Fig. 14.4.

14.7 Learning Objectives, Quiz and Application

You should be able to define the degree of exactness of a quadrature formula and derive abstract error estimates based on it. You should be able to specify the Newton–Cotes formulas and apply them to examples and to describe the construction of the Gauss quadrature and name the properties of the method.

Quiz 14.1 Decide for each of the following statements whether it is true or false. You should be able to justify your answer.

For every quadrature formula, $Q(\alpha f + \beta g) = \alpha Q(f) + \beta Q(g)$ holds.	
If a quadrature formula is exact of degree $r \geq 1$, then the weights of the quadrature formula are positive.	
Every Newton-Cotes formula with $n + 1 = 2q$ nodes is exact of degree $n + 2$.	
The Gauss quadrature uses the $n + 1$ zeros of an orthogonal polynomial $\pi_n \in \mathscr{P}_n$ as quadrature points.	
The trapezoidal rule on the interval $[-1, 1]$ approximates the integral of the function f by $\bigl[f(-1) + f(1)\bigr]/2$.	

Application 14.1 Based on samples and statistical considerations, the weight of a hen's egg can be approximated as a normally distributed random variable X with expected value $\mu = 57$ g and standard deviation $\sigma = 7$ g. The probability that the weight of an egg lies in the interval $[m_1, m_2]$ is thus given by

$$P(m_1 \leq X \leq m_2) = \frac{1}{\sqrt{2\pi\sigma^2}} \int_{m_1}^{m_2} e^{-(x-\mu)^2/(2\sigma^2)} \, dx.$$

(i) Determine, with a composite quadrature formula as well as the identity

$$\int_0^\infty e^{-t^2/2} \, dt = \sqrt{\pi/2},$$

the probability that an egg weighs more than 63 g.

(ii) Compare your result with an approach to calculating the probability without numerical integration using the identity

$$e^{-t^2} = 1 - t^2 + \frac{t^4}{2!} - \frac{t^6}{3!} + \cdots$$

and the exact integration of some monomials. In which situations is this approach useful?

(iii) Specify numerically with four decimal places accuracy the so-called 68-95-99.7 rule, which gives the probabilities for the deviation by one, two or three standard deviations from the expected value, i.e. the quantities $P(|X - \mu| \leq j\sigma)$ for $j = 1, 2, 3$.

Chapter 15
Nonlinear Problems

15.1 Root Finding and Minimisation Problems

For an open set $U \subset \mathbb{R}^n$ and mappings $f : U \to \mathbb{R}^n$ and $g : U \to \mathbb{R}$, we consider the following problems:

(N) Find $x^* \in U$ such that $f(x^*) = 0$.

(M) Find $x^* \in U$ such that $g(x^*) = \min_{x \in U} g(x)$.

These problems are connected via the optimality condition $\nabla g(x^*) = 0$ or via the minimisation of $x \mapsto \|f(x)\|^2$. Furthermore, root finding is equivalent to determining a fixed point of the mapping $\Phi(x) = f(x) + x$. In general, it is neither possible nor sensible to determine a solution exactly, and therefore sequences $(x_k)_{k=0,1,\ldots}$ are constructed iteratively, which under suitable conditions converge to a solution. The following terms are used to classify the convergence behaviour.

Definition 15.1 A numerical method that defines a sequence $(x_k)_{k=0,1,\ldots}$ of approximations for a numerical problem is called

(i) *globally convergent*, if the sequence $(x_k)_{k=0,1,\ldots}$ for every starting vector $x_0 \in U$ converges to a solution $x^* \in U$, and
(ii) *locally convergent*, if for every solution $x^* \in U$ there exists a number $\varepsilon > 0$ such that the sequence $(x_k)_{k=0,1,\ldots}$ for every starting vector $x_0 \in B_\varepsilon(x^*) \cap U$ converges to x^*.

Obviously, every globally convergent method is also locally convergent. To characterise the convergence speed of methods, we assume that $x_k \neq x^*$ for all $k \in \mathbb{N}_0$.

Definition 15.2 A locally convergent method is called *convergent of order* $\alpha \geq 1$, if a $q \in \mathbb{R}$ exists, such that for every solution $x^* \in U$, every starting vector $x_0 \in B_\varepsilon(x^*) \cap U$ and the sequence $(x_k)_{k \in \mathbb{N}_0}$ generated by the method for the approximation errors $\delta_k = \|x^* - x_k\|$, there holds

$$\limsup_{k \to \infty} \frac{\delta_{k+1}}{\delta_k^\alpha} = q$$

and in the case $\alpha = 1$ we additionally have $q \leq 1$. A method that is convergent of order α is called *linearly convergent* if $\alpha = 1$ and $q < 1$ and *quadratically convergent* if $\alpha = 2$ and $q \in \mathbb{R}_{\geq 0}$ holds. It is called *superlinear* or *sublinearly convergent* if $\alpha = 1$ and $q = 0$ or $\alpha = 1$ and $q = 1$ hold, respectively.

Examples 15.1

(i) If $\Phi : \mathbb{R}^n \to \mathbb{R}^n$ is a contraction, then the method $x_{k+1} = \Phi(x_k)$ for approximating a fixed point of Φ is globally and linearly convergent.

(ii) If $\Phi \in C^1(\mathbb{R})$, then the fixed point iteration $x_{k+1} = \Phi(x_k)$ for determining a fixed point x^* of Φ is locally linearly convergent, provided that $|\Phi'(x^*)| < 1$ by the mean value theorem. If $|\Phi'(x^*)| > 1$, then the method is divergent. If $\Phi'(x_*) = 0$ and $\Phi \in C^2(\mathbb{R})$ then local quadratic convergence occurs, which is an immediate consequence of a Taylor approximation argument, i.e.,

$$x_{k+1} - x^* = \Phi(x_k) - \Phi(x^*) = \frac{1}{2}\Phi''(\xi)(x_k - x^*)^2.$$

The cases of linear and quadratic convergence are illustrated in Fig. 15.1.

Remarks 15.1

(i) In the so-called *asymptotic region*, that is, after a sufficient number of iterations, linear convergence results in an error reduction by the factor q, while quadratic convergence doubles the number of correct decimal places in each step.

(ii) Convergent fixed-point iterations are robust with respect to rounding errors since they are *self-stabilizing* in the sense that every iterate can be interpreted as a new starting value.

Fig. 15.1 Linear (left) and quadratic (right) convergence of fixed-point iterations

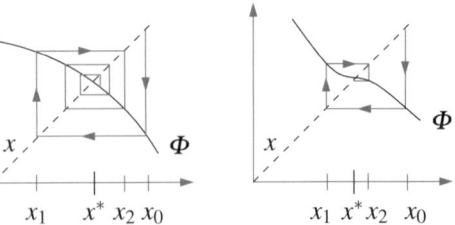

15.2 Approximation of Roots

The bisection method is based on the fact that any continuous function $f \in C^0([a, b])$ with the property $f(a)f(b) \leq 0$ has a root in the interval $[a, b]$. If then $c \in (a, b)$ is arbitrary, it follows

$$f(a)f(c) \leq 0 \quad \text{or} \quad f(c)f(b) \leq 0$$

and the subinterval $[a, c]$ or $[c, b]$ contains at least one root of f, see Fig. 15.2.

Algorithm 15.1 (Bisection Method) *Let $f \in C^0([a, b])$ with $f(a)f(b) \leq 0$ and $\varepsilon_{stop} > 0$. Set $(a_0, b_0) = (a, b)$ and $k = 0$.*

(1) Define $c_k = (a_k + b_k)/2$.
(2) Set

$$(a_{k+1}, b_{k+1}) = \begin{cases} (a_k, c_k) & \text{if } f(a_k)f(c_k) \leq 0, \\ (c_k, b_k) & \text{otherwise.} \end{cases}$$

(3) Stop if $b_{k+1} - a_{k+1} \leq \varepsilon_{stop}$; otherwise increase $k \to k+1$ and repeat step (1).

Since the current interval is halved at each step, the following statement is immediately apparent.

Proposition 15.1 *The bisection method is linearly convergent with the approximations $x_k = c_k$ for $k = 0, 1, 2, \ldots$ with $q = 1/2$. It stops after $J \leq 1 + \log_2\left((b-a)/\varepsilon_{stop}\right)$ steps and the interval $[a_J, b_J]$ contains a root.*

While the bisection method is only meaningful in one dimension, the secant method approximates a derivative, which is also possible in multiple dimensions. The easily determined root of the secant defines the new reference point, see Fig. 15.3.

Algorithm 15.2 (Secant Method) *Let $f \in C^0([a, b])$ and $\varepsilon_{stop} > 0$. Set $x_0 = a$, $x_1 = b$ and $k = 1$.*

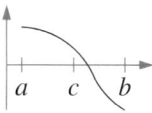

Fig. 15.2 The sign change of a continuous function implies the existence of a root

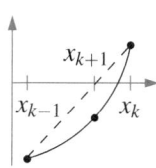

Fig. 15.3 The root of the secant serves as an approximation of a root

Fig. 15.4 The root of the tangent serves in the Newton method as an approximation of a root

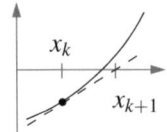

(1) If $f(x_k) \neq f(x_{k-1})$, then define

$$x_{k+1} = x_k - \frac{x_k - x_{k-1}}{f(x_k) - f(x_{k-1})} f(x_k).$$

(2) *Stop if* $|x_{k+1} - x_k| \leq \varepsilon_{stop}$; *otherwise increase* $k \to k+1$ *and repeat step (1).*

Remarks 15.2

(i) When implementing the secant method, cancellation effects can occur and rounding errors can become significant.
(ii) An alternative termination criterion is $|f(x_{k+1})| \leq \varepsilon_{stop}$.
(iii) The *regula-falsi* method combines the bisection method with the secant method, so that the interval $[x_{k-1}, x_k]$ always contains a root.

The quantity $(f(x_k) - f(x_{k-1}))/(x_k - x_{k-1})$ that appears in the secant method is an approximation of the derivative $f'(x_k)$. This slope defines a tangent to f at the point $(x_k, f(x_k))$, which is used in the Newton method to determine the new approximation x_{k+1}, see Fig. 15.4. This is especially easier to implement in multi-dimensional cases, provided the Jacobian matrix can be easily determined.

A Taylor approximation of the C^1 mapping $f : U \to \mathbb{R}^n$ around the point x shows

$$0 = f(x^*) = f(x) + Df(x)(x^* - x) + \varphi(x^* - x).$$

If the approximation $x = x_k$ is close to x^*, neglecting the term $\varphi(x^* - x)$ implies that

$$f(x_k) + Df(x_k)(x^* - x_k) \approx 0$$

and this motivates that by

$$x_{k+1} = x_k - Df(x_k)^{-1} f(x_k) \approx x^*$$

an improved approximation $x_{k+1} \approx x^*$ is defined, provided $Df(x_k)$ is regular.

Algorithm 15.3 (Newton Method) *Let* $f \in C^1(U; \mathbb{R}^n)$, $x_0 \in U$ *and* $\varepsilon_{stop} > 0$. *Set* $k = 0$.

(1) If $Df(x_k)$ is regular, then define

$$x_{k+1} = x_k - Df(x_k)^{-1} f(x_k).$$

15.2 Approximation of Roots

(2) Stop if $\|x_{k+1} - x_k\| \leq \varepsilon_{stop}$; otherwise increase $k \to k+1$ and repeat step (1).

The Newton method is locally quadratically convergent.

Proposition 15.2 *Let $f \in C^2(U; \mathbb{R}^n)$ and $x^* \in U$ be a root of f in U, such that $Df(x^*)$ is regular. Then there exists a number $\varepsilon > 0$, such that for every initial value $x_0 \in B_\varepsilon(x^*) \cap U$ the Newton method is executable and convergent. For the iterates $(x_k)_{k=0,1,\ldots}$ we have*

$$\|x^* - x_{k+1}\| \leq c\|x^* - x_k\|^2$$

with a constant $c \geq 0$.

Proof Since $\det Df(x^*) \neq 0$ and the mapping $x \mapsto \det Df(x)$ is continuous, there exists a number $\tilde{\varepsilon} > 0$, such that $\det Df(x) \neq 0$ and $\|Df(x)^{-1}\| \leq c_1$ for all $x \in B_{\tilde{\varepsilon}}(x^*) \subset U$. Let $x_k \in B_{\tilde{\varepsilon}}(x^*)$ for a $k \geq 0$. The Taylor expansion

$$0 = f(x^*) = f(x_k) + Df(x_k)(x^* - x_k) + \varphi(x^* - x_k)$$

with a function $\varphi : \mathbb{R}^n \to \mathbb{R}$, which fulfils $|\varphi(z)| \leq c_2 |z|^2$ for all $|z| \leq c_3$, implies

$$\|f(x_k) + Df(x_k)(x^* - x_k)\| \leq c_2 \|x^* - x_k\|^2$$

if $\|x^* - x_k\| \leq c_3$. With the iteration rule we get

$$x^* - x_{k+1} = x^* - x_k + Df(x_k)^{-1} f(x_k)$$
$$= Df(x_k)^{-1} \big(f(x_k) + Df(x_k)(x^* - x_k)\big).$$

Hence it follows

$$\|x^* - x_{k+1}\| \leq \|Df(x_k)^{-1}\| \|f(x_k) + Df(x_k)(x^* - x_k)\| \leq c_1 c_2 \|x^* - x_k\|^2.$$

With $\varepsilon \leq \min\{1/(c_1 c_2), \tilde{\varepsilon}, c_3\}$ it follows, provided $x_k \in B_\varepsilon(x^*)$, that

$$\|x^* - x_{k+1}\| < c_1 c_2 \varepsilon \|x^* - x_k\| \leq \|x^* - x_k\| < \varepsilon \leq \tilde{\varepsilon}$$

and thus $x_{k+1} \in B_{\tilde{\varepsilon}}(x^*)$. The iteration is therefore well-defined and convergent, provided $x_0 \in B_\varepsilon(x^*)$. □

Remark 15.3 If x_0 is not close enough to x^*, divergence can occur, see Fig. 15.5. This can be avoided in many cases by introducing damping, i.e. by the modification $x_{k+1} = x_k - \omega Df(x_k)^{-1} f(x_k)$ with $0 < \omega < 1$, However, in general, no quadratic convergence is then guaranteed.

Fig. 15.5 The Newton method is generally only locally convergent

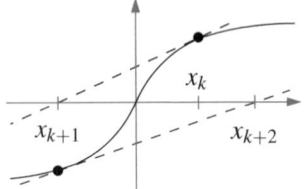

15.3 One-Dimensional Minimisation

The global minimisation of a continuous function on a compact set is rarely achievable without further additional conditions on the function. Therefore, one usually restricts oneself to determining local minima. In the case of convex functions, these are already global minima.

Algorithm 15.4 (Discrete Search) *Let $a = x_0 < x_1 < \cdots < x_n = b$ be a partitioning into $n \geq 3$ subintervals and $g \in C^0([a,b])$. Determine x_k with $g(x_k) = \min\{g(x_1), g(x_2), \ldots, g(x_{n-1})\}$, see Fig. 15.6.*

Proposition 15.3 *If x_k is the point determined by the discrete search, then the interval $[x_{k-1}, x_{k+1}]$ contains a local minimum $x^*_{loc} \in [a,b]$, i.e. there exists a $\delta > 0$ with $g(x^*_{loc}) \leq g(x)$ for all $x \in B_\delta(x^*_{loc}) \cap [a,b]$.*

Proof On the compact interval $[x_{k-1}, x_{k+1}]$, the function g attains its minimum. □

In interval reduction methods, the discrete search is applied to intervals of decreasing lengths, see Fig. 15.7.

Algorithm 15.5 (Interval Reduction) *Let $g \in C^0([a,b])$ and $\varepsilon_{stop} > 0$. Set $a_0 = a$, $b_0 = b$ and $k = 0$.*

(1) Choose $c_k, d_k \in (a_k, b_k)$ with $a_k < c_k < d_k < b_k$ and set

$$(a_{k+1}, b_{k+1}) = \begin{cases} (a_k, d_k) & \text{if } g(c_k) \leq g(d_k), \\ (c_k, b_k) & \text{otherwise.} \end{cases}$$

(2) Stop if $b_{k+1} - a_{k+1} \leq \varepsilon_{stop}$; otherwise increase $k \to k+1$ and repeat step (1).

Remark 15.4 An optimized choice of points c_k and d_k leads to a uniform reduction of the intervals and a minimal number of function evaluations.

Fig. 15.6 The minimum of a finite set of function values provides an approximation of a local minimum

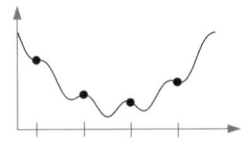

15.4 Multidimensional Minimisation

Fig. 15.7 Reduction of the search region based on multiple function values

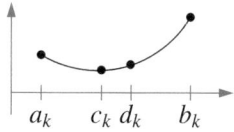

In multiple dimensions, successive one-dimensional minimisations are usually performed along suitable search directions. A canonical choice of the respective search direction is the direction of steepest descent, which is given by the negative gradient of the function to be minimised. We follow the presentations in [7, 8].

Algorithm 15.6 (Gradient Method) *Let $g \in C^1(\mathbb{R}^n)$, $x_0 \in \mathbb{R}^n$, $\sigma \in (0, 1)$ and $\varepsilon_{stop} > 0$. Set $k = 0$.*

(1) *Define $d_k = -\nabla g(x_k)$ and determine the maximum number $\alpha_k \in \{2^{-\ell} : \ell \in \mathbb{N}_0\}$, for which the* Armijo condition

$$g(x_k + \alpha_k d_k) \leq g(x_k) - \sigma \alpha_k \|d_k\|^2$$

is fulfilled, see Fig. 15.8.
(2) *Set $x_{k+1} = x_k + \alpha_k d_k$.*
(3) *Stop if $\|\alpha_k d_k\| \leq \varepsilon_{stop}$; otherwise increase $k \to k+1$ and repeat step (1).*

Remark 15.5 The method performs a discrete search in each iteration step to determine the step size α_k. The existence of an admissible step size follows from an exercise. If $g \in C^2(\mathbb{R}^n)$ and an upper bound for $\|D^2 g\|$ is explicitly available, a fixed step size can be chosen and the Armijo search can be omitted.

In the analysis of the method, the termination criterion is ignored and the convergence of the search directions $(d_k)_{k \in \mathbb{N}_0}$ to zero is proven.

Proposition 15.4 *Let $g \in C^2(\mathbb{R}^n)$ and $x_0 \in \mathbb{R}^n$ be such that the sublevel set*

$$N_g^-(x_0) = \{x \in \mathbb{R}^n : g(x) \leq g(x_0)\}$$

is bounded, i.e. $N_g^-(x_0) \subset K_R(x_0)$ for some $R > 0$, see Fig. 15.9. Then for the iterates of the gradient method, it follows that $\nabla g(x_k) \to 0$ as $k \to \infty$.

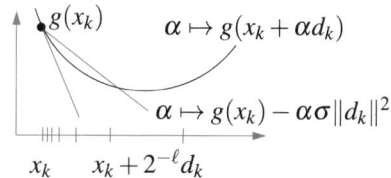

Fig. 15.8 The Armijo condition guarantees a predetermined relative reduction of the function value

Fig. 15.9 Sublevel set $N_g^-(x_0)$ (hatched) of a function g at the level $g(x_0)$

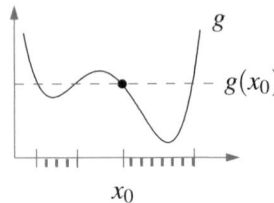

and $\alpha_k > (1-\sigma)/\gamma$ for all $k \in \mathbb{N}_0$ with $\gamma = \max_{x \in K_{R+m}(x_0)} \|D^2 g(x)\|$ and $m = \max_{x \in K_R(x_0)} \|\nabla g(x)\|$.

Proof The sequence $(g(x_k))_{k \in \mathbb{N}_0}$ is monotonically decreasing, so that $(x_k)_{k \in \mathbb{N}_0} \subset N_g^-(x_0)$ and $g(x_k) \geq c_0 = \min_{x \in N_g^-(x_0)} g(x)$ for all $k \in \mathbb{N}_0$. From the Armijo condition it follows

$$g(x_0) \geq g(x_1) + \sigma \alpha_0 \|\nabla g(x_0)\|^2$$
$$\geq g(x_2) + \sigma \alpha_1 \|\nabla g(x_1)\|^2 + \sigma \alpha_0 \|\nabla g(x_0)\|^2$$
$$\geq \cdots \geq g(x_{\ell+1}) + \sigma \sum_{k=0}^{\ell} \alpha_k \|\nabla g(x_k)\|^2.$$

Hence, $\sum_{k=0}^{\infty} \alpha_k \|\nabla g(x_k)\|^2 \leq (g(x_0) - c_0)/\sigma$ and it follows that $\alpha_k \|\nabla g(x_k)\|^2 \to 0$. It is therefore sufficient to show that $\alpha_k \geq \delta > 0$ for all $k \in \mathbb{N}_0$ and a number $\delta > 0$ applies. For each $k \in \mathbb{N}_0$, either $\alpha_k = 1$ or the Armijo condition is violated for $2\alpha_k$. The latter means

$$2\sigma \alpha_k \|\nabla g(x_k)\|^2 > g(x_k) - g(x_k + 2\alpha_k d_k).$$

A Taylor approximation implies that a $\xi \in K_{R+m}(x_0)$ exists with

$$g(x_k + 2\alpha_k d_k) = g(x_k) + \nabla g(x_k) \cdot (2\alpha_k d_k) + \frac{1}{2}(2\alpha_k)^2 D^2 g(\xi)[d_k, d_k].$$

With $d_k = -\nabla g(x_k)$ and $D^2 g(\xi)[d_k, d_k] \leq \gamma \|d_k\|^2$ it follows

$$2\sigma \alpha_k \|d_k\|^2 > 2\alpha_k \|d_k\|^2 - 2\gamma \alpha_k^2 \|d_k\|^2$$

or $(1-\sigma)\alpha_k < \gamma \alpha_k^2$ and thus $\alpha_k > (1-\sigma)/\gamma$ for all $k \in \mathbb{N}_0$. □

Remarks 15.6

(i) The sequence $(x_k)_{k \in \mathbb{N}_0}$ is generally not convergent. If $x_s \in N_g^-(x_0)$ is an accumulation point of a subsequence $(x_{k_n})_{n \in \mathbb{N}_0}$, then x_s is a stationary point of g, that is $\nabla g(x_s) = 0$, and x_s can be a local minimum or maximum or a saddle point. However, local maxima and saddle points are unstable with respect to perturbations, so the gradient method usually converges to a local minimum in practice.

(ii) From the estimates $\sum_{k=0}^{\ell} \alpha_k \|\nabla g(x_k)\|^2 \leq (g(x_0) - c_0)/\sigma$ and $\alpha_k \geq (1-\sigma)/\gamma$ of the proof it follows that

$$(\ell + 1) \min_{k=0,\ldots,\ell} \|\nabla g(x_k)\|^2 \leq \gamma (g(x_0) - c_0)/(\sigma(1-\sigma))$$

or $\min_{k=0,\ldots,n} \|\nabla g(x_k)\| = \mathcal{O}(n^{-1/2})$ applies. For uniformly convex functions, improved convergence properties can be established.

15.5 Learning Objectives, Quiz and Application

You should be familiar with different methods for the approximate calculation of zeros and minima. You should be able to motivate the methods and explain their properties.

Quiz 15.1 Decide for each of the following statements whether it is true or false. You should be able to justify your answer.

If the series $\sum_{k=0}^{\infty} \delta_k$ converges, then the sequence $(\delta_k)_{k \geq 0}$ is linearly convergent.	
The sequence $\delta_k = \sin^2(1/k)$, $k \in \mathbb{N}$, is quadratically convergent to zero.	
The gradient method with a function g defines a convergent sequence $(x_k)_{k \in \mathbb{N}_0}$, whose limit is a critical point of g.	
Sufficient for the convergence of the gradient method is that $g \in C^2(\mathbb{R}^n)$ applies and g is convex.	
If the Newton method converges, then $\|f(x_k)\| \leq c \|x^* - x_k\|$ with a constant $c \geq 0$ for all $k \geq 0$.	

Application 15.1 In the shape optimisation of a rotationally symmetric drinking glass, whose base is circular with a diameter of 3 cm, is 10 cm high and has a volume of about 0.21 ℓ, the surface area should be minimised. The shape of the glass should be described by a cubic curve $s : [0, 10] \to \mathbb{R}$ such that the surface of the glass is given by

$$A(s) = 2\pi \int_0^{10} s(1 + |s'|^2)^{1/2} \, dx + \pi(3/2)^2$$

Use the nodes $0 = x_0 < x_1 < x_2 < x_3 = 10$ to describe the desired curve with the values $y_0 = 1.5$ and y_1, y_2, y_3. For the choice $x_1 = 5$ and the approximation of the volume

$$V(s) = \pi \int_0^{10} (s(x))^2 \, dx$$

using the Simpson rule, the value y_1 can be eliminated. Formulate the surface as a function of y_2 and y_3 and minimise the resulting expression numerically by discretising $A(s)$ appropriately.

Chapter 16
Conjugate Gradient Method

16.1 Quadratic Minimisation

If $A \in \mathbb{R}^{n \times n}$ is symmetric and positive definite, then the solution $x^* \in \mathbb{R}^n$ of the system of equations $Ax = b$ is the unique minimum point of the function

$$\phi(x) = \frac{1}{2}\|b - Ax\|^2_{A^{-1}} = \frac{1}{2}\left(A^{-1}(b - Ax)\right) \cdot (b - Ax) \geq 0,$$

because for every symmetric and positive definite matrix $B \in \mathbb{R}^{n \times n}$ a norm in $\mathbb{R}^n$ is defined by $v \mapsto \|v\|_B = \sqrt{(Bv) \cdot v}$. With a variant of the descent method, for an approximate solution or an initial value $\tilde{x} \in \mathbb{R}^n$ and a search direction $\tilde{d} \in \mathbb{R}^n$, a new approximation $\tilde{x} + \tilde{\alpha}\tilde{d}$ is obtained by minimising $\tilde{\psi} : t \mapsto \phi(\tilde{x} + t\tilde{d})$. We have that

$$\tilde{\psi}(t) = \frac{1}{2}\|b - A\tilde{x}\|^2_{A^{-1}} - t(b - A\tilde{x}) \cdot \tilde{d} + \frac{t^2}{2}(A\tilde{d}) \cdot \tilde{d}$$

and differentiating with respect to t shows that the minimum is given by

$$\tilde{\alpha} = \frac{(b - A\tilde{x}) \cdot \tilde{d}}{(A\tilde{d}) \cdot \tilde{d}}.$$

If the search direction is chosen as the negative gradient of ϕ at $\tilde{x}$, that is

$$\tilde{d} = -\nabla\phi(\tilde{x}) = b - A\tilde{x},$$

then for the new approximate solution we have that

$$\tilde{x}^{new} = \tilde{x} + \tilde{\alpha}\tilde{d} = \tilde{x} + \tilde{\alpha}(b - A\tilde{x}),$$

which corresponds exactly to a step of a Richardson procedure. The repeated execution of this strategy defines a sequence of approximate solutions $(x_k)_{k=0,1,\ldots}$ for which an exercise in the case of symmetric, positive definite matrices shows the convergence behaviour

$$\|x_k - x^*\|_A \leq \left(\frac{\kappa - 1}{\kappa + 1}\right)^k \|x_0 - x^*\|_A$$

with $\kappa = \text{cond}_2(A)$. For large condition numbers, therefore, only a small improvement is generally achieved in each iteration step.

16.2 Conjugate Search Directions

The search directions occurring in the descent method are successively orthogonal to each other, but only lead to slow convergence. A strong acceleration is achieved by using so-called A-conjugate search directions. In the following, it is always assumed that $A \in \mathbb{R}^{n \times n}$ is symmetric and positive definite. In this chapter, we follow the presentation in [8].

Definition 16.1 The vectors $x, y \in \mathbb{R}^n$ are called *A-conjugate*, if $x \cdot (Ay) = 0$ holds.

The concept of A-conjugacy generalises the concept of orthogonality, because orthogonal vectors are A-conjugate with respect to $A = I_n$.

Lemma 16.1 *Assume that the vectors $d_0, d_1, \ldots, d_k \in \mathbb{R}^n \setminus \{0\}$ are pairwise A-conjugate, that is $d_i \cdot A d_j = 0$ for all $0 \leq i, j \leq k$ with $i \neq j$. If $x_0 \in \mathbb{R}^n$ and x_{j+1} is obtained from x_j by successively minimising ϕ in the direction of d_j, that is*

$$x_{j+1} = x_j + \alpha_j d_j = x_0 + \sum_{\ell=0}^{j} \alpha_\ell d_\ell,$$

$$\alpha_j = \frac{d_j \cdot (b - Ax_j)}{d_j \cdot Ad_j} = \frac{d_j \cdot (b - Ax_0)}{d_j \cdot Ad_j}$$

for $j = 1, 2, \ldots, k$, then x_{j+1} is the minimum of ϕ in the set

$$x_0 + \text{span}\{d_0, d_1, \ldots, d_j\}.$$

Proof For $j = 1, 2, \ldots, k+1$, $x_j \in x_0 + \text{span}\{d_0, d_1, \ldots, d_{j-1}\}$ and with the A-conjugacy of the vectors $d_0, d_1, \ldots, d_{j-1}$ it follows

$$d_j \cdot A(x_j - x_0) = 0$$

and thus $d_j \cdot (b - Ax_j) = d_j \cdot (b - Ax_0)$, which proves the second representation of α_j. From this it follows

$$\phi(x_j + \alpha_j d_j) = \phi(x_j) + \frac{\alpha_j^2}{2} d_j \cdot Ad_j - \alpha_j d_j \cdot (b - Ax_j)$$

$$= \phi(x_j) + \frac{\alpha_j^2}{2} d_j \cdot Ad_j - \alpha_j d_j \cdot (b - Ax_0)$$

$$= \phi(x_j) + \psi_j(\alpha_j)$$

with the quadratic function $\psi_j(t) = (t^2/2) d_j \cdot Ad_j - t d_j \cdot (b - Ax_0)$. Inductively, it follows

$$\phi\left(x_0 + \sum_{\ell=0}^{j} \alpha_\ell d_\ell\right) = \phi(x_0) + \sum_{\ell=0}^{j} \psi_\ell(\alpha_\ell).$$

A necessary and sufficient condition for a minimum point of the convex function ϕ in the set $x_0 + \text{span}\{d_0, \ldots, d_j\}$ is the vanishing of the partial derivatives with respect to the coefficients α_i, $i = 0, 1, \ldots, j$, that is

$$\frac{\partial}{\partial \alpha_i} \phi\left(x_0 + \sum_{\ell=0}^{j} \alpha_\ell d_\ell\right) = \psi_i'(\alpha_i) = 0$$

for $i = 0, 1, \ldots, j$. This corresponds exactly to the choice of coefficients and thus the statement of the lemma is proven. □

Remark 16.1 The lemma shows that the coefficients $\alpha_1, \ldots, \alpha_{n-1}$ can be determined independently of each other, provided the A-conjugate vectors are given.

16.3 Calculation of Conjugate Directions

The determination of A-conjugate search directions is carried out simultaneously with the step-by-step improvement of the approximate solutions. For an approximation x_k, the *residual of x_k* is defined by

$$r_k = b - Ax_k.$$

If $r_k = 0$, then x_k solves the system of equations $Ax = b$ and if $x_{k+1} = x_k + \alpha_k d_k$, then obviously $r_{k+1} = r_k - \alpha_k A d_k$.

Lemma 16.2 *For any vector $x_0 \in \mathbb{R}^n$ and $r_0 = b - Ax_0$ as well as $d_0 = r_0$, the recursion*

$$r_{k+1} = r_k - \alpha_k A d_k, \quad d_{k+1} = r_{k+1} - \beta_k d_k,$$

$$\alpha_k = \frac{d_k \cdot r_k}{d_k \cdot A d_k}, \quad \beta_k = \frac{d_k \cdot A r_{k+1}}{d_k \cdot A d_k}$$

determines a sequence of non-vanishing A-conjugate vectors $d_0, d_1, \ldots, d_k$ until $r_{k+1} = 0$ holds. For the Krylov *space defined by $\mathcal{K}_k(A, r_0) = \text{span}\{r_0, A r_0, \ldots, A^{k-1} r_0\}$, we have*

$$\mathcal{K}_k(A, r_0) = \text{span}\{d_0, d_1, \ldots, d_{k-1}\} = \text{span}\{r_0, r_1, \ldots, r_{k-1}\}$$

and r_k is orthogonal to these spaces.

Proof We use the abbreviations $\mathcal{K}_k = \mathcal{K}_k(A, r_0)$, $\mathcal{D}_k = \text{span}\{d_0, d_1, \ldots, d_{k-1}\}$, and $\mathcal{R}_k = \text{span}\{r_0, r_1, \ldots, r_{k-1}\}$. Assume that the equality $\mathcal{K}_k = \mathcal{D}_k = \mathcal{R}_k$ and the A-conjugacy of the vectors $d_0, d_1, \ldots, d_{k-1}$ are proven for some $k \geq 1$. For $k = 1$ the statements are obviously correct and we infer them for $k+1$ in four steps.

(i) We have that $r_k \in \mathcal{K}_{k+1}$. Because $r_{k-1} \in \mathcal{K}_k \subset \mathcal{K}_{k+1}$ as well as $d_{k-1} \in \mathcal{K}_k$ and thus $A d_{k-1} \in \mathcal{K}_{k+1}$ it follows

$$r_k = r_{k-1} - \alpha_{k-1} A d_{k-1} \in \mathcal{K}_{k+1}.$$

(ii) We have that $r_k \perp \mathcal{D}_k$. The A-conjugacy of $d_0, d_1, \ldots, d_{k-1}$, the identity $x_k = x_0 + \sum_{i=0}^{k-1} \alpha_i d_i$ and the choice of α_ℓ show for $0 \leq \ell \leq k-1$, that

$$d_\ell \cdot r_k = d_\ell \cdot (b - A x_k) = d_\ell \cdot (b - A x_0) - \alpha_\ell d_\ell \cdot A d_\ell = 0.$$

(iii) We have that $\mathcal{K}_{k+1} = \mathcal{D}_{k+1} = \mathcal{R}_{k+1}$. If $r_k \neq 0$ then

$$\mathcal{K}_k = \mathcal{R}_k = \mathcal{D}_k \subsetneq \mathcal{R}_{k+1} = \text{span}\{r_0, r_1, \ldots, r_k\} \subset \mathcal{K}_{k+1},$$

it follows by dimension that $\mathcal{R}_{k+1} = \mathcal{K}_{k+1}$. With $d_k = r_k - \beta_{k-1} d_{k-1}$ we also get $\mathcal{D}_{k+1} = \mathcal{R}_{k+1}$.

(iv) We have that $d_\ell \cdot A d_k = 0$ for $\ell = 0, 1, \ldots, k-1$. From $d_k = r_k - \beta_{k-1} d_{k-1}$ it follows

$$d_\ell \cdot A d_k = d_\ell \cdot A r_k - \beta_{k-1} d_\ell \cdot A d_{k-1}.$$

If $\ell = k-1$, the choice of β_{k-1} implies that $d_{k-1} \cdot A d_k = 0$ holds. If on the other hand $\ell \leq k-2$, the orthogonality $r_k \perp \mathcal{K}_k$, the inclusion $A d_\ell \in \mathcal{K}_k$ and the A-conjugacy of the vectors $d_0, d_1, \ldots, d_{k-1}$ imply that $d_\ell \cdot A d_k = 0$ holds. □

16.4 CG Method

For the efficient implementation of the iterative method with A-conjugated search directions, we note that the orthogonality of r_k and d_{k-1}, that is $r_k \cdot d_{k-1} = 0$, implies the equation

$$\alpha_k = \frac{d_k \cdot r_k}{d_k \cdot Ad_k} = \frac{(r_k - \beta_{k-1}d_{k-1}) \cdot r_k}{d_k \cdot Ad_k} = \frac{\|r_k\|^2}{d_k \cdot Ad_k}$$

From $r_k \in \mathcal{K}_{k+1}(A, r_0) = \text{span}\{d_0, d_1, \ldots, d_k\} \perp r_{k+1}$ it follows

$$d_k \cdot Ar_{k+1} = (Ad_k) \cdot r_{k+1} = \frac{1}{\alpha_k}(r_k - r_{k+1}) \cdot r_{k+1} = -\frac{d_k \cdot Ad_k}{\|r_k\|^2}\|r_{k+1}\|^2$$

and thus

$$\beta_k = \frac{d_k \cdot Ar_{k+1}}{d_k \cdot Ad_k} = -\frac{\|r_{k+1}\|^2}{\|r_k\|^2}.$$

With these identities, the *conjugate gradient method* is implemented as follows.

Algorithm 16.1 (CG Method) Let $A \in \mathbb{R}^{n \times n}$ be symmetric and positive definite, $b \in \mathbb{R}^n$, $x_0 \in \mathbb{R}^n$ and $\varepsilon_{stop} > 0$. Define $d_0 = r_0 = b - Ax_0$ and $k = 0$.

(1) Set $x_{k+1} = x_k + \alpha_k d_k$, $r_{k+1} = r_k - \alpha_k Ad_k$ and $d_{k+1} = r_{k+1} - \beta_k d_k$ with

$$\alpha_k = \frac{\|r_k\|^2}{d_k \cdot Ad_k}, \quad \beta_k = -\frac{\|r_{k+1}\|^2}{\|r_k\|^2}.$$

(2) Stop if $\|r_{k+1}\|/\|b\| \leq \varepsilon_{stop}$ holds; otherwise increase $k \to k+1$ and repeat step (1).

Remark 16.2 The algorithm terminates after at most n steps, since otherwise we have $r_n \perp \text{span}\{d_0, d_1, \ldots, d_{n-1}\}$ holds and the vectors $d_0, d_1, \ldots, d_{n-1}$ are linearly independent, unless $r_k = 0$ for a $0 \leq k \leq n-1$ already holds. In particular, the exact solution of the linear system is obtained with a maximum of n steps.

The difference between the CG method and the descent method is schematically illustrated in Fig. 16.1.

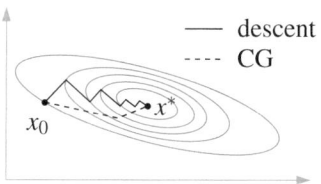

Fig. 16.1 The CG method often requires fewer iteration steps than the descent method

16.5 Convergence of the CG Method

In many cases, the CG method provides a good approximation of the solution of the system of equations after a few steps.

Proposition 16.1 *For the iterates $x_0, x_1, \ldots$ of the CG method and the solution x^* of the system of equations $Ax = b$, we have with $\kappa = \mathrm{cond}_2(A)$*

$$\|x^* - x_k\|_A \le 2\left(\frac{\kappa^{1/2}-1}{\kappa^{1/2}+1}\right)^k \|x^* - x_0\|_A.$$

Proof The minimality property of the iterates from Lemma 16.1 and the representations of the Krylov spaces from Lemma 16.2 imply

$$\phi(x_k) = \min_{y \in x_0 + \mathrm{span}\{d_0,\ldots,d_{k-1}\}} \phi(y) = \min_{y \in x_0 + \mathcal{K}_k(A, r_0)} \phi(y).$$

Since $b - Ay = A(x^* - y)$ and $\|Av\|_{A^{-1}} = \|v\|_A$, we have

$$\phi(x_k) = \frac{1}{2}\|x^* - x_k\|_A^2, \quad \phi(y) = \frac{1}{2}\|x^* - y\|_A^2.$$

This leads to

$$\|x^* - x_k\|_A^2 = \min_{y \in x_0 + \mathcal{K}_k(A, r_0)} \|x^* - y\|_A^2.$$

For every $y \in x_0 + \mathcal{K}_k(A, r_0)$ there exists a vector $c = [c_1, c_2, \ldots, c_k]^T \in \mathbb{R}^k$ with

$$y = x_0 + c_1 A^0 r_0 + c_2 A^1 r_0 + \cdots + c_k A^{k-1} r_0$$
$$= x_0 + c_1 A(x^* - x_0) + \cdots + c_k A^k (x^* - x_0),$$

where we have used $r_0 = A(x^* - x_0)$. If $\mathscr{P}_k$ denotes the space of polynomials of maximum degree k, then

$$\|x^* - x_k\|_A^2 = \min_{c \in \mathbb{R}^k} \|x^* - x_0 - c_1 A(x^* - x_0) - \cdots - c_k A^k(x^* - x_0)\|_A^2$$
$$= \min_{c \in \mathbb{R}^k} \|(I_n - c_1 A - \ldots - c_k A^k)(x^* - x_0)\|_A^2 = \min_{p \in \mathscr{P}_k,\, p(0)=1} \|p(A)(x^* - x_0)\|_A^2.$$

Since A is symmetric and positive definite, there exist eigenvalues $0 < \lambda_1 \le \lambda_2 \le \cdots \le \lambda_n$ and corresponding orthonormal eigenvectors $v_1, v_2, \ldots, v_n \in \mathbb{R}^n$. With suitable coefficients $\gamma_1, \gamma_2, \ldots, \gamma_n$ we have

$$x^* - x_0 = \sum_{i=1}^n (v_i \cdot (x^* - x_0)) v_i = \sum_{i=1}^n \gamma_i v_i$$

16.5 Convergence of the CG Method

and

$$\|x^* - x_0\|_A^2 = \left(A(x^* - x_0)\right) \cdot (x^* - x_0) = \left(\sum_{i=1}^n \lambda_i \gamma_i v_i\right) \cdot \left(\sum_{j=1}^n \gamma_j v_j\right) = \sum_{i=1}^n \lambda_i \gamma_i^2.$$

For each polynomial $p \in \mathscr{P}_k$ with $p(0) = 1$ using $p(A)v_i = p(\lambda_i)v_i$, it follows that

$$\|p(A)(x^* - x_0)\|_A^2 = \Big\|\sum_{i=1}^n \gamma_i p(A) v_i\Big\|_A^2 = \Big\|\sum_{i=1}^n \gamma_i p(\lambda_i) v_i\Big\|_A^2$$

$$= \left(\sum_{i=1}^n \gamma_i p(\lambda_i) A v_i\right) \cdot \left(\sum_{j=1}^n \gamma_j p(\lambda_j) v_j\right) = \sum_{i=1}^n \gamma_i^2 |p(\lambda_i)|^2 \lambda_i$$

$$\leq \max_{i=1,\ldots,n} |p(\lambda_i)|^2 \sum_{i=1}^n \gamma_i^2 \lambda_i = \max_{i=1,\ldots,n} |p(\lambda_i)|^2 \|x^* - x_0\|_A^2.$$

In the case $\lambda_1 = \lambda_2 = \cdots = \lambda_n$ we can find a polynomial $p \in \mathscr{P}_k$ with $p(0) = 1$, such that $p(\lambda_i) = 0$ for $i = 1, 2, \ldots, n$ and the statement of the proposition is proven. We assume in the following that $\lambda_1 < \lambda_n$ holds. With the k-th Chebyshev polynomial $T_k \in \mathscr{P}_k$ whose roots are contained in the interval $[-1, 1]$ and in this interval is given by $T_k(s) = \cos(k \arccos(s))$, we set

$$q(t) = T_k\left(\frac{\lambda_n + \lambda_1 - 2t}{\lambda_n - \lambda_1}\right) \Big/ T_k\left(\frac{\lambda_n + \lambda_1}{\lambda_n - \lambda_1}\right).$$

Then $q \in \mathscr{P}_k$ with $q(0) = 1$ holds. If $t \in [\lambda_1, \lambda_n]$ then $(\lambda_n + \lambda_1 - 2t)/(\lambda_n - \lambda_1) \in [-1, 1]$ and from $\max_{s \in [-1,1]} |T_k(s)| \leq 1$ it follows

$$\max_{i=1,\ldots,n} |q(\lambda_i)| \leq \left[T_k\left(\frac{\lambda_n + \lambda_1}{\lambda_n - \lambda_1}\right)\right]^{-1} = \left[T_k\left(\frac{\lambda_n/\lambda_1 + 1}{\lambda_n/\lambda_1 - 1}\right)\right]^{-1}.$$

An exercise shows

$$T_k\left(\frac{s+1}{s-1}\right) \geq \frac{1}{2} \frac{(s^{1/2} + 1)^k}{(s^{1/2} - 1)^k}$$

for $s > 1$ and thus with $\kappa = \lambda_n/\lambda_1$, it follows that

$$\|x^* - x_k\|_A \leq \max_{i=1,\ldots,n} |q(\lambda_i)| \|x^* - x_0\|_A \leq 2 \frac{(\kappa^{1/2} - 1)^k}{(\kappa^{1/2} + 1)^k} \|x^* - x_0\|_A.$$

This proves the claim. $\square$

Example 16.1 If $\kappa = 100$, then the CG method yields an error reduction by $q \approx 0.8$ in each iteration step and about 20 steps are required to reach 1% of the initial error. In the descent method, $q \approx 0.98$ is obtained and more than 200 steps are required.

Remark 16.3 The condition number is relevant for two aspects of numerical mathematics. On the one hand, it describes the effects of perturbations on the solution of a system of linear equations and on the other hand, it indicates how many steps are required in the approximate iterative solution of a system of equations. In the first case, the choice of norms that determine the condition number is usually dictated by the application, while in the second case, the condition number induced by the spectral norm is of interest.

16.6 Learning Objectives, Quiz and Application

You should be able to explain the concept of conjugate search directions and their significance in the iterative solution of systems of linear equations. You should be able to state sufficient conditions for the convergence of the CG method. Furthermore, you should be able to carry out comparative effort considerations.

Quiz 16.1 Decide for each of the following statements whether it is true or false. You should be able to justify your answer.

With the Cholesky decomposition $A = LL^T$ of the matrix A, $\|x\|_A = \|Lx\|$.	
If $x_0 = x^*$ is the solution of $Ax = b$, then the Krylov spaces are trivial, that is, $\mathcal{K}_k = \{0\}$ for $k = 1, 2, \ldots, n$.	
For $x \in \mathbb{R}^n$ and $A \in \mathbb{R}^{n \times n}$, $\|x\|_A = \|Ax\|_{A^{-1}}$.	
One iteration step of the CG method requires an effort of $\mathcal{O}(n^2)$ operations.	
Non-vanishing, pairwise A-conjugate vectors are linearly independent.	

Application 16.1 For a simple mathematical description of a two-dimensional diffusion process we consider a grid on the domain $[0, 1]^2$ with grid points $x_{ij} = (i, j)h$, $i, j = 0, 1, \ldots, n$ and grid width $h = 1/n$. Let u_{ij}^k denote the concentration of a substance near the grid point x_{ij} at time t_k, that is, the quotient of the amount of particles of the considered substance in the region $x_{ij} + [-h/2, h/2]^2$ and the area h^2. The probability that a particle within the time interval $[t_k, t_{k+1}]$ of length τ jumps from the vicinity of a grid point to the vicinity of a neighbouring point is denoted by p. This is proportional to the length h of the interface, to the length τ of the time interval, inversely proportional to the area h^2 and inversely proportional to the average distance h, that is, with a diffusion constant $c > 0$ we have

$$p = c\frac{\tau}{h^2}.$$

16.6 Learning Objectives, Quiz and Application

Obviously, $p \leq 1/4$ should hold. At the boundary points the concentration is kept at zero by removing or adding substance. The quantity f_{ij}^{k+1} denotes the amount added or removed in the vicinity of an inner grid point x_{ij} in the time interval $[t_k, t_{k+1}]$ relative to the volume h^2. For the concentration at time t_{k+1}, we thus have

$$u_{ij}^{k+1} = (1-4p)u_{ij}^k - p\left(u_{i-1,j}^k + u_{i+1,j}^k + u_{i,j-1}^k + u_{i,j+1}^k\right) + \tau f_{ij}^{k+1}$$

for inner grid points and $u_{ij}^{k+1} = 0$ for grid points on the sides of $[0, 1]^2$. If the grid function f_{ij}^k is constant over time, an equilibrium will be established after a certain period of time, that is $u_{ij}^{k+1} \approx u_{ij}^k$ for all $0 \leq i, j \leq n$ and all $k \geq K$.

(i) Show that the equilibrium state of the diffusion process can be determined as the solution of a system of linear equations with a symmetric and positive definite system matrix.
(ii) Experimentally investigate the dependence of the condition number of the matrix A on h and determine the necessary number of iteration steps of the descent and CG methods to achieve an accuracy $\varepsilon_{stop} = h$.
(iii) Solve the system of linear equations approximately with the CG method and present the approximate solution with the help of the MATLAB commands meshgrid and surf for the case $f_{ij} = 1$ graphically.

Chapter 17
Sparse Matrices and Preconditioning

17.1 Sparse Matrices

The CG method requires a matrix-vector multiplication in each iteration step and is therefore particularly efficient when this is associated with low effort. This is the case when only a few entries of the system matrix are different from zero. In the following, the matrix $A \in \mathbb{R}^{n \times n}$ always represents a sequence $(A_\ell)_{\ell \in \mathbb{N}}$ with $A_\ell \in \mathbb{R}^{n_\ell \times n_\ell}$ with $n_\ell \to \infty$ for $\ell \to \infty$.

Definition 17.1 The matrix $A \in \mathbb{R}^{n \times n}$ is called *sparse* if for the number of entries different from zero $N_{nz} = |\{(i,j) : 1 \le i, j \le n, a_{ij} \ne 0\}|$ we have that $N_{nz} = \mathcal{O}(n)$. The index nz stands for *not zero*.

Example 17.1 Band matrices $A \in \mathbb{R}^{n \times n}$ with a number $k \in \mathbb{N}$ of non-vanishing subdiagonals independent of n, i.e. $a_{ij} \ne 0$ implies $|i - j| \in \{d_1, d_2, \ldots, d_k\}$ with numbers $d_r \in \mathbb{N}_0$, $r = 1, 2, \ldots, k$, are sparse. The *bandwidth* is given by $w = \max_{r=1,\ldots,k} d_r$, see Fig. 17.1.

To save memory, sparse matrices are not stored as $n \times n$-arrays. Instead, lists $I, J \in \mathbb{N}^{N_{nz}}$ and $X \in \mathbb{R}^{N_{nz}}$ are used, which contain the positions and values of the entries of A different from zero, i.e. we have

$$a_{ij} \ne 0 \iff \exists 1 \le k \le N_{nz}, (i,j) = (I_k, J_k), a_{ij} = X_k.$$

This representation is called *coordinate representation*. More generally, if a position (i, j) appears repeatedly in the index lists, the corresponding values are usually summed up. The memory requirement is further reduced in the *compressed-column-storage (CCS)* format by defining I and X as above and the ℓ-th entry of a list $\widetilde{J} \in \mathbb{N}_0^n$ specifies from which position in I and X the entries of the ℓ-th column begin.

Fig. 17.1 Schematic representation of a band matrix with few entries different from zero

Example 17.2 For the following matrix $A \in \mathbb{R}^{4\times 4}$, the lists I, J, X and $\tilde{J}$ result:

$$A = \begin{bmatrix} 7 & 0 & 1 & 0 \\ 0 & 8 & 0 & 3 \\ 4 & 0 & 5 & 0 \\ 2 & 0 & 0 & 3 \end{bmatrix}, \quad \begin{aligned} I &= [1, 3, 4, 2, 1, 3, 2, 4]^T, \\ J &= [1, 1, 1, 2, 3, 3, 4, 4]^T, \\ X &= [7, 4, 2, 8, 1, 5, 3, 3]^T, \end{aligned} \quad \tilde{J} = [1, 4, 5, 7]^T.$$

The matrix-vector multiplication with a matrix in the coordinate format can be easily implemented.

Remark 17.1 The vector $y = Az$ is calculated by:

$$y = 0; \quad \text{for } \ell = 1 : N_{nz}; \ y_{I(\ell)} = y_{I(\ell)} + X(\ell) z_{J(\ell)}; \quad \text{end}$$

17.2 Preconditioned CG Method

The number of required iterations of the CG method for the approximate solution of the linear system $Ax = b$ depends on the condition number of the symmetric and positive definite system matrix A. By choosing a suitable invertible matrix $C \in \mathbb{R}^{n\times n}$, however, it is attractive to consider the equivalent system

$$(CA)x = Cb.$$

If $\mathrm{cond}(CA) \ll \mathrm{cond}(A)$, it can be expected that this system can be solved faster and more robustly, provided the matrix C has a simple structure, so that the multiplication with C can be implemented efficiently. In terms of the condition number, the choice $C = A^{-1}$ would be optimal, but then the multiplication with C would be equivalent to solving the original problem $Ax = b$. Therefore, an approximate inverse is chosen in the sense of the following definition.

Definition 17.2 Let $A \in \mathbb{R}^{n\times n}$ be regular. A regular matrix $C \in \mathbb{R}^{n\times n}$ is called *preconditioning matrix* for A, if $\mathrm{cond}(CA) \leq \mathrm{cond}(A)$ holds and the computational effort of the matrix-vector multiplication $z \mapsto Cz$ is less than the direct solution of the linear system $Ax = b$.

A simple type of preconditioning is the *row equilibration*.

17.2 Preconditioned CG Method

Proposition 17.1 *Let $A \in \mathbb{R}^{n \times n}$ be regular and the diagonal matrix $C \in \mathbb{R}^{n \times n}$ for $i = 1, 2, \ldots, n$ defined by*

$$C_{ii} = \Big(\sum_{j=1}^{n} |a_{ij}|\Big)^{-1}.$$

Then C is a preconditioning matrix for A with respect to the row sum norm.

Proof The matrix $B = CA$ satisfies $\sum_{j=1}^{n} |b_{ij}| = 1$ for all $i = 1, 2, \ldots, n$, and consequently $\|B\|_\infty = 1$. For any diagonal matrix $T \in \mathbb{R}^{n \times n}$ it follows

$$\|TB\|_\infty = \max_{1 \le i \le n} |t_{ii}| \sum_{j=1}^{n} |b_{ij}| = \max_{1 \le i \le n} |t_{ii}| = \|T\|_\infty$$

and thus we get

$$\operatorname{cond}_\infty(B) = \|B^{-1}\|_\infty = \|(TB)^{-1}T\|_\infty \le \|(TB)^{-1}\|_\infty \|T\|_\infty$$
$$= \|(TB)^{-1}\|_\infty \|TB\|_\infty = \operatorname{cond}_\infty(TB).$$

Since the estimate also applies for $T = C^{-1}$ and the matrix-vector multiplication $z \mapsto Cz$ can be realised with n operations, the statement follows. □

In general, the preconditioned system matrix CA is neither symmetric nor positive definite, even if A and C have these properties, and therefore the convergence of the CG method for the preconditioned system is not immediately guaranteed. However, this can be circumvented by using the Cholesky decomposition $C = VV^\mathsf{T}$, because we have

$$Ax = b \iff V^\mathsf{T} A V \widetilde{x} = V^\mathsf{T} b, \quad \widetilde{x} = V^{-1} x$$

and the matrix $V^\mathsf{T} A V$ is symmetric and positive definite. The preconditioned CG method solves this transformation, without using the Cholesky factorisation explicitly. To demonstrate this, we apply the CG method to the preconditioned system of equations

$$\widetilde{A}\widetilde{x} = \widetilde{b}, \quad \widetilde{A} = V^\mathsf{T} A V, \quad \widetilde{b} = V^\mathsf{T} b,$$

so that the sought solution x is given by $x = V\widetilde{x}$. In the iteration rules of the CG method

$$\widetilde{x}_{k+1} = \widetilde{x}_k + \alpha_k \widetilde{d}_k, \quad \widetilde{r}_{k+1} = \widetilde{r}_k - \alpha_k \widetilde{A} \widetilde{d}_k, \quad \widetilde{d}_{k+1} = \widetilde{r}_{k+1} - \beta_k \widetilde{d}_k,$$
$$\alpha_k = \|\widetilde{r}_k\|^2 / (\widetilde{d}_k \cdot \widetilde{A}\widetilde{d}_k), \quad \beta_k = -\|\widetilde{r}_{k+1}\|^2 / \|\widetilde{r}_k\|^2,$$

the explicit use of the product $V^T A V$ should be avoided. To do this, the first and third equation are multiplied by V and the second by V^{-T}. This gives rise to the quantities $x_k = V\tilde{x}_k$, $r_k = V^{-T}\tilde{r}_k$ and $d_k = V\tilde{d}_k$ which satisfy

$$x_{k+1} = x_k + \alpha_k d_k, \quad r_{k+1} = r_k - \alpha_k A d_k \quad d_{k+1} = Cr_{k+1} - \beta_k d_k,$$

where the equations $V^{-T}\tilde{A}\tilde{d}_k = Ad_k$ and $V\tilde{r}_{k+1} = Cr_{k+1}$ were exploited. For the calculation of the coefficients the following results

$$\alpha_k = \frac{V^T r_k \cdot V^T r_k}{V^{-1} d_k \cdot V^T AV(V^{-1} d_k)} = \frac{Cr_k \cdot r_k}{d_k \cdot Ad_k},$$

$$\beta_k = -\frac{V^T r_{k+1} \cdot V^T r_{k+1}}{V^T r_k \cdot V^T r_k} = -\frac{Cr_{k+1} \cdot r_{k+1}}{Cr_k \cdot r_k}.$$

By introducing the variable $z_k = Cr_k$ the following procedure is obtained.

Algorithm 17.1 (Preconditioned CG Method) *Let $A, C \in \mathbb{R}^{n \times n}$ be symmetric and positive definite, $b \in \mathbb{R}^n$, $x_0 \in \mathbb{R}^n$ and $\varepsilon_{stop} > 0$. Define $r_0 = b - Ax_0$, $k = 0$ and set $d_0 = z_0 = Cr_0$.*

(1) *Set $x_{k+1} = x_k + \alpha_k d_k$ and $r_{k+1} = r_k - \alpha_k A d_k$ as well as $z_{k+1} = Cr_{k+1}$ and define $d_{k+1} = z_{k+1} - \beta_k d_k$ with*

$$\alpha_k = \frac{z_k \cdot r_k}{d_k \cdot A d_k}, \quad \beta_k = -\frac{z_{k+1} \cdot r_{k+1}}{z_k \cdot r_k}.$$

(2) *Stop if $\|r_{k+1}\|/\|b\| \leq \varepsilon_{stop}$; otherwise increase $k \to k+1$ and repeat (1).*

Remarks 17.2

(i) The reformulation of the system of equations requires to impose the property $\text{cond}(V^T AV) \leq \text{cond}(A)$ on a preconditioning matrix $C = VV^T$.
(ii) The construction of suitable preconditioning matrices is usually based on particular properties of the underlying application.

17.3 Further Preconditioning Matrices

Stationary iteration methods of the form

$$x_{k+1} = x_k - R(Ax_k - b) = (I_n - RA)x_k + Rb$$

can be interpreted as fixed point iterations of the system of equations

$$RAx = Rb$$

17.3 Further Preconditioning Matrices

They are convergent, provided $\varrho(I_n - RA) < 1$ holds, and motivate the choice of $C = R$ as a preconditioning matrix, since then in rough approximation $CA \approx I_n$, so we can expect $\operatorname{cond}(CA) \approx 1$. Whether a preconditioning matrix is actually well-defined must be checked in each individual case.

Examples 17.3

(i) With the decomposition $A = L + D + R$ into the diagonal part D and the strict lower and upper part L and R the Jacobi method is defined by

$$Dx_{k+1} = -(L + R)x_k + b = (D - A)x_k + b$$

or

$$x_{k+1} = x_k - D^{-1}(Ax_k - b),$$

which motivates the preconditioning matrix $C_J = D^{-1}$.

(ii) The Gauss-Seidel method leads to the matrix $C_{GS} = D + L$, which is generally not symmetric. The symmetric Gauss-Seidel preconditioning matrix of a symmetric matrix $A = L + D + L^\mathsf{T} \in \mathbb{R}^{n \times n}$ is defined by

$$C_{SGS} = \left[(D + L)D^{-1}(D + L)^\mathsf{T}\right]^{-1}.$$

The direct solution of a sparse system of equations using an LU or Cholesky decomposition can be inefficient, as the factors of the decomposition are generally not sparse. This effect is referred to as *fill-in*. The incomplete calculation of an LU or Cholesky decomposition can, however, lead to a suitable preconditioning matrix. A population structure $\mathscr{B} \subset \{1, 2, \ldots, n\} \times \{1, 2, \ldots, n\}$ is specified for the factors and it is required that

$$(LU)_{ij} = a_{ij}, \ (i, j) \in \mathscr{B}, \quad \ell_{ij} = u_{ij} = 0, \ (i, j) \notin \mathscr{B}.$$

For certain classes of matrices, the existence of the incomplete LU or Cholesky decomposition can be proven. The calculation is done by ignoring the entries in the null pattern in the algorithms for the complete factorisations. It should be noted that the factors of the incomplete Cholesky decompositions may not be regular, which is referred to as *pivot breakdown*.

Algorithm 17.2 (Incomplete Cholesky Decomposition) *Let $A \in \mathbb{R}^{n \times n}$ be symmetric and positive definite and $\mathscr{B} \subset \{1, 2, \ldots, n\} \times \{1, 2, \ldots, n\}$ be symmetric.*

The non-trivial entries of L are calculated by:

for $k = 1 : n$

$$\ell_{kk} = \left(a_{kk} - \sum_{j=1,\ldots,k-1,\,(j,k)\in\mathscr{B}} \ell_{kj}^2 \right)^{1/2}$$

    for $i = k+1 : n$

        if $(i,k) \in \mathscr{B}$; $\ell_{ik} = \left(a_{ik} - \sum_{\substack{j=1,\ldots,k-1,\\ (j,k)\in\mathscr{B},\,(i,j)\in\mathscr{B}}} \ell_{ij}\ell_{kj} \right)/\ell_{kk}$;   end

    end

end

With an incomplete factorisation, a preconditioning matrix can be defined.

Example 17.4 If the incomplete Cholesky decomposition $A = LL^\mathsf{T} + E$ exists, then $C = (LL^\mathsf{T})^{-1}$ defines a possible preconditioning matrix. Typical definitions for the occupancy structure are those of the given matrix A, that is, $\mathscr{B} = \{(i,j) : a_{ij} \neq 0\}$, which is referred to as *zero-fill-in*, or a bandwidth $w \in \mathbb{N}_0$ is specified and $\mathscr{B} = \{(i,j) : |i - j| \leq w\}$ is defined.

17.4 Learning Objectives, Quiz and Application

You should be able to explain the concept of a sparse matrix and illustrate it with examples. Furthermore, you should be familiar with the basic ideas of using a preconditioning matrix in the CG method and be able to name some examples.

Quiz 17.1 Decide for each of the following statements whether it is true or false. You should be able to justify your answer.

If for the number of non-zero entries $N_z = |\{(i,j) : 1 \leq i, j \leq n,\, a_{ij} = 0\}| = \mathscr{O}(n^2)$, then A is sparse.
A sparse matrix $A \in \mathbb{R}^{n \times n}$ is specified in the CCS format by $\mathscr{O}(n)$ pieces of information.
The product of two sparse matrices is a sparse matrix.
Every row-equilibrated matrix $A \in \mathbb{R}^{n \times n}$, that is $\sum_{j=1}^{n} |a_{ij}| = 1$, satisfies $\mathrm{cond}_\infty(A) = 1$.
The preconditioning of a system of linear equations leads to a system that can be solved with the effort $\mathscr{O}(n)$.

Application 17.1 To illustrate Google's PageRank algorithm, a model internet with N pages is considered. Let n_i be the number of links leading from the i-th page to

17.4 Learning Objectives, Quiz and Application

Fig. 17.2 Links in a model internet

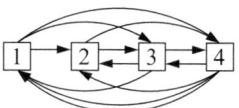

other pages. The variable $x_i \geq 0$ is supposed to indicate the relevance of the i-th page and for each link leading from the j-th to the i-th page, it increases by the value x_j/n_j. In the sketch shown in Fig. 17.2, for example, we have

$$x_1 = \frac{0}{2}x_2 + \frac{1}{3}x_3 + \frac{2}{4}x_4.$$

Overall, a system of linear equations for determining the vector $x = [x_1, x_2, \ldots, x_N]^T$, which describes a balance state of the proportional page accesses of a group of users, is established when they repeatedly switch between pages at random.

(i) Show that the determination of a solution of the system of equations can be formulated as an eigenvalue problem $\lambda x = Ax$ with $\lambda = 1$.

(ii) Determine the Gerschgorin circles for A^T, to show that $|\lambda| \leq 1$ for all eigenvalues of A, and prove that $\lambda = 1$ is an eigenvalue of A^T or A.

(iii) Determine with the help of MATLAB an eigenvector x of the matrix A for the eigenvalue 1 with $x_i \geq 0$, $i = 1, 2, \ldots, N$, and $\|x\|_1 = 1$ for the model internet shown in Fig. 17.2.

(iv) Perform 5 steps of the power method with the starting vector $x_0 = [1, 1, 1, 1]^T/4$ and normalise with respect to the norm $\|\cdot\|_1$.

(v) Discuss whether the matrix A can be assumed to be sparse in reality and whether the effort can be reduced by using suitable storage formats and algorithms for matrix-vector multiplication.

Chapter 18
Multidimensional Approximation

18.1 Grids and Triangulations

There are various approaches to approximating functions and integrals in multiple dimensions, which depend on the properties of the underlying domain. A domain is defined as an open and connected set $\Omega \subset \mathbb{R}^d$ with $d \in \mathbb{N}$, which is also always assumed to be bounded in the following. The simplest situation arises when Ω is the product of intervals, i.e. when Ω is a right-angled, axis-parallel parallelepiped of the form

$$\Omega = (a_1, b_1) \times (a_2, b_2) \times \cdots \times (a_d, b_d) = \prod_{i=1}^{d}(a_i, b_i)$$

In this case, one-dimensional arguments can be transferred to the multidimensional case using tensor product approaches.

Definition 18.1 A *(tensor product) grid* of the domain $\Omega = \prod_{i=1}^{d}(a_i, b_i)$ is a set of points

$$\mathcal{G}_h = \{x = (a_1, a_2, \ldots, a_d) + (j_1 h_1, j_2 h_2, \ldots, j_d h_d) :$$
$$0 \leq j_i \leq n_i,\ i = 1, 2, \ldots, d\}$$

with *grid fineness* $h_i = (b_i - a_i)/n_i$, $n_i \in \mathbb{N}$, $i = 1, 2, \ldots, d$, see Fig. 18.1. The grid is called *uniform* if $h_1 = h_2 = \cdots = h_d = h$.

In the case of a more general bounded domain $\Omega \subset \mathbb{R}^d$, we assume that it has a polygonal boundary, i.e. there exist affine-linear subspaces $H_k = \{x \in \mathbb{R}^d : d_k \cdot x = c_k\}$

Fig. 18.1 Tensor product grid of a rectangle with grid fineness h_1 and h_2

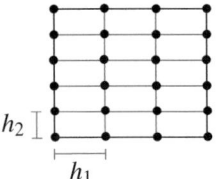

Fig. 18.2 Triangulation of a two-dimensional domain into triangles

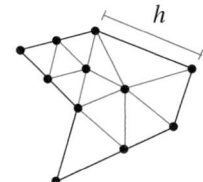

with $d_k \in \mathbb{R}^d$ and $c_k \in \mathbb{R}$, so that

$$\partial \Omega = \bigcup_{k=1}^{K} (\partial \Omega \cap H_k).$$

Domains of this type can be divided into simple subdomains. A *simplex* in $\mathbb{R}^d$ is a closed subset $T \subset \mathbb{R}^d$, which is given as the convex hull of $d+1$ points $z_0, z_1, \ldots, z_d \in \mathbb{R}^d$, i.e.

$$T = \text{conv}\{z_0, z_1, \ldots, z_d\} = \left\{ x \in \mathbb{R}^d : x = \sum_{i=0}^{d} \theta_i z_i, \ \theta_i \geq 0, \ \sum_{i=0}^{d} \theta_i = 1 \right\},$$

so that T is non-degenerate, i.e. it has a non-empty interior or a positive d-dimensional volume. For $d = 1, 2, 3$, simplices are intervals, triangles or tetrahedra, respectively, see Fig. 18.2.

Definition 18.2 A *(regular) triangulation* of the polygonal domain Ω is a set $\mathscr{T}_h = \{T_1, T_2, \ldots, T_J\}$ of simplices $T_j \subset \mathbb{R}^d$, $j = 1, 2, \ldots, J$, so that

$$\overline{\Omega} = \bigcup_{j=1}^{J} T_j$$

and the intersection $T_j \cap T_k$ of two different simplices is either empty or a common subsimplex, i.e. a common corner, edge or side surface. The simplices of a triangulation are also referred to as *elements* and the set $\mathscr{N}_h$ of the corners of elements as *nodes*. The triangulation is called *uniform* if all elements are congruent. It has the *(maximum) mesh width* $h > 0$, if $\text{diam}(T) \leq h$ holds for all $T \in \mathscr{T}_h$.

Different polynomial spaces are used on parallelepipeds and simplices.

18.2 Approximation on Tensor Product Grids

Definition 18.3 Let $A \subset \mathbb{R}^d$ be a closed set and $k \in \mathbb{N}_0$. The set of polynomials of *partial degree k* and of *total degree k* on A are defined by

$$\mathcal{Q}_k(A) = \left\{ q(x) = \sum_{0 \leq i_1, i_2, \ldots, i_d \leq k} a_{i_1 i_2 \ldots i_d} x_1^{i_1} x_2^{i_2} \ldots x_d^{i_d} : a_{i_1 i_2 \ldots i_d} \in \mathbb{R} \right\},$$

$$\mathcal{P}_k(A) = \left\{ p(x) = \sum_{\substack{0 \leq i_1, i_2, \ldots, i_d \leq k, \\ i_1 + i_2 + \cdots + i_d \leq k}} a_{i_1 i_2 \ldots i_d} x_1^{i_1} x_2^{i_2} \ldots x_d^{i_d} : a_{i_1 i_2 \ldots i_d} \in \mathbb{R} \right\}.$$

Remarks 18.1

(i) Polynomials of partial degree k are linear combinations of tensor products of one-dimensional polynomials of degree k.
(ii) We have $\dim \mathcal{Q}_1(A) = 2^d$ and $\dim \mathcal{P}_1(A) = d+1$, which corresponds exactly to the number of corners of parallelepipeds and simplices in $\mathbb{R}^d$.

Example 18.1 The polynomial $q(x_1, x_2) = x_1^2 x_2^3$ is of total degree 5 and partial degree 3.

18.2 Approximation on Tensor Product Grids

By means of suitable linear transformations, every right-angled parallelepiped can be mapped onto the set $\Omega = (0, 1)^d$ and in the following, this case is always considered together with a uniform tensor product grid of grid size $h > 0$.

Definition 18.4 For a given function $f \in C^0([0, 1]^d)$ and a given grid size $h = 1/n$, the *tensor product interpolation task* consists in determining a polynomial $q \in \mathcal{Q}_n([0, 1]^d)$ with

$$q(x) = f(x)$$

for all $x \in \mathcal{G}_h = \{h(i_1, i_2, \ldots, i_d) : 0 \leq i_1, i_2, \ldots, i_d \leq n\}$.

Proposition 18.1 *The tensor product interpolation task is uniquely solvable.*

Proof To illustrate the idea of the proof, we consider the case $d = 2$. Let $E : \mathcal{Q}_n([0, 1]^2) \to \mathbb{R}^{(n+1)^2}$ be the linear mapping $q \mapsto (q(x) : x \in \mathcal{G}_h)$. Let $q \in \mathcal{Q}_n([0, 1]^2)$ have the property $Eq = 0$. For $(s, t) \in [0, 1]^2$ the expression $q(s, t)$ has the representations

$$q(s, t) = \sum_{0 \leq \ell, m \leq n} a_{\ell m} s^\ell t^m = \sum_{0 \leq \ell \leq n} \Big(\sum_{0 \leq m \leq n} a_{\ell m} t^m \Big) s^\ell = \sum_{0 \leq \ell \leq n} b_\ell(t) s^\ell.$$

For each fixed $t_j = jh$, $j = 0, 1, \ldots, n$, the polynomial $s \mapsto q(s, t_j)$ has the zeros $s_i = ih$, $i = 0, 1, \ldots, n$, and it follows $b_\ell(t_j) = 0$ for all $j, \ell = 0, 1, \ldots, n$.

For each $\ell = 0, 1, \ldots, n$ the polynomial $t \mapsto b_\ell(t)$ therefore has the roots t_j, $j = 0, 1, \ldots, n$ and it follows $b_\ell(t) = 0$ for all $t \in [0, 1]$ and thus $a_{\ell m} = 0$ for all $\ell, m = 0, 1, \ldots, n$ respectively $q = 0$. Thus E is injective and due to $\dim \mathcal{Q}_n([0, 1]^2) = (n + 1)^2$ also bijective. □

The numerical integration of a function $f \in C^0([0, 1]^d)$ is reduced to the approximation of one-dimensional integrals by means of the iteration formula based on Fubini's theorem

$$I^d(f) = \int_{[0,1]^d} f(x)\, dx = \int_0^1 \int_0^1 \cdots \int_0^1 f(x_1, x_2, \ldots, x_d)\, dx_1\, dx_2 \ldots dx_d.$$

Proposition 18.2 *If $Q : C^0([0, 1]) \to \mathbb{R}$ is a quadrature formula with non-negative weights and points $(w_i, t_i)_{i=0,\ldots,n}$ with degree of exactness $k \geq 0$, then*

$$Q^d(f) = \sum_{i_1=0}^n \sum_{i_2=0}^n \cdots \sum_{i_d=0}^n w_{i_1} w_{i_2} \ldots w_{i_d} f(t_{i_1}, t_{i_2}, \ldots, t_{i_d})$$

defines an iterated quadrature formula *$Q^d : C^0([0, 1]^d) \to \mathbb{R}$ that is exact for all $p \in \mathcal{Q}_{kd}([0, 1]^d)$. Furthermore, we have*

$$\left| I^d(f) - Q^d(f) \right| \leq \sum_{i=1}^d \sup_{\widehat{x}_i \in [0,1]^{d-1}} \left| I(f_{\widehat{x}_i}) - Q(f_{\widehat{x}_i}) \right|,$$

where $f_{\widehat{x}_i}$ for $\widehat{x}_i = (x_1, \ldots, x_{i-1}, x_{i+1}, \ldots, x_d) \in [0, 1]^{d-1}$ denotes the mapping

$$t \mapsto f(x_1, \ldots, x_{i-1}, t, x_{i+1}, \ldots, x_d).$$

Proof We consider the case $d = 2$. Then we have

$$I^2(f) - Q^2(f) = \int_0^1 \int_0^1 f(x_1, x_2)\, dx_1\, dx_2 - \sum_{i_1=0}^n \sum_{i_2=0}^n w_{i_1} w_{i_2} f(t_{i_1}, t_{i_2})$$

$$= \int_0^1 \left[\int_0^1 f(x_1, x_2)\, dx_1 - \sum_{i_1=0}^n w_{i_1} f(t_{i_1}, x_2) \right] dx_2$$

$$+ \int_0^1 \sum_{i_1=0}^n w_{i_1} f(t_{i_1}, x_2)\, dx_2 - \sum_{i_1=0}^n \sum_{i_2=0}^n w_{i_1} w_{i_2} f(t_{i_1}, t_{i_2})$$

$$= \int_0^1 \left(If(\cdot, x_2) - Qf(\cdot, x_2) \right) dx_2$$

$$+ \sum_{i_1=0}^n w_{i_1} \left(If(t_{i_1}, \cdot) - Qf(t_{i_1}, \cdot) \right).$$

Together with the property $\sum_{i=0}^{n} w_i = 1$ the asserted statement is obtained by taking the absolute value. □

Remark 18.2 The effort of the iterated quadrature formula grows exponentially with respect to d, that is $(n+1)^d$ function evaluations are required. The error order, on the other hand, is independent of the dimension and is determined by the one-dimensional degree of exactness.

18.3 Two-Dimensional Fourier Transform

Based on the observation that with a basis $(\omega^k)_{k=0,1,\ldots,n-1}$ of $\mathbb{C}^n$ through the matrices $(\omega^k(\omega^\ell)^\mathsf{T})_{k,\ell=0,\ldots,n-1}$ a basis of the vector space $\mathbb{C}^{n \times n}$ is defined, the discrete Fourier transform can be generalised to the two-dimensional case.

Proposition 18.3 *For every matrix $Y \in \mathbb{C}^{n \times n}$ there exist uniquely determined coefficients $B = (b_{k\ell})_{k,\ell=0,\ldots,n-1} \in \mathbb{C}^{n \times n}$, such that*

$$Y = \sum_{k,\ell=0}^{n-1} b_{k\ell} E^{k\ell}$$

with the orthogonal basis defined by the matrices $E^{k\ell} = \left(e^{\mathrm{i}(j_1 k + j_2 \ell) 2\pi/n}\right)_{j_1,j_2=0,\ldots,n-1} \in \mathbb{C}^{n \times n}$ for $k,\ell = 0,1,\ldots,n-1$ with respect to the scalar product $E : F = \sum_{j,m=0}^{n-1} E_{jm} \overline{F}_{jm}$. With $T_n \in \mathbb{C}^{n \times n}$ defined by $(T_n)_{jk} = e^{\mathrm{i}jk2\pi/n}$, $j,k = 0,1,\ldots,n-1$, we have

$$Y = \frac{1}{n^2} \overline{T}_n B \overline{T}_n, \quad B = T_n Y T_n.$$

Proof Exercise. □

Remarks 18.2

(i) The matrix multiplications required for the transformation can be performed with $\mathcal{O}(n^2 \log n)$ operations. To do this, the one-dimensional fast Fourier transformation is first applied to the columns of Y and then to the rows of the resulting matrix.
(ii) The two-dimensional Fourier transform is the basis for image compression techniques such as the *jpeg* format.

18.4 Approximation on Triangulations

Spline spaces can be generalised using triangulations.

Definition 18.5 For $k, m \geq 0$ and a triangulation $\mathcal{T}_h$ of a domain $\Omega \subset \mathbb{R}^d$, denote

$$\mathscr{S}^{m,k}(\mathcal{T}_h) = \{v_h \in C^k(\overline{\Omega}) : v_h|_T \in \mathscr{P}_m(T) \text{ for all } T \in \mathcal{T}_h\}$$

the *spline space of degree m and order k with respect to* $\mathcal{T}_h$.

By using affine-linear transformations investigations of the spline spaces can be reduced to the case of the standard simplex

$$\widehat{T} = \mathrm{conv}\{\widehat{z}_0, \widehat{z}_1, ..., \widehat{z}_d\}$$

where $\widehat{z}_0 = 0$ and $\widehat{z}_i = e_i$ for $i = 1, 2, \ldots, d$ with the canonical basis $(e_1, e_2, \ldots, e_d) \subset \mathbb{R}^d$.

Lemma 18.1 *For* $i = 0, 1, \ldots, d$ *let* $\widehat{\varphi}_i \in \mathscr{P}_1(\widehat{T})$ *be the uniquely defined hat function satisfying the conditions* $\widehat{\varphi}_i(\widehat{z}_j) = \delta_{ij}$, $j = 0, 1, \ldots, d$. *If* $T = \mathrm{conv}\{z_0, z_1, \ldots, z_d\} \in \mathbb{R}^d$ *is a non-degenerate simplex, then by*

$$\widehat{x} \mapsto \Phi_T(\widehat{x}) = \sum_{i=0}^d \widehat{\varphi}_i(\widehat{x}) z_i$$

an affine-linear diffeomorphism $\Phi_T : \widehat{T} \to T$ *is defined with the property* $\Phi_T(\widehat{z}_i) = z_i$, $i = 0, 1, \ldots, d$, *see Fig. 18.3. The volume of* T *is given by* $|\det D\Phi_T|/d!$.

Proof The hat functions on $\widehat{T}$ are given by $\widehat{\varphi}_i(\widehat{x}) = \widehat{x}_i$, $i = 1, 2, \ldots, d$, and $\widehat{\varphi}_0(\widehat{x}) = 1 - \widehat{x}_1 - \cdots - \widehat{x}_d$ for $\widehat{x} = (\widehat{x}_1, \widehat{x}_2, \ldots, \widehat{x}_d) \in \widehat{T}$ and the mapping Φ_T fulfils $\Phi_T(\widehat{z}_i) = z_i$, $i = 0, 1, \ldots, d$. For all $\widehat{x} \in \widehat{T}$ we have

$$\Phi_T(\widehat{x}) = z_0 + Q_T \widehat{x} = z_0 + [z_1 - z_0, z_2 - z_0, \ldots, z_d - z_0]\widehat{x}.$$

The determinant of Q_T is defined as the volume of the image of the unit cube $[0, 1]^d$ under the linear mapping Q_T, with which the volume of the image of the standard simplex is given by $|\det Q_T|/d!$. Since this coincides with the volume of T and is therefore positive, it follows that Φ_T is a diffeomorphism. $\square$

The hat functions from the proof can be transformed with the diffeomorphism Φ_T onto the elements and lead to the concept of the *nodal basis*, with which the spline

Fig. 18.3 The diffeomorphism Φ_T maps the standard simplex $\widehat{T}$ bijectively onto the simplex T

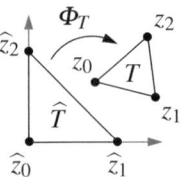

18.4 Approximation on Triangulations

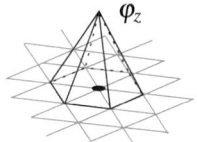

Fig. 18.4 Hat function φ_z associated with a node z in a triangulation

interpolation task in the space $\mathscr{S}^{1,0}(\mathscr{T}_h)$ can be solved. A typical hat function is shown in Fig. 18.4.

Proposition 18.4 *There exists a uniquely determined basis $(\varphi_z : z \in \mathscr{N}_h)$ of the space $\mathscr{S}^{1,0}(\mathscr{T}_h)$ with the property $\varphi_z(y) = \delta_{zy}$ for all $z, y \in \mathscr{N}_h$. For $f \in C^0(\overline{\Omega})$ it is defined by*

$$\mathscr{I}_h f = \sum_{z \in \mathscr{N}_h} f(z) \varphi_z$$

the nodal interpolant $\mathscr{I}_h f \in \mathscr{S}^{1,0}(\mathscr{T}_h)$ with the property $\mathscr{I}_h f(z) = f(z)$ for all $z \in \mathscr{N}_h$.

Proof Let $z \in \mathscr{N}_h$ and $T \in \mathscr{T}_h$. If $z \notin T$, then define $\varphi_z|_T = 0$. Otherwise, let $i \in \{0, 1, \ldots, d\}$, such that $\Phi_T(\widehat{z}_i) = z$ holds and define $\varphi_z|_T = \widehat{\varphi}_i \circ \Phi_T^{-1}$. In this way, functions $(\varphi_z : z \in \mathscr{N}_h) \subset \mathscr{S}^1(\mathscr{T}_h)$ with the properties $\varphi_z(y) = \delta_{zy}$ for $z, y \in \mathscr{N}_h$ are defined. To prove that this is a basis, let $s_h \in \mathscr{S}^{1,0}(\mathscr{T}_h)$ be arbitrary. By

$$\widetilde{s}_h = \sum_{z \in \mathscr{N}_h} s_h(z) \varphi_z$$

a function $\widetilde{s}_h \in \mathscr{S}^{1,0}(\mathscr{T}_h)$ is defined with $\widetilde{s}_h(z) = s_h(z)$ for all $z \in \mathscr{N}_h$. For each $T \in \mathscr{T}_h$ the function $\widehat{e} = (\widetilde{s}_h - s_h) \circ \Phi_T$ is affine-linear on $\widehat{T}$ with $\widehat{e}(0) = 0$ and $\widehat{e}(e_i) = 0$, $i = 0, 1, \ldots, d$. From this follows $\widehat{e} = 0$ and overall $s_h = \widetilde{s}_h$. □

The interpolation error can be bounded as in the one-dimensional case.

Proposition 18.5 *Let $f \in C^2(\overline{\Omega})$ and $\mathscr{T}_h$ be a regular triangulation of Ω. Then we have*

$$\|f - \mathscr{I}_h f\|_{C^0(\overline{\Omega})} \leq \frac{h^2}{2} \|D^2 f\|_{C^0(\overline{\Omega})}.$$

Proof We define $e = f - \mathscr{I}_h f$ and let $x_m \in \overline{\Omega}$ and $T \in \mathscr{T}_h$, such that $x_m \in T$ and $|e(x_m)| = \|e\|_{C^0(\overline{\Omega})}$ holds. Obviously, we have $e|_T \in C^2(T)$. If x_m is in the interior of T, then $\nabla e(x_m) = 0$. If x_m is a corner of T, then $e|_T = 0$ follows. If x_m is on a side of T, then there exists a corner $z \in \mathscr{N}_h \cap T$, such that the derivative of the mapping $t \mapsto e(z + t(x_m - z))$ at the point $t = 1$ vanishes, that is $\nabla e(x_m) \cdot (x_m - z) = 0$. In all three cases there exists a $z \in \mathscr{N}_h \cap T$, such that with a Taylor approximation for a $\xi \in T$ we have

$$0 = e(z) = e(x_m) + \frac{1}{2}(z - x_m)^\mathsf{T} D^2 e(\xi)(z - x_m).$$

Since $|z - x_m| \le h$ and $D^2 \mathscr{I}_h f|_T = 0$ hold, the assertion follows. □

Composite quadrature formulas on triangulated domains are defined using the reference element.

Definition 18.6 Let $\widehat{Q} : C^0(\widehat{T}) \to \mathbb{R}$ be a quadrature formula on $\widehat{T}$, defined by quadrature points and weights $(\widehat{\xi}_i, \widehat{w}_i)_{i=0,\ldots,n}$, that is $\widehat{Q} f = \sum_{i=0}^n \widehat{w}_i \widehat{f}(\widehat{\xi}_i)$. A corresponding *composite quadrature formula* $Q_{\mathscr{T}_h} : C^0(\overline{\Omega}) \to \mathbb{R}$ is defined by

$$Q_{\mathscr{T}_h}(f) = \sum_{T \in \mathscr{T}_h} \sum_{i=0}^n |\det D\Phi_T| \widehat{w}_i f(\Phi_T(\widehat{\xi}_i))$$

Remark 18.4 If the quadrature formula $\widehat{Q} : C^0(\widehat{T}) \to \mathbb{R}$ is exact of total degree $m \ge 0$, that is, the integrals of all polynomials $q \in \mathscr{P}_m(\widehat{T})$ are exactly reproduced, then the composite quadrature formula $Q_{\mathscr{T}_h}$ is exact for all $f \in \mathscr{S}^{m,0}(\mathscr{T}_h)$.

Example 18.2 Gaussian quadrature formulas with one, three or seven quadrature points on $\widehat{T} \subset \mathbb{R}^2$ are defined by $\widehat{\xi} \in \mathbb{R}^{n \times 2}$ and $\widehat{w} \in \mathbb{R}^n$ with

$$\widehat{\xi} = \frac{1}{3}\begin{bmatrix} 1 \\ 1 \end{bmatrix}^\mathsf{T}, \quad \widehat{w} = \frac{1}{2},$$

and

$$\widehat{\xi} = \frac{1}{6}\begin{bmatrix} 1 & 4 & 1 \\ 1 & 1 & 4 \end{bmatrix}^\mathsf{T}, \quad \widehat{w} = \frac{1}{6}[1, 1, 1]^\mathsf{T},$$

or with $s = \sqrt{15}$

$$\widehat{\xi} = \frac{1}{21}\begin{bmatrix} 6-s & 9+2s & 6-s & 6+s & 6+s & 9-2s & 7 \\ 6-s & 6-s & 9+2s & 9-2s & 6+s & 6+s & 7 \end{bmatrix}^\mathsf{T},$$

$$\widehat{w} = \frac{1}{2400}[155-s, 155-s, 155-s, 155+s, 155+s, 155+s, 270]^\mathsf{T}.$$

These quadrature formulas are exact for the polynomial spaces $\mathscr{P}_1(\widehat{T})$, $\mathscr{Q}_2(\widehat{T})$ or $\mathscr{P}_5(\widehat{T})$, respectively; they are schematically shown in Fig. 18.5.

Fig. 18.5 Schematic representation of Gaussian quadrature formulas on the reference triangle

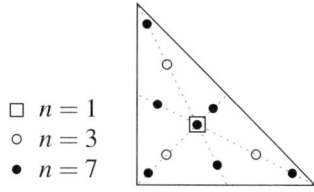

☐ $n = 1$
○ $n = 3$
● $n = 7$

18.5 Learning Objectives, Quiz and Application

You should be familiar with approaches to interpolation and quadrature of functions in several variables. You should be able to state interpolation estimates and explain the problems of quadrature in high-dimensional spaces.

Quiz 18.1 Decide for each of the following statements whether it is true or false. You should be able to justify your answer.

The polynomial $q(x, y) = x^2 y^3 z^4 + 3x^5$ has the partial degree 4 and the total degree 5.
We have that $\dim \mathscr{Q}_k(\mathbb{R}^d) = (k+1)^d$.
We have that $\dim \mathscr{P}_k(\mathbb{R}^d) = (d+1)k$.
We have that $\dim \mathscr{S}^1(\mathscr{T}_h) = |\mathscr{N}_h|$, where $|\mathscr{N}_h|$ denotes the cardinality of $\mathscr{N}_h$.
If $\mathscr{T}_h$ is a triangulation of a region $\Omega \subset \mathbb{R}^2$ with edges $\mathscr{E}_h$ and nodes $\mathscr{N}_h$, then the cardinalities of the sets satisfy $|\mathscr{N}_h| - |\mathscr{E}_h| + |\mathscr{T}_h| = 1$.

Application 18.1 At a narrow point of a river, its width d_F is to be determined. To this end, sighting marks are placed on the opposite banks. At some distance from the spot on the river, there is a town with a church and a water tower, the distance between which is known with high accuracy and denoted by d_T, see Fig. 18.6. A sighting device, which can measure the angle between two sighting points, and the possibility of installing additional sighting marks are available. Use a triangulation to determine the size d_F. What error influences must be taken into account and how can these be minimised? How should it be interpreted if the sum of the angles at an inner node deviates from 2π, but this cannot be attributed to measurement errors? How should geographical peculiarities be taken into account?

Fig. 18.6 Determination of an unknown from a known distance

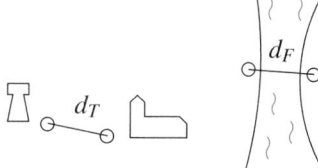

Part III
Numerics for Differential Equations

Chapter 19
Ordinary Differential Equations

19.1 Fundamentals

Many time-varying processes can be described by so-called *ordinary differential equations*. In this case, a differentiable function $y : [0, T) \to \mathbb{R}$ is sought, which for a given mapping $f : (0, T) \times \mathbb{R} \to \mathbb{R}$ satisfies the equation

$$y'(t) = f(t, y(t))$$

for all $t \in (0, T)$ as well as the *initial condition* $y(0) = y_0$ for a given number $y_0 \in \mathbb{R}$. One refers to t as the *independent* and y as the *dependent variable* of the *initial value problem*. The differential equation is often written in the form $y' = f(t, y)$, that is the argument t is omitted in the function y and its derivatives. A differential equation is *linear*, if the mapping $s \mapsto f(t, s)$ is linear for all $t \in (0, T)$.

Example 19.1 For $k \in \mathbb{R}$ we consider the linear differential equation $y'(t) = ky(t)$, that is $f(t, s) = ks$ is independent of t. For each $c \in \mathbb{R}$ the function

$$y(t) = ce^{kt}$$

is a solution of the differential equation on any interval $(0, T)$. An initial condition $y(0) = y_0$ determines $c = y_0$.

Remarks 19.1

(i) The initial value problem $y' = ky$, $y(0) = y_0$ describes the development of an account with initial capital y_0 at a fixed interest rate k per unit of time and immediate consideration of compound interest.
(ii) According to *Newton's law of cooling*, the change in temperature θ of a body is proportional to the difference to the surrounding temperature θ_s, that is $\theta'(t) = -k(\theta(t) - \theta_s)$.

(iii) The identity $y'(t) = ky(t)$ means that the change of y at time t is proportional to the value of y at this time.

(iv) The differential equation $y' = ky$ also describes the development of a population, where $k > 0$ applies when the birth rate is higher than the mortality rate.

In many applications several relevant quantities are considered simultaneously, whose values influence each other. This leads to *systems of differential equations*, where functions $y_1, y_2, \ldots, y_n : [0, T) \to \mathbb{R}$ are sought with the property that

$$y_1'(t) = f_1(t, y_1(t), y_2(t), \ldots, y_n(t)),$$
$$\vdots$$
$$y_n'(t) = f_n(t, y_1(t), y_2(t), \ldots, y_n(t))$$

hold for all $t \in (0, T)$. Such systems can be written in vector notation as $y'(t) = f(t, y(t))$, where $y = [y_1, y_2, \ldots, y_n]^\mathsf{T}$ and

$$f(t, s) = \begin{bmatrix} f_1(t, s_1, s_2, \ldots, s_n) \\ \vdots \\ f_n(t, s_1, s_2, \ldots, s_n) \end{bmatrix}$$

for $s = [s_1, s_2, \ldots, s_n]^\mathsf{T} \in \mathbb{R}^n$. An initial condition is then defined by a vector $y_0 \in \mathbb{R}^n$.

19.2 The Predator-Prey Model

The predator-prey model according to Lotka–Volterra describes the development of the number of predators and prey, such as raptors and mice, where it is assumed that the predators feed exclusively on the prey. Let $y_1(t)$ and $y_2(t)$ be the number of prey and predators, respectively, at time t in suitable units, so that for $y_1 = y_2 = 1$ a state of equilibrium occurs, that is, in this case, the increase in y_2 exactly corresponds to the decrease in y_1 due to death and being eaten. The change in the number of prey y_1 is then proportional to their number, with the proportionality factor depending on the number of predators and is positive if $y_2 < 1$ applies, and negative if $y_2 > 1$ applies, that is, for example

$$y_1'(t) = \alpha\bigl(1 - y_2(t)\bigr)y_1(t).$$

Similarly, the change in the number of predators y_2 is proportional to their number, with the proportionality factor being positive if more prey are available than in the

Fig. 19.1 Typical periodic solution in the predator-prey model

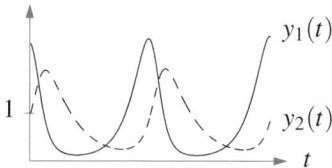

state of equilibrium, that is, for example

$$y_2'(t) = \beta(y_1(t) - 1)y_2(t).$$

A typical development of the populations for the case $y_1(0) > 1$ and $y_2(0) = 1$ is shown in Fig. 19.1 and demonstrates that a large number of prey leads to an increase in predators until a critical value is reached, and a low number of predators leads to an increase in prey.

19.3 Higher Order Equations

The equations considered so far only contained first-order derivatives. More generally, one can consider ordinary differential equations of *m-th order*, which can be abstractly written as

$$y^{(m)}(t) = f(t, y(t), y'(t), y''(t), \ldots, y^{(m-1)}(t))$$

with a function $f : (0, T) \times \mathbb{R}^m \to \mathbb{R}$. However, higher order differential equations can be written as a system of first order differential equations by introducing auxiliary variables. To this end, $z = [z_1, z_2, \ldots, z_m]^\mathsf{T}$ is defined by

$$z_1 = y, \quad z_2 = y', \quad z_3 = y'', \quad \ldots, \quad z_m = y^{(m-1)}$$

and the system

$$z_1'(t) = z_2(t),$$
$$\vdots$$
$$z_{m-1}'(t) = z_m(t),$$
$$z_m'(t) = f(t, z_1(t), z_2(t), \ldots, z_m(t))$$

is considered, which can be written in an obvious way as a vectorial differential equation $z' = \widetilde{f}(t, z)$. For higher order differential equations, it is generally not sufficient to only prescribe the function value at $t = 0$. In addition, the derivatives

Fig. 19.2 Oscillation behaviour of a damped spring pendulum

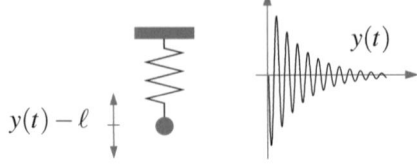

up to order $m-1$ must be given as initial data, that is

$$y(0) = y_{0,0}, \quad y'(0) = y_{0,1}, \quad \ldots, \quad y^{(m-1)}(0) = y_{0,m-1}$$

or with z defined above, the condition $z(0) = z_0$ with $z_0 = [y_{0,0}, y_{0,1}, \ldots, y_{0,m-1}]^T \in \mathbb{R}^m$.

Example 19.2 The differential equation $y'' = -c^2 y$ has the solutions $y(t) = \alpha \sin(ct)$ with the property $y(0) = 0$ for any choice of $\alpha \in \mathbb{R}$. By prescribing $y'(0)$, α is uniquely determined.

Remark 19.2 The deflection of a spring pendulum, which is fixed at the upper end, loaded at the lower end of the spring with the point mass m and in the resting position has the length ℓ, satisfies the force equilibrium

$$m\,y''(t) + r y'(t) + D\bigl(y(t) - \ell\bigr) = 0$$

from inertial force, frictional force and restoring force. The resting position is given by the weight force through $\ell = mg/D$. To predict the oscillation behaviour for $t > 0$ in addition to the initial deflection $y(0)$ the initial velocity $y'(0)$ must also be known. A typical solution is shown in Fig. 19.2.

19.4 Autonomous Equations

Differential equations $y'(t) = f(t, y(t))$, in which the function f does not depend on t, i.e. $f(t, s) = \tilde{f}(s)$ applies, are called *autonomous* differential equations. By adding the equation $z'(t) = 1$ it is shown that every differential equation can be written as a system of autonomous differential equations.

Remark 19.3 A solution of a system of autonomous differential equations $y' = f(y)$ with $f : \mathbb{R}^n \to \mathbb{R}^n$ is also referred to as an *integral curve* of the vector field f, because y can be geometrically interpreted as a curve in $\mathbb{R}^n$, whose tangent at each point is just prescribed by f, see Fig. 19.3. This is also referred to as a *phase diagram*. From it, qualitative properties of solutions such as periodicity or damping can be read off.

Fig. 19.3 Solutions of autonomous differential equations are integral curves of the vector field f

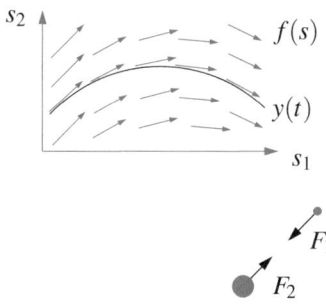

Fig. 19.4 Attractive gravitational forces act between bodies

19.5 Two-Body Problems

Attractive gravitational forces act between bodies, which are proportional to the product of the masses and inversely proportional to the square of the distance. With Newton's second law, which states that the change in momentum of a body or the product of mass and acceleration corresponds to the sum of the forces acting, this allows motion equations to be formulated. If the functions $y_1, y_2 : [0, T) \to \mathbb{R}^3$ describe the positions of the centres of two bodies of masses m_1, m_2, then it follows,

$$m_1 y_1'' = F_1(y_1, y_2) = \gamma \frac{m_1 m_2}{\|y_1 - y_2\|^2} \frac{y_2 - y_1}{\|y_1 - y_2\|},$$

$$m_2 y_2'' = F_2(y_1, y_2) = \gamma \frac{m_1 m_2}{\|y_1 - y_2\|^2} \frac{y_1 - y_2}{\|y_1 - y_2\|},$$

where $\gamma \approx 6.673 \cdot 10^{-11} \mathrm{m}^3/(\mathrm{kg\,s})$ is the gravitational constant. Note the opposite directions of the forces acting. With the initial positions $y_1(0)$ and $y_2(0)$ as well as the initial velocity vectors $y_1'(0)$ and $y_2'(0)$, the positions of the bodies can then be predicted, as long as they have a positive distance, see Fig. 19.4.

19.6 Explicit Solutions

In special situations, ordinary differential equations can be explicitly solved. For *separated* equations of the form $y' = f(t)g(y)$, the formal equivalence

$$\frac{dy}{dt} = f(t)g(y) \quad \Longleftrightarrow \quad \frac{dy}{g(y)} = f(t)dt \quad \Longleftrightarrow \quad \int \frac{1}{g(y)} = \int f(t)$$

with antiderivatives $G(y)$ of $1/g(y)$ and $F(t) + c$ of $f(t)$ leads to the identities

$$G(y) = F(t) + c \quad \Longleftrightarrow \quad y(t) = G^{-1}(F(t) + c).$$

This procedure is referred to as *separation of variables*. The method of *variation of constants* allows the solution of equations of the form $y' = f(t)y + h(t)$. First, the homogeneous equation $z' = f(t)z$ is solved and then a function φ is sought such that $y = \varphi z$ holds. With the product rule, we get

$$f(t)\varphi z + h(t) = y' = \varphi' z + \varphi z' = \varphi' z + \varphi f(t)z,$$

thus the condition $\varphi' = h/z$.

Examples 19.3

(i) For the equation $y' = y^2$, with $F(t) = t$ and $G(y) = -1/y$, the solutions $y(t) = -1/(t+c)$ are obtained.
(ii) In the case $y' = ky + h(t)$, $z(t) = ce^{kt}$ satisfies the equation $z' = kz$ and with $\varphi(t) = \int_0^t h(s)c^{-1}e^{-ks}\,ds$, a general solution is obtained.

19.7 Learning Objectives, Quiz and Application

You should be able to explain ordinary differential equations and initial value problems and illustrate them with examples. For some special cases, you should be able to construct explicit solutions.

Quiz 19.1 Decide for each of the following statements whether it is true or false. You should be able to justify your answer.

The necessary number of initial data for the well-posedness of an ordinary differential equation corresponds to the order of the differential equation.	
If y is a solution of the autonomous differential equation $y' = f(y)$, then $y(t+c)$ is also a solution for every $c \in \mathbb{R}$.	
The identity $y' = y(y(t))$ defines an ordinary differential equation.	
The differential equation $my' = ky$ describes the conservation of momentum of a body of mass m.	
If $f(s) = 0$ for some $s \in \mathbb{R}^n$, then the constant mapping $y(t) = s$ is a solution of the autonomous differential equation $y' = f(y)$.	

Application 19.1 The growth of a population is only meaningfully described by the differential equation $y' = ky$ within a certain range. When a capacity limit y_{max} is reached, no further increase in the population will occur. Explain why this effect can be described by the equation $y' = k(1 - y/y_{max})y$ and sketch solutions of this differential equation.

Chapter 20
Existence, Uniqueness and Stability

20.1 Existence and Uniqueness

A central existence result is based on Banach's fixed point theorem. For this, let X be a Banach space, i.e. X is a vector space on which a norm $\|\cdot\| : X \to \mathbb{R}$ is defined, with respect to which every Cauchy sequence in X converges.

Proposition 20.1 *If $\Psi : X \to X$ is a contraction on the Banach space X, i.e. there exists a constant $K < 1$, such that*

$$\|\Psi(u) - \Psi(v)\| \leq K \|u - v\|$$

for all $u, v \in X$, then Ψ has a unique fixed point $y \in X$, i.e. it holds that $\Psi(y) = y$.

The resulting existence statement uses an equivalent representation of an ordinary differential equation as an integral equation.

Lemma 20.1 *Let $f \in C^0([0, T] \times \mathbb{R})$. The function $y \in C^1([0, T])$ satisfies*

$$y'(t) = f(t, y(t)), \ t \in (0, T), \quad y(0) = y_0$$

if and only if $y \in C^0([0, T])$ and

$$y(t) = y_0 + \int_0^t f(s, y(s)) \, ds$$

for all $t \in [0, T]$.

© The Author(s), under exclusive license to Springer-Verlag GmbH, DE, part of Springer Nature 2025
S. Bartels, *Numerical Mathematics 3x9*, La Matematica per il 3+2 160,
https://doi.org/10.1007/978-3-662-70890-3_20

Proof

(i) First, let $y \in C^1([0,T])$ be a solution of the differential equation, which we write with s instead of t. The fundamental theorem of calculus yields

$$y(t) - y(0) = \int_0^t y'(s)\,ds = \int_0^t f(s, y(s))\,ds.$$

With the initial condition $y(0) = y_0$, the integral equation follows.

(ii) Conversely, let $y \in C^0([0,T])$ satisfy the integral equation. The fundamental theorem of calculus implies that the right-hand side of the integral equation defines a differentiable mapping with derivative $f(t, y(t))$ and value y_0 for $t = 0$. This implies $y \in C^1([0,T])$ and y solves the initial value problem. □

The integral representation shows that y is a solution of the fixed point equation $y = \Psi[y]$ if $\Psi : C^0([0,T]) \to C^0([0,T])$ is defined by

$$\Psi[y](t) = y_0 + \int_0^t f(s, y(s))\,ds$$

In the following *Picard-Lindelöf theorem*, a norm is constructed on the space $C^0([0,T])$ with respect to which Ψ is a contraction. For the sake of clarity, scalar equations are considered.

Proposition 20.2 *Assume that the mapping* $f \in C^0([0,T] \times \mathbb{R})$ *is uniformly Lipschitz continuous in the second argument, i.e. there exists an* $L \geq 0$, *such that*

$$|f(t,v) - f(t,w)| \leq L|v - w|$$

for all $t \in [0,T]$ *and all* $v, w \in \mathbb{R}$. *Then the initial value problem*

$$y'(t) = f(t, y(t)), \quad t \in (0,T), \quad y(0) = y_0$$

has a unique solution $y \in C^1([0,T])$.

Proof The operator Ψ is defined as above. For each $u \in C^0([0,T])$ the conditions on f imply that $\Psi[u] \in C^0([0,T])$ holds. On $C^0([0,T])$ we consider the weighted norm

$$\|u\|_L = \sup_{t \in [0,T]} e^{-2Lt}|u(t)|.$$

With this norm, $C^0([0,T])$ is complete and it suffices to show that Ψ is a contraction with respect to $\|\cdot\|_L$. For $u, v \in C^0([0,T])$ and $t \in [0,T]$ the following holds

20.1 Existence and Uniqueness

$$e^{-2Lt}\left|\Psi[u](t) - \Psi[v](t)\right| = e^{-2Lt}\left|\int_0^t f(s, u(s)) - f(s, v(s))\,ds\right|$$

$$\leq Le^{-2Lt}\int_0^t |u(s) - v(s)|\,ds$$

$$= Le^{-2Lt}\int_0^t e^{2Ls}e^{-2Ls}|u(s) - v(s)|\,ds$$

$$\leq Le^{-2Lt}\|u - v\|_L \int_0^t e^{2Ls}\,ds$$

$$= Le^{-2Lt}\frac{1}{2L}\left(e^{2Lt} - 1\right)\|u - v\|_L$$

$$\leq \frac{1}{2}\|u - v\|_L.$$

By forming the supremum on the left side we obtain

$$\|\Psi[u] - \Psi[v]\|_L \leq \frac{1}{2}\|u - v\|_L,$$

that is $\Psi : C^0([0, T]) \to C^0([0, T])$ is a contraction and Banach's fixed point theorem implies the existence of a unique fixed point $y \in C^0([0, T])$. According to the definition of Ψ and the previous lemma, this is equivalent to y being a solution of the initial value problem. □

The constructive proof of Banach's fixed point theorem shows that the fixed point $y \in C^0([0, T])$ is given as the limit of the recursively defined sequence

$$y^{k+1} = \Psi[y^k]$$

with any starting function $y^0 \in C^0([0, T])$. This observation can be used for the construction of numerical methods for solving initial value problems, however, functions must be suitably interpolated and integrated.

Remark 20.1 The condition of uniform Lipschitz continuity on the function f is a restrictive assumption. If f is merely continuous, then with the *Peano's theorem* the existence of a local solution can be proven, that is there exist $0 < T_* \leq T$ and $y \in C^1([0, T_*))$, such that y solves the initial value problem on the interval $(0, T_*)$.

Examples 20.1

(i) The initial value problem $y' = ky$, $y(0) = y_0$, has a unique solution on any interval $(0, T]$ and for any $k \in \mathbb{R}$.
(ii) The initial value problem $y' = y^2$, $y(0) = y_0$, with $y_0 > 0$ has the unique solution $y(t) = (T_* - t)^{-1}$ on the interval $[0, T_*)$ with $T_* = 1/y_0$.

(iii) The initial value problem $y' = y^{1/2}$, $y(0) = 0$, has the solutions $y(t) = 0$ as well as $y(t) = t^2/4$.

Frequently, the solution of an initial value problem possesses higher regularity properties than just differentiability.

Proposition 20.3 *If $f \in C^m([0, T] \times \mathbb{R}^n)$, then $y \in C^{m+1}([0, T])$ follows. In the case $m \geq 1$, solutions of corresponding initial value problems are unique.*

Proof Exercise. □

20.2 Gronwall's Lemma

Gronwall's lemma controls the growth of the solution of a differential equation.

Lemma 20.2 *Let $u \in C^0([0, T])$ and $\alpha, \beta \in \mathbb{R}$ with $\beta \geq 0$, such that*

$$u(t) \leq \alpha + \beta \int_0^t u(s)\, ds$$

for all $t \in [0, T]$. Then for all $t \in [0, T]$, it follows that

$$u(t) \leq \alpha e^{\beta t}.$$

Proof Let $v \in C^1([0, T])$ be defined by

$$v(t) = e^{-\beta t} \int_0^t \beta u(s)\, ds.$$

The product rule and the assumptions of the lemma imply

$$v'(t) = -\beta e^{-\beta t} \int_0^t \beta u(s)\, ds + e^{-\beta t} \beta u(t) \leq \beta e^{-\beta t} \alpha.$$

With $v(0) = 0$ it follows

$$e^{-\beta t} \int_0^t \beta u(s)\, ds = v(t) = \int_0^t v'(s)\, ds \leq \beta \alpha \int_0^t e^{-\beta s}\, ds = \alpha\left(1 - e^{-\beta t}\right).$$

Multiplication with $e^{\beta t}$ leads to

$$u(t) \leq \alpha + \int_0^t \beta u(s)\, ds \leq \alpha + \alpha e^{\beta t}\left(1 - e^{-\beta t}\right) = \alpha e^{\beta t}$$

and proves the lemma. □

Remark 20.2 Gronwall's lemma is often given in differential form. The condition then reads $u'(t) \leq \beta u(t)$ and from the resulting inequality $(\log u)' = u'/u \leq \beta$ it is evident that u grows at most exponentially.

20.3 Stability

The *stability of an initial value problem* refers to the conditioning of the associated mathematical operation, i.e. the effects of perturbations on solutions of the initial value problem. We assume that $y \in C^1([0, T])$ is the unique solution of the initial value problem

$$y'(t) = f(t, y(t)), \quad y(0) = y_0$$

and that for perturbations $\tilde{f}$ and $\tilde{y}_0$ of the function f and the initial data y_0, the function $\tilde{y} \in C^1([0, T])$ is the unique solution of the associated perturbed initial value problem

$$\tilde{y}'(t) = \tilde{f}(t, \tilde{y}(t)), \quad \tilde{y}(0) = \tilde{y}_0$$

Assuming that the perturbations are small, it can be shown that y and $\tilde{y}$ are close to each other for certain times.

Proposition 20.4 *Let $f, \tilde{f} \in C^0([0, T] \times \mathbb{R})$, such that a $\delta > 0$ exists with*

$$|f(t, v) - \tilde{f}(t, v)| \leq \delta$$

for all $t \in [0, T]$ and $v \in \mathbb{R}$, and let f be uniformly Lipschitz-continuous with respect to the second argument, i.e. there exists a number $L \geq 0$, such that

$$|f(t, v) - f(t, w)| \leq L|v - w|$$

for all $t \in [0, T]$ and all $v, w \in \mathbb{R}$. Furthermore, let $y_0, \tilde{y}_0 \in \mathbb{R}$ with $|y_0 - \tilde{y}_0| \leq \delta_0$ for a $\delta_0 > 0$. Let $y, \tilde{y} \in C^1([0, T] \times \mathbb{R})$ be solutions of the initial value problems

$$y'(t) = f(t, y(t)), \quad y(0) = y_0,$$
$$\tilde{y}'(t) = \tilde{f}(t, \tilde{y}(t)), \quad \tilde{y}(0) = \tilde{y}_0$$

in $[0, T]$. Then we have

$$\sup_{t \in [0,T]} |y(t) - \tilde{y}(t)| \leq (\delta_0 + \delta T)e^{LT}.$$

Proof The difference $y - \tilde{y}$ satisfies the integral equation

$$y(t) - \tilde{y}(t) = y_0 - \tilde{y}_0 + \int_0^t f(s, y(s)) - \tilde{f}(s, \tilde{y}(s)) \, ds$$

and this implies that

$$|y(t) - \tilde{y}(t)| \leq |y_0 - \tilde{y}_0| + \int_0^t |f(s, y(s)) - \tilde{f}(s, \tilde{y}(s))| \, ds.$$

The triangle inequality and the assumptions on f show that

$$|f(s, y(s)) - \tilde{f}(s, \tilde{y}(s))| \leq |f(s, y(s)) - f(s, \tilde{y}(s))| + |f(s, \tilde{y}(s)) - \tilde{f}(s, \tilde{y}(s))|$$
$$\leq L|y(s) - \tilde{y}(s)| + \delta$$

holds, and with $|y_0 - \tilde{y}_0| \leq \delta_0$ it follows

$$|y(t) - \tilde{y}(t)| \leq \delta_0 + \delta t + L \int_0^t |y(s) - \tilde{y}(s)| \, ds.$$

For the function $u(t) = |y(t) - \tilde{y}(t)|$ with $\alpha = \delta_0 + \delta T$ and $\beta = L$ it follows

$$u(t) \leq \alpha + \beta \int_0^t u(s) \, ds.$$

The lemma of Gronwall implies $u(t) \leq \alpha e^{\beta t} \leq \alpha e^{\beta T}$, from which the statement of the proposition follows. □

Remark 20.3 The error in the solution of the differential equation is proportional to δ_0 and δ, however, the proportionality factor is exponentially dependent on T and L. The initial value problem is therefore well conditioned or stable, provided LT is sufficiently small.

Example 20.2 Considering two spring pendulums with spring constants D and $\tilde{D}$, the solutions y and $\tilde{y}$ get out of phase and the solutions differ greatly from each other for large times, see Fig. 20.1. This reflects the exponential dependence on the time horizon T.

Fig. 20.1 Small perturbations of initial data can become noticeable in the long-term behaviour

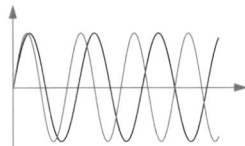

20.4 Learning Objectives, Quiz and Application

You should be able to reformulate an initial value problem as an equivalent integral equation. Based on this, you should be able to explain the ideas leading to the Picard-Lindelöf theorem. You should be able to derive the Gronwall lemma and explain its importance for the conditioning of initial value problems.

Quiz 20.1 Decide for each of the following statements whether it is true or false. You should be able to justify your answer.

Every solution of the differential equation $y'' = c^2 y$ is of the form $y(t) = \alpha \sin(ct) + \beta \cos(ct)$.
If $f \in C^1(\mathbb{R})$, then the initial value problem $y' = f(y)$, $y(0) = y_0$, has a solution $y \in C^1([0, T])$ for all $y_0 \in \mathbb{R}$ and $T > 0$.
Every contraction $\Psi : \mathbb{R}^n \to \mathbb{R}^n$ is continuously differentiable.
There exist autonomous differential equations $y' = f(y)$ that have solutions $y \in C^1([0, T])$ with the property $y \notin C^2([0, T])$.
If $y, \widetilde{y} \in C^1([0, T])$ are solutions of the differential equation $y' = f(y)$ with a Lipschitz-continuous mapping $f : \mathbb{R} \to \mathbb{R}$, then $|y(t) - \widetilde{y}(t)| \leq |y(0) - \widetilde{y}(0)|$ for all $t \in [0, T]$.

Application 20.1 The flight path of a rocket in the Earth's gravitational field can be described by a simplification of the two-body problem, assuming that the centre of the Earth remains unchanged and can be set as $y_{Earth} = 0$. Furthermore, it is assumed that the rocket flies perpendicular to the Earth's surface. and the fuel is depleted, so that no further acceleration occurs. Show that the height z of the rocket is described by the equation

$$z''(t) = \frac{a}{(z(t))^2}$$

with a suitable constant a and determine the solution for different initial velocities by using the approach $z(t) = \alpha(t - t_0)^\beta$. Discuss sufficient conditions for the global existence of the solution.

Chapter 21
Single-Step Methods

21.1 Euler Method

A simple method for the numerical approximation of solutions to ordinary differential equations of the form

$$y'(t) = f(t, y(t)), \quad y(0) = y_0$$

arises from the approximation of the derivative by a *(forward) difference quotient*, that is from

$$y'(t) \approx \frac{y(t+\tau) - y(t)}{\tau}$$

with a fixed *step size* $\tau > 0$. If $y \in C^1([0, T])$, the right-hand side converges to $y'(t)$ as $\tau \to 0$. The approximation leads to

$$y(t+\tau) \approx y(t) + \tau f(t, y(t))$$

and means that, as long as an approximation of y at time t is known, an approximation at time $t + \tau$ can be directly calculated. Starting with the initial data at $t_0 = 0$ the approximations at the *time steps* $t_k = k\tau$, $k = 1, 2, \ldots, K$, are obtained, where K is the largest natural number with the property $K\tau \leq T$, denoted by $K = \lfloor T/\tau \rfloor$.

Algorithm 21.1 (Explicit Euler Method) *Let $f \in C^0([0, T] \times \mathbb{R})$, $y_0 \in \mathbb{R}$ and $\tau > 0$. Set $k = 0$ and $K = \lfloor T/\tau \rfloor$.*

© The Author(s), under exclusive license to Springer-Verlag GmbH, DE, part of Springer Nature 2025
S. Bartels, *Numerical Mathematics 3x9*, La Matematica per il 3+2 160,
https://doi.org/10.1007/978-3-662-70890-3_21

Fig. 21.1 Euler methods approximate solutions through polygonal chains

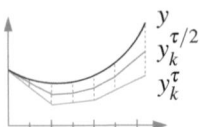

(1) Compute

$$y_{k+1} = y_k + \tau f(t_k, y_k).$$

(2) Stop if $k+1 > K$; otherwise increase $k \to k+1$ and repeat step (1).

Geometrically, the curve $t \mapsto y(t)$ is approximated by a polygonal chain that connects the values $(y_k)_{k=0,\ldots,K}$, see Fig. 21.1. Therefore, the method is also referred to as *Euler's polygonal chain method*.

Remark 21.1 In general, the approximations y_k do not coincide with the exact solution $y(t_k)$ at the times t_k, $k = 1, 2, \ldots, K$.

Definition 21.1 A method of the form

$$y_{k+1} = y_k + \tau \Phi(t_k, y_k, y_{k+1}, \tau), \quad k = 0, 1, \ldots, K-1,$$

is called a *single-step method* with *increment function* $\Phi : [0, T] \times \mathbb{R} \times \mathbb{R} \times \mathbb{R}_{\geq 0} \to \mathbb{R}$. If Φ is independent of y_{k+1}, the method is referred to as *explicit* and otherwise as *implicit*.

The *implicit Euler method* results from the use of the *backward difference quotient*

$$y'(t) \approx \frac{y(t) - y(t - \tau)}{\tau}$$

and the evaluation of the differential equation at t_{k+1}.

Algorithm 21.2 (Implicit Euler Method) *Let $f \in C^0([0, T] \times \mathbb{R})$, $y_0 \in \mathbb{R}$ and $\tau > 0$. Set $k = 0$ and $K = \lfloor T/\tau \rfloor$.*

(1) Determine $y_{k+1} \in \mathbb{R}$ as the solution of the equation

$$y_{k+1} = y_k + \tau f(t_{k+1}, y_{k+1}).$$

(2) Stop if $k+1 > K$; otherwise, increase $k \to k+1$ and repeat step (1).

Remarks 21.2

(i) In contrast to the explicit method, the implicit Euler method requires solving a system of equations at each iteration step. The solvability of this system must be ensured in each case.

(ii) For the explicit and implicit Euler methods, the increment functions Φ are given by

$$\Phi_{expl}(t_k, y_k, y_{k+1}, \tau) = f(t_k, y_k), \quad \Phi_{impl}(t_k, y_k, y_{k+1}, \tau) = f(t_k + \tau, y_{k+1}).$$

Higher accuracy is achieved by using the two approximations y_k and y_{k+1}.

Example 21.1 The *midpoint method* is defined by

$$\Phi(t_k, y_k, y_{k+1}, \tau) = f\bigl(t_k + \tau/2, (y_k + y_{k+1})/2\bigr).$$

21.2 Consistency

If the function values $y(t_k)$ of the exact solution of a differential equation at the time steps t_k, $k = 0, 1, \ldots, K$, are inserted into a numerical method, it can be assessed how accurate the method is. In the case of the explicit Euler method, using the differential equation evaluated at t_k, we have

$$\frac{y(t_{k+1}) - y(t_k)}{\tau} - f\bigl(t_k, y(t_k)\bigr) = \frac{y(t_{k+1}) - y(t_k)}{\tau} - y'(t_k).$$

A Taylor approximation shows

$$\left| \frac{y(t_{k+1}) - y(t_k)}{\tau} - y'(t_k) \right| \leq \frac{\tau}{2} \sup_{t \in [t_k, t_{k+1}]} |y''(t)|.$$

The values of the exact solution thus fulfil the numerical method up to the *consistency term* $(\tau/2)\|y''\|_{C^0([0,T])}$. To generalise this approach, for a given value z_k at time step t_k, we consider the *local initial value problem*

$$z'(t) = f\bigl(t, z(t)\bigr), \ t \in [t_k, t_{k+1}], \quad z(t_k) = z_k.$$

The deviation of the solution $z(t_{k+1})$ at time t_{k+1} from the approximation defined by the single-step method

$$z_{k+1} = z_k + \tau \Phi(t_k, z_k, z_{k+1}, \tau)$$

is given by

$$z(t_{k+1}) - z_{k+1} = z(t_{k+1}) - z_k - \tau \Phi(t_k, z_k, z_{k+1}, \tau)$$

see Fig. 21.2.

Fig. 21.2 The discretisation error of a time step defines the consistency of a method

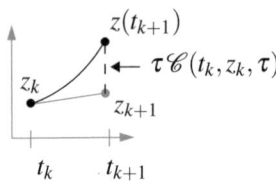

Definition 21.2 The *local discretisation error* $\mathscr{C}(t_k, z_k, \tau)$ of the increment function Φ is defined by

$$\mathscr{C}(t_k, z_k, \tau) = \frac{z(t_{k+1}) - z_k}{\tau} - \Phi(t_k, z_k, z_{k+1}, \tau).$$

The method defined by Φ is called *consistent of order* $p \geq 0$, if for all functions $f \in C^p([0, T] \times \mathbb{R})$, which are uniformly Lipschitz continuous in the second argument, and $k = 0, 1, \ldots, K - 1$ as well as $z_k \in \mathbb{R}$ we have that

$$\mathscr{C}(t_k, z_k, \tau) = \mathcal{O}(\tau^p)$$

for $\tau \to 0$, that is, if there exist $c_1, c_2 > 0$, so that $|\mathscr{C}(t_k, z_k, \tau)| \leq c_1 \tau^p$ holds for all $0 < \tau \leq c_2$.

Remarks 21.3

(i) If Φ is Lipschitz continuous in the third argument, then z_{k+1} can be replaced by $z(t_{k+1})$ to determine the consistency order, that is, using $z_k = z(t_k)$,

$$\widetilde{\mathscr{C}}(t_k, z_k, \tau) = \frac{z(t_{k+1}) - z(t_k)}{\tau} - \Phi\big(t_k, z(t_k), z(t_{k+1}), \tau\big).$$

In the case $z_k = y(t_k)$, this corresponds to the substitution of the function values of the exact solution into the numerical scheme.

(ii) For $z_k = y(t_k)$, the local solution z matches y on the interval $[t_k, t_{k+1}]$ and we have

$$\widetilde{\mathscr{C}}\big(t_k, y(t_k), \tau\big) = \frac{y(t_{k+1}) - y(t_k)}{\tau} - \Phi\big(t_k, y(t_k), y(t_{k+1}), \tau\big).$$

We will mostly use this expression in the following.

Examples 21.1

(i) For the explicit Euler method it follows from $\Phi(t_k, z_k, z_{k+1}, \tau) = f(t_k, z_k)$, $z_k = z(t_k)$ and $z'(t_k) = f(t_k, z(t_k))$, that $\mathscr{C}(t_k, z_k, \tau) = \widetilde{\mathscr{C}}(t_k, z_k, \tau)$ with

$$\mathscr{C}(t_k, z_k, \tau) = \frac{z(t_{k+1}) - z(t_k)}{\tau} - z'(t_k)$$

holds. With a Taylor approximation it follows

$$|\mathscr{C}(t_k, z_k, \tau)| \leq \frac{\tau}{2} \sup_{t \in [t_k, t_{k+1}]} |z''(t)|,$$

so that the explicit Euler method is consistent of order $p = 1$.
(ii) An analogous argument shows that the implicit Euler method also has the consistency order $p = 1$.
(iii) Also from Taylor approximations, the consistency order $p = 2$ of the midpoint method is derived.

21.3 Discrete Gronwall Lemma and Convergence

The consistency of a single-step method is a measure of the exactness of a method. Based on this, we will show that the approximations $(y_k)_{k=0,\ldots,K}$ are close to the exact function values $(y(t_k))_{k=0,\ldots,K}$. For this purpose, a single-step method

$$y_{k+1} = y_k + \tau \Phi(t_k, y_k, y_{k+1}, \tau)$$

with consistency order p is given, that is, we have

$$\widetilde{\mathscr{C}}(t_k, y(t_k), \tau) = \frac{y(t_{k+1}) - y(t_k)}{\tau} - \Phi(t_k, y(t_k), y(t_{k+1}), \tau) = \mathscr{O}(\tau^p).$$

The following error estimate is based on the interpretation of the exact solution values $(y(t_k))_{k=0,\ldots,K}$ as a solution of the numerical method perturbed by terms of order $\mathscr{O}(\tau^p)$. For this, the following *discrete version of the Gronwall lemma* is needed.

Lemma 21.1 *Let $(u_k)_{k=0,\ldots,K}$ be a sequence of non-negative, real numbers and $\alpha, \beta \in \mathbb{R}$ with $\beta \geq 0$, such that*

$$u_\ell \leq \alpha + \tau \sum_{k=0}^{\ell-1} \beta u_k$$

for all $\ell = 0, 1, \ldots, K$. Then it follows for all $\ell = 0, 1, \ldots, K$, that

$$u_\ell \leq \alpha \exp(\ell \tau \beta).$$

Proof Exercise. □

If one interprets the sum as a quadrature formula, the relationship to the continuous Gronwall lemma becomes apparent.

Definition 21.3 A single-step method is called *convergent of order* $p \geq 0$, if for all functions $f \in C^p([0, T] \times \mathbb{R})$, which are uniformly Lipschitz continuous in the second argument, initial data $y_0 \in \mathbb{R}$ and the exact solution $y \in C^{p+1}([0, T])$ as well as the approximations $(y_\ell)_{\ell=0,\ldots,K}$ we have

$$\max_{\ell=0,\ldots,K} |y(t_\ell) - y_\ell| = \mathcal{O}(\tau^p).$$

With the help of the discrete Gronwall lemma, the consistency of order p of a method leads to the convergence of order p of the method, that is, we obtain a *general error estimate for single-step methods*.

Proposition 21.1 *Let the single-step method defined by Φ be well-defined and consistent of order p. Assume that the increment function Φ is uniformly Lipschitz continuous in the second and third argument, that is, there exists $M \geq 0$, such that*

$$|\Phi(t, a_1, b_1, \tau) - \Phi(t, a_2, b_2, \tau)| \leq M(|a_1 - a_2| + |b_1 - b_2|)$$

for all $t \in [0, T]$, $a_1, a_2, b_1, b_2 \in \mathbb{R}$ and $\tau > 0$. If $\tau \leq 1/(2M)$, then it follows

$$\max_{\ell=0,\ldots,K} |y(t_\ell) - y_\ell| \leq 2cT\tau^p \exp(4MT)$$

with a constant $c \geq 0$ independent of τ.

Proof For the function values $(y(t_k))_{k=0,\ldots,K}$, according to the definition of the consistency term, we have

$$y(t_{k+1}) = y(t_k) + \tau\Phi(t_k, y(t_k), y(t_{k+1}), \tau) + \tau\widetilde{\mathscr{C}}(t_k, y(t_k), \tau),$$

while the approximations $(y_k)_{k=0,\ldots,K}$ are defined by

$$y_{k+1} = y_k + \tau\Phi(t_k, y_k, y_{k+1}, \tau)$$

Subtracting the two identities leads to

$$y(t_{k+1}) - y_{k+1} = y(t_k) - y_k + \tau[\Phi(t_k, y(t_k), y(t_{k+1}), \tau) - \Phi(t_k, y_k, y_{k+1}, \tau)]$$
$$+ \tau\widetilde{\mathscr{C}}(t_k, y(t_k), \tau).$$

With the triangle inequality, the Lipschitz continuity of Φ and the consistency order p, it follows

$$|y(t_{k+1}) - y_{k+1}| \leq |y(t_k) - y_k| + \tau M(|y(t_k) - y_k| + |y(t_{k+1}) - y_{k+1}|) + c\tau^{p+1}.$$

With the definition $u_k = |y(t_k) - y_k|$ we get

$$(1 - \tau M)u_{k+1} \leq (1 + \tau M)u_k + c\tau^{p+1},$$

21.3 Discrete Gronwall Lemma and Convergence

or

$$u_{k+1} \le \frac{1+\tau M}{1-\tau M} u_k + \frac{c}{1-\tau M} \tau^{p+1}.$$

Subtracting u_k on both sides yields using $1 - \tau M \ge 1/2$

$$\begin{aligned}
u_{k+1} - u_k &\le \left(\frac{1+\tau M}{1-\tau M} - 1\right) u_k + \frac{c}{1-\tau M} \tau^{p+1} \\
&= \tau \frac{2M}{1-\tau M} u_k + \frac{c\tau}{1-\tau M} \tau^p \\
&\le \tau 4 M u_k + 2c\tau \tau^p.
\end{aligned}$$

A summation over $k = 0, 1, \ldots, \ell - 1$ with $0 \le \ell \le K$ leads to

$$u_\ell - u_0 \le 4M\tau \sum_{k=0}^{\ell-1} u_k + 2cK\tau \tau^p.$$

Hence, the sequence $(u_k)_{k=0,\ldots,K}$ satisfies the conditions of the discrete Gronwall lemma with

$$\alpha = u_0 + 2c(K\tau)\tau^p, \quad \beta = 4\tau M$$

and with $u_0 = 0$ and $\ell\tau \le K\tau \le T$ the assertion follows. □

Remarks 21.4

(i) Similar to the stability estimate, the constant in the error estimate critically depends on the product MT.
(ii) The proof of the proposition shows that it is sufficient to approximate the initial data with an accuracy $|y_0 - y(0)| = \mathcal{O}(\tau^p)$.

In some special cases the convergence results can be improved and reveal an important difference between explicit and implicit schemes.

Example 21.3 We consider the equation $y' = \lambda y$ with a number $\lambda < 0$. For the explicit Euler method we obtain for $e_k = y(t_k) - y_k$ the error equation

$$e_{k+1} = (1+\tau\lambda) e_k + \tau \widetilde{\mathscr{C}}_{expl}(t_k, y(t_k), \tau).$$

Unless $|1 + \tau\lambda| \le 1$ or $\tau \le 1/(2|\lambda|)$, a strong error amplification is to be expected in the time stepping. This explains the step size condition in the previous result. For the implicit Euler method we obtain the error equation

$$(1-\tau\lambda) e_{k+1} = e_k + \tau \widetilde{\mathscr{C}}_{impl}(t_k, y(t_k), \tau).$$

Since $1 - \tau\lambda \geq 1$, a damping effect for the error takes place. In particular, noting $|1 - \tau\lambda|^{-1} \leq 1$ we obtain by taking absolute values that for $u_k = |e_k|$ we have

$$u_{k+1} \leq u_k + c\tau^2.$$

A summation over $k = 0, 1, \ldots, \ell - 1$ with $0 \leq \ell \leq K$ shows that

$$\max_{k=0,\ldots,K} |y(t_k) - y_k| \leq cT\tau.$$

This estimate holds without a step size condition and without an exponential dependence on T and $|\lambda|$.

21.4 Higher-Order Methods

The consistency order of the Euler methods is given by the Taylor formulas used in the derivation. This motivates the use of a higher accuracy approximation such as

$$\frac{y(t_{k+1}) - y(t_k)}{\tau} = y'(t_k) + \frac{\tau}{2} y''(t_k) + \mathcal{O}(\tau^2),$$

provided $y \in C^3([0, T])$. Based on this identity, there are various ways to construct an increment function.

Examples 21.4 ([8, 9])

(i) Differentiating the differential equation $y' = f(t, y)$ with respect to t, one gets with the partial derivatives $\partial_t f$ and $\partial_y f$ of f that

$$y''(t) = \partial_t f(t, y(t)) + \partial_y f(t, y(t)) y'(t)$$
$$= \partial_t f(t, y(t)) + \partial_y f(t, y(t)) f(t, y(t)).$$

The use of this identity in the above Taylor formula shows that the expression

$$\frac{y(t_{k+1}) - y(t_k)}{\tau} - \frac{\tau}{2} \Big(\partial_t f(t_k, y(t_k)) + \partial_y f(t_k, y(t_k)) f(t_k, y(t_k)) \Big)$$

approximates the derivative $y'(t_k)$ up to an error $\mathcal{O}(\tau^2)$ and motivates the use of the explicit increment function

$$\Phi(t_k, y_k, y_{k+1}, \tau) = f(t_k, y_k) + \frac{\tau}{2} \Big(\partial_t f(t_k, y_k) + \partial_y f(t_k, y_k) f(t_k, y_k) \Big).$$

The calculations imply $\mathscr{C}(t_k, y_k, \tau) = \mathcal{O}(\tau^2)$.

(ii) With coefficients $a, b, c, d \in \mathbb{R}$ to be determined, the approach

$$\Phi(t, y, \tau) = af(t, y) + bf(t + c\tau, y + \tau df(t, y))$$

is considered and inserted into the definition of the consistency term, where for brevity t and y stand for t_k and y_k. Using the Taylor approximation

$$f(t+c\tau, y+d\tau f(t, y)) = f(t, y) + \partial_t f(t, y)c\tau + \partial_y f(t, y) d\tau f(t, y) + \mathcal{O}(\tau^2)$$

the conditions $a + b = 1$, $bc = 1/2$ and $bd = 1/2$ arise for the parameters a, b, c, d for the consistency of order $p = 2$. The solution $a = b = 1/2$, $c = d = 1$ defines *Heun's method*

$$\Phi(t, y, \tau) = \frac{1}{2}\big(f(t, y) + f(t + \tau, y + \tau f(t, y))\big)$$

and the solution $a = 0$, $b = 1$, $c = d = 1/2$ defines the *Euler–Collatz method*

$$\Phi(t, y, \tau) = f\Big(t + \frac{\tau}{2}, y + \frac{\tau}{2}f(t, y)\Big).$$

Remark 21.5 The terms appearing in the methods

$$y_k + \theta\tau f(t_k, y_k) \approx y_k + \theta\tau y'(t_k)$$

approximate the unknown values $y(t_{k+1})$ in the case $\theta = 1$ and $y(t_{k+1/2}) = (y_k + y_{k+1})/2$ in the case $\theta = 1/2$, where $t_{k+1/2} = t_k + \tau/2$.

21.5 Learning Objectives, Quiz and Application

You should be able to derive particular single-step methods and show their differences. You should be able to motivate and define the concept of consistency and explain its use in deriving error estimates.

Quiz 21.1 Decide for each of the following statements whether it is true or false. You should be able to justify your answer.

The increment function $\Phi(t, a, b, \tau) = \alpha(a+b)/2$ defines a single-step method for the differential equation $y' = \alpha y$.
Explicit single-step methods are always well-defined.
The local discretisation error of the explicit Euler method for the differential equation $y' = \lambda y$ is given by $(z(t_{k+1}) - z_k)/\tau - \lambda z_k$.
The increment function $\Phi(t, a, b, \tau) = f(t + \tau/2, a + \tau f(t, a)/2)$ defines a method of consistency order $p = 2$.
In general, the error $|y_\ell - y(t_\ell)|$ increases with each time step.

Application 21.1

(i) The speed of the chemical reaction of two substances A and B with product $2B$ is determined by a reaction coefficient α and the differential equations

$$C'_A = -\alpha C_A C_B, \quad C'_B = \alpha C_A C_B$$

where $C_A, C_B : [0, T] \to [0, 1]$ indicate the respective concentrations. In the reaction equation, this is taken into account by the notation $A + B \xrightarrow{\alpha} 2B$. Show that the sum of the concentrations $C_A + C_B$ is constant.

(ii) We consider the reaction scheme

$$A \xrightarrow{\alpha} B, \quad 2B \xrightarrow{\beta} B + C, \quad B + C \xrightarrow{\gamma} A + C,$$

with the reaction coefficients $\alpha = 0.04$, $\beta = 3 \cdot 10^7$, $\gamma = 10^4$, that is for example, that the substance B is very quickly converted into the substance C. Formulate a system of differential equations to describe the reaction scheme, show that the sum of the concentrations is constant and determine numerically the maximum concentration of the substance B, if at the beginning of the process only the substance A is present.

(iii) Test various MATLAB routines for the numerical solution of the problem in the time interval $[0, 1/2]$ and comment on the results.

Chapter 22
Runge-Kutta Methods

22.1 Motivation

The construction of numerical methods based on Taylor approximations with higher consistency order usually leads to schemes in which many function evaluations, especially of derivatives, are required in each step. This is generally associated with very high effort. The starting point for the development of methods that avoid the evaluation of derivatives of f is a local integral representation of the differential equation $y' = f(t, y)$. We have

$$y(t_{k+1}) = y(t_k) + \int_{t_k}^{t_{k+1}} y'(s)\,ds = y(t_k) + \int_{t_k}^{t_{k+1}} f\bigl(s, y(s)\bigr)\,ds.$$

If the integral is approximated by the value of the integrand at the point t_k multiplied by the length of the integration range $t_{k+1} - t_k = \tau$, the result is

$$y(t_{k+1}) \approx y(t_k) + \tau f\bigl(t_k, y(t_k)\bigr),$$

and this motivates the explicit Euler method. It is obvious to apply more exact quadrature formulas to obtain methods of higher accuracy. In the case of the midpoint rule, with $t_{k+1/2} = (k + 1/2)\tau$, we get

$$y(t_{k+1}) \approx y(t_k) + \tau f\bigl(t_{k+1/2}, y(t_{k+1/2})\bigr).$$

The function value $y(t_{k+1/2})$ can be approximated using an approximation of $y(t_k)$, because a Taylor approximation and the evaluation of the differential equation at t_k

show for a $\xi \in [t_k, t_{k+1/2}]$

$$y(t_{k+1/2}) = y(t_k) + \frac{\tau}{2}y'(t_k) + \frac{\tau^2}{8}y''(\xi)$$

$$= y(t_k) + \frac{\tau}{2}f(t_k, y(t_k)) + \frac{\tau^2}{8}y''(\xi).$$

Overall, this leads to the Euler–Collatz method

$$y_{k+1} = y_k + \tau f\left(t_k + \frac{\tau}{2}, y_k + \frac{\tau}{2}f(t_k, y_k)\right).$$

With a Taylor approximation of the right side, it is shown that this method has the consistency order $p = 2$.

22.2 Runge-Kutta Methods

If $(\alpha_\ell, \gamma_\ell)_{\ell=1,\ldots,m}$ is a quadrature formula on the interval $[0, 1]$, then $(t_k + \tau\alpha_\ell, \tau\gamma_\ell)_{\ell=1,\ldots,m}$ defines a quadrature formula on $[t_k, t_{k+1}]$ and we obtain

$$y(t_{k+1}) = y(t_k) + \int_{t_k}^{t_{k+1}} y'(s)\,ds \approx y(t_k) + \tau \sum_{\ell=1}^{m} \gamma_\ell \eta_\ell^k$$

with the approximations

$$\eta_\ell^k \approx y'(t_k + \tau\alpha_\ell) = f(t_k + \tau\alpha_\ell, y(t_k + \tau\alpha_\ell)).$$

The right side is approximated using

$$y(t_k + \tau\alpha_\ell) \approx y(t_k) + \tau \sum_{j=1}^{m} \beta_{\ell j} y'(t_k + \tau\alpha_j) \approx y(t_k) + \tau \sum_{j=1}^{m} \beta_{\ell j} \eta_j^k$$

and thus the quantities η_ℓ^k can be determined as the solution of the nonlinear system of equations

$$\eta_\ell^k = f\left(t_k + \tau\alpha_\ell, y_k + \tau \sum_{j=1}^{m} \beta_{\ell j} \eta_j^k\right)$$

for $\ell = 1, 2, \ldots, m$. This leads to the following definition.

Definition 22.1 For $\alpha_\ell, \beta_{\ell j}, \gamma_\ell \in \mathbb{R}$, $j, \ell = 1, 2, \ldots, m$, an *m-stage Runge-Kutta method* is defined by

22.2 Runge-Kutta Methods

Fig. 22.1 Runge-Kutta methods use implicitly defined intermediate steps

Table 22.1 Butcher tableau of a Runge-Kutta method

α_1	β_{11}	$\ldots$	β_{1m}
$\vdots$	$\vdots$		$\vdots$
α_m	β_{m1}	$\ldots$	β_{mm}
	γ_1	$\ldots$	γ_m

$$y_{k+1} = y_k + \tau \sum_{\ell=1}^{m} \gamma_\ell \eta_\ell^k, \quad \eta_\ell^k = f\left(t_k + \tau\alpha_\ell, y_k + \tau \sum_{j=1}^{m} \beta_{\ell j} \eta_j^k\right).$$

Intuitively, a Runge-Kutta method uses intermediate steps $t_{k+s/m} = t_k + \tau\alpha_s$, $s = 1, 2, \ldots, m$, and corresponding approximations $y_{k+s/m} = y_k + \tau \sum_{\ell=1}^{s} \gamma_\ell \eta_\ell^k$ to determine y_{k+1}, see Fig. 22.1.

Remark 22.1 Runge-Kutta methods are single-step methods, where the increment function $\Phi(t_k, y_k, \tau) = \sum_{\ell=1}^{m} \gamma_\ell \eta_\ell^k$ is defined by the solution of a possibly nonlinear problem. In terms of single-step methods, Runge-Kutta methods are explicit, however, this view is not meaningful.

Remark 22.2 A Runge-Kutta method is determined by the associated *Butcher tableau*, in which the coefficients are schematically arranged as in Table 22.1.

Examples 22.1

(i) For $m = 1$, $\alpha_1 = 0$, $\beta_{11} = 0$, $\gamma_1 = 1$ we get the explicit Euler method defined by

$$\eta_1^k = f(t_k, y_k), \quad y_{k+1} = y_k + \tau \eta_1^k.$$

(ii) The Euler–Collatz method and the Heun method are defined by the respective Butcher tableau shown in Table 22.2.

(iii) The *trapezoidal method* results from the use of the trapezoidal rule, that is

$$y_{k+1} = y_k + \frac{\tau}{2}\left(f(t_k, y_k) + f(t_{k+1}, y_{k+1})\right) = y_k + \frac{\tau}{2} \sum_{\ell=1}^{2} \eta_\ell^k$$

with $\eta_1^k = f(t_k, y_k)$ and $\eta_2^k = f(t_k + \tau, y_{k+1})$.

Table 22.2 Butcher tableau of the explicit Euler method, the Euler-Collatz and Heun methods, as well as the trapezoidal method

0	0
	1

0	0	0
1/2	1/2	0
	0	1

0	0	0
1	1	0
	1/2	1/2

0	0	0
1	1/2	1/2
	1/2	1/2

22.3 Well-Posedness

The execution of an iteration step with a Runge-Kutta method requires the solution of a system of equations. If β is a strict lower triangular matrix, this can be solved explicitly.

Definition 22.2 A Runge-Kutta method is called *explicit* if $\beta_{\ell j} = 0$ holds for all $1 \leq \ell \leq j \leq n$. It is called *implicit* otherwise.

Remark 22.3 If a Runge-Kutta method is explicit, the expressions η_ℓ^k can be determined successively. For $\ell = 1, 2, \ldots, m$ the following then applies

$$\eta_\ell^k = f\left(t_k + \tau\alpha_\ell, y_k + \tau \sum_{j=1}^{\ell-1} \beta_{\ell j}\eta_j^k\right).$$

Example 22.1 Examples of explicit, four-stage Runge-Kutta methods are the classic Runge-Kutta method and the 3/8 rule, which are defined by the respective Butcher tableau shown in Table 22.3.

In the implicit case, a fixed point equation must be solved. With the abbreviations $t = t_k$, $y = y_k$ and $\eta_\ell = \eta_\ell^k$, a vector $\eta = [\eta_1, \eta_2, \ldots, \eta_m]^\mathsf{T}$ is to be determined such that

$$\eta_1 = f(t + \tau\alpha_1, y + \tau\beta_{11}\eta_1 + \tau\beta_{12}\eta_2 + \cdots + \tau\beta_{1m}\eta_m),$$

$$\vdots$$

$$\eta_m = f(t + \tau\alpha_m, y + \tau\beta_{m1}\eta_1 + \tau\beta_{m2}\eta_2 + \cdots + \tau\beta_{mm}\eta_m)$$

holds, which can be written abstractly as a vectorial equation $\eta = \Psi(\eta)$.

Table 22.3 Butcher tableau of the classic Runge-Kutta method and the 3/8 rule

0				
1/2	1/2			
1/2	0	1/2		
1	0	0	1	
	1/6	1/3	1/3	1/6

0				
1/3	1/3			
2/3	−1/3	1		
1	1	−1	1	
	1/8	3/8	3/8	1/8

22.4 Consistency

Table 22.4 Butcher tableau for the implicit Euler, the midpoint and the Radau-3 methods

$$
\begin{array}{c|c} 1 & 1 \\ \hline & 1 \end{array} \qquad \begin{array}{c|c} 1/2 & 1/2 \\ \hline & 1 \end{array} \qquad \begin{array}{c|cc} 1/3 & 5/12 & -1/12 \\ 1 & 3/4 & 1/4 \\ \hline & 3/4 & 1/4 \end{array}
$$

Proposition 22.1 *If $f \in C^0([0, T] \times \mathbb{R})$ is uniformly Lipschitz continuous in the second argument with Lipschitz constant $L \geq 0$ and if*

$$L\tau \|\beta\|_\infty < 1,$$

with the row sum norm $\|\beta\|_\infty = \max_{\ell=1,\ldots,m} \sum_{j=1}^m |\beta_{\ell j}|$, then Ψ is a contraction with respect to the maximum norm on $\mathbb{R}^m$ and there exists a unique fixed point $\eta \in \mathbb{R}^m$ of Ψ.

Proof Let $\xi, \zeta \in \mathbb{R}^m$. Then we have that

$$\|\Psi(\xi) - \Psi(\zeta)\|_\infty$$

$$= \max_{\ell=1,\ldots,m} \left| f\left(t + \tau\alpha_\ell, y + \tau \sum_{j=1}^m \beta_{\ell j}\xi_j\right) - f\left(t + \tau\alpha_\ell, y + \tau \sum_{j=1}^m \beta_{\ell j}\zeta_j\right) \right|$$

$$\leq \max_{\ell=1,\ldots,m} L\tau \sum_{j=1}^m |\beta_{\ell j}| \max_{j=1,\ldots,m} |\xi_j - \zeta_j| = L\tau \|\beta\|_\infty \|\xi - \zeta\|_\infty.$$

The Banach fixed point theorem implies in the case $L\tau\|\beta\|_\infty < 1$ the existence of a unique fixed point. □

Remark 22.4 A fixed point of the contraction $\Psi : \mathbb{R}^m \to \mathbb{R}^m$ can be approximated with any initial value $\xi^0 \in \mathbb{R}^m$ through the iteration $\xi^{i+1} = \Psi(\xi^i)$. Under suitable conditions, the nonlinear system of equations can be approximately solved using the Newton method. An initial value can be defined using the approximation to the previous time step.

Example 22.3 Examples of implicit Runge-Kutta methods are the implicit Euler method, the midpoint method and the Radau-3 method, which are defined by the respective Butcher tableau shown in Table 22.4.

22.4 Consistency

Since Runge-Kutta methods are based on quadrature formulas, the exactness of the underlying quadrature formula is decisive for the consistency of the method. The quadrature formula $(\alpha_\ell, \gamma_\ell)_{\ell=1,\ldots,m}$ is called *exact of degree r* on the interval $[0, \widehat{a}]$,

if
$$\int_0^{\widehat{a}} q(s)\,ds = \sum_{\ell=1}^{m} \gamma_\ell q(\alpha_\ell)$$

for all polynomials q up to degree r. In this case, for every function $\phi \in C^{r+1}([0,T])$, we have that

$$\int_t^{t+\tau\widehat{a}} \phi(s)\,ds = \tau \sum_{\ell=1}^{m} \gamma_\ell \phi(t + \tau\alpha_\ell) + \mathcal{O}(\tau^{r+2}).$$

In the sense of this statement, the trivial quadrature formula, which approximates every integral by the value 0, is exact of degree $r = -1$. The exactness of degree $p-1$ is necessary for the consistency of order p.

Lemma 22.1 *Assume that the Runge-Kutta method defined by the coefficients $\alpha_\ell, \beta_{\ell j}, \gamma_\ell$, $j, \ell = 1, 2, \ldots, m$, is consistent of order $p \geq 0$. Then the quadrature formula defined by $(\alpha_\ell, \gamma_\ell)_{\ell=1,\ldots,m}$ is exact of degree $p-1$ on $[0, 1]$.*

Proof For $0 \leq n \leq p-1$, let $y : [0, 1] \to \mathbb{R}$ be the solution of the differential equation $y'(t) = f(t, y(t))$, $y(0) = 0$, with $f(t, z) = t^n$. Obviously $y(t) = t^{n+1}/(n+1)$. The consistency of order p of the Runge-Kutta method implies that for all $\tau > 0$ the estimate

$$\left|\widetilde{\mathscr{C}}(0, 0, \tau)\right| = \left|\frac{y(\tau) - y(0)}{\tau} - \Phi(0, 0, \tau)\right| \leq c\tau^p$$

holds and hence

$$\left|\frac{1}{\tau}\frac{\tau^{n+1}}{n+1} - \sum_{\ell=1}^{m} \gamma_\ell (\tau\alpha_\ell)^n\right| \leq c\tau^p.$$

A division of this inequality by τ^n and the limit $\tau \to 0$ imply, that the quadrature formula is exact for the monomials $t \mapsto t^n$, $0 \leq n \leq p-1$, in the interval $[0, 1]$. By its linearity this is true for all polynomials of degree $p-1$. □

The reversal of the statement holds under additional conditions on the coefficients $\beta_{\ell j}$ and leads to a sufficient condition for a consistency statement, which follows the presentation in [8].

Proposition 22.2 *Let $\alpha_\ell, \beta_{\ell j}, \gamma_\ell$, $j, \ell = 1, 2, \ldots, m$ be such that the quadrature formula defined by $(\alpha_\ell, \gamma_\ell)_{\ell=1,\ldots,m}$ on the interval $[0, 1]$ is exact of degree $p-1$ and for $\ell = 1, 2, \ldots, m$ the quadrature formula defined by $(\alpha_j, \beta_{\ell j})_{j=1,\ldots,m}$ is exact on $[0, \alpha_\ell]$ of degree $p-2$. Then the Runge-Kutta method defined by $\alpha_\ell, \beta_{\ell j}, \gamma_\ell$ has the consistency order p.*

22.4 Consistency

Proof From local integral representations of the differential equation $y' = f(s, y)$ and the assumed exactness of the quadrature formulas, it follows that

$$y(t+\tau) - y(t) = \int_t^{t+\tau} f(s, y(s))\, ds = \tau \sum_{\ell=1}^m \gamma_\ell f\big(t+\tau\alpha_\ell, y(t+\tau\alpha_\ell)\big) + \mathcal{O}(\tau^{p+1})$$

and for $\ell = 1, 2, \ldots, m$

$$y(t+\tau\alpha_\ell) - y(t) = \int_t^{t+\tau\alpha_\ell} f(s, y(s))\, ds = \tau \sum_{j=1}^m \beta_{\ell j} f\big(t+\tau\alpha_j, y(t+\tau\alpha_j)\big) + \mathcal{O}(\tau^p).$$

For the consistency term, with the first equation, the abbreviations $t = t_k$ and $t_{k,\ell} = t_k + \tau\alpha_\ell$ and

$$\eta_\ell = f\bigg(t + \tau\alpha_\ell,\, y(t) + \tau \sum_{s=1}^m \beta_{\ell s} \eta_s\bigg)$$

and $\Phi(t, y(t), \tau) = \sum_{\ell=1}^m \gamma_\ell \eta_\ell$, it follows that

$$\big|\widetilde{\mathscr{C}}(t, y(t), \tau)\big| = \bigg|\frac{y(t+\tau) - y(t)}{\tau} - \sum_{\ell=1}^m \gamma_\ell \eta_\ell\bigg|$$

$$= \bigg|\sum_{\ell=1}^m \gamma_\ell \bigg[f\big(t_{k,\ell}, y(t_{k,\ell})\big) - f\bigg(t_{k,\ell}, y(t) + \tau \sum_{j=1}^m \beta_{\ell j}\eta_j\bigg)\bigg]\bigg| + \mathcal{O}(\tau^p)$$

$$\leq L \sum_{\ell=1}^m |\gamma_\ell|\bigg|y(t+\tau\alpha_\ell) - y(t) - \tau \sum_{j=1}^m \beta_{\ell j}\eta_j\bigg| + \mathcal{O}(\tau^p) = L \sum_{\ell=1}^m |\gamma_\ell| r_\ell + \mathcal{O}(\tau^p).$$

Here, using the second equation,

$$r_\ell = \bigg|y(t+\tau\alpha_\ell) - y(t) - \tau \sum_{j=1}^m \beta_{\ell j}\eta_j\bigg|$$

$$= \bigg|\tau \sum_{j=1}^m \beta_{\ell j}\bigg[f\big(t_{k,j}, y(t_{k,j})\big) - f\bigg(t_{k,j}, y(t) + \tau \sum_{n=1}^m \beta_{jn}\eta_n\bigg)\bigg]\bigg| + \mathcal{O}(\tau^p)$$

$$\leq \tau L \sum_{j=1}^m |\beta_{\ell j}|\bigg|y(t_{k,j}) - y(t) + \tau \sum_{n=1}^m \beta_{jn}\eta_n\bigg| + \mathcal{O}(\tau^p)$$

$$= \tau L \sum_{j=1}^m |\beta_{\ell j}| r_j + \mathcal{O}(\tau^p).$$

From this estimate it follows

$$\|r\|_\infty \leq \tau L \|\beta\|_\infty \|r\|_\infty + c\tau^p$$

or $\|r\|_\infty \leq 2c\tau^p$, provided $\tau L\|\beta\|_\infty \leq 1/2$ holds. Overall, we get

$$\left|\widetilde{\mathscr{C}}(t, y(t), \tau)\right| \leq c\tau^p$$

and this proves the assertion. $\square$

Remark 22.5 Alternatively, the consistency order of a Runge-Kutta method can be investigated using Taylor approximations. For example, using the abbreviation y for $y(t)$, we have

$$\frac{y(t+\tau) - y(t)}{\tau} = y'(t) + \frac{\tau}{2} y''(t) + \mathcal{O}(\tau^2)$$

$$= f(t, y) + \frac{\tau}{2}\left[\partial_t f(t, y) + \partial_y f(t, y) f(t, y)\right] + \mathcal{O}(\tau^2),$$

where $y' = f(t, y)$ and the identity $y'' = \partial_t f(t, y) + \partial_y f(t, y)y'$ resulting by differentiation were exploited. For the increment function $\Phi(t, y(t), \tau) = \sum_{\ell=1}^{m} \gamma_\ell \eta_\ell$ the Taylor approximations

$$\eta_\ell = f\left(t + \tau\alpha_\ell, y + \tau \sum_{j=1}^{m} \beta_{\ell j} \eta_j\right)$$

$$= f(t, y) + \partial_t f(t, y)\tau\alpha_\ell + \partial_y f(t, y)\left(\tau \sum_{j=1}^{m} \beta_{\ell j} \eta_j\right) + \mathcal{O}(\tau^2)$$

as well as $\eta_j = f(t, y) + \mathcal{O}(\tau)$, imply that

$$\Phi(t, y, \tau) = \sum_{\ell=1}^{m} \gamma_\ell \left[f(t, y) + \partial_t f(t, y)\tau\alpha_\ell + \partial_y f(t, y)\tau \sum_{j=1}^{m} \beta_{\ell j} \eta_j \right] + \mathcal{O}(\tau^2)$$

$$= \sum_{\ell=1}^{m} \gamma_\ell \left[f(t, y) + \partial_t f(t, y)\tau\alpha_\ell + \partial_y f(t, y)\tau \sum_{j=1}^{m} \beta_{\ell j} f(t, y) \right] + \mathcal{O}(\tau^2).$$

A comparison of the coefficients in the resulting identity for

$$\frac{y(t+\tau) - y(t)}{\tau} - \Phi(t, y(t), \tau)$$

22.5 Learning Objectives, Quiz and Application

implies the sufficient conditions

$$\sum_{\ell=1}^{m} \gamma_\ell = 1, \quad \sum_{\ell=1}^{m} \gamma_\ell \alpha_\ell = \frac{1}{2}, \quad \sum_{\ell=1}^{m} \sum_{j=1}^{m} \gamma_\ell \beta_{\ell j} = \frac{1}{2}$$

for second order consistency. The last condition can be replaced by $\alpha_\ell = \sum_{j=1}^{m} \beta_{\ell j}$.

Examples 22.4 The explicit Euler method has the consistency order $p = 1$, the midpoint method, the Euler–Collatz method and Heun's method have the order $p = 2$, the Radau-3 method the consistency order $p = 3$ and the classic Runge–Kutta method as well as the 3/8 rule have the order $p = 4$.

Remarks 22.6

(i) Explicit m-stage Runge–Kutta methods have at most the consistency order $p = m$.
(ii) By using Gaussian quadrature formulas, which have the degree of exactness $2m - 1$ at m quadrature points, implicit Runge–Kutta methods with consistency order $p = 2m$ can be constructed.

22.5 Learning Objectives, Quiz and Application

You should be familiar with the approach to deriving Runge–Kutta methods and you should be able to create a Butcher tableau. You should moreover be able to describe sufficient criteria for determining the consistency of a Runge–Kutta method.

Quiz 22.1 Decide for each of the following statements whether it is true or false. You should be able to justify your answer.

The midpoint method results from an application of the midpoint quadrature formula.
An m-stage Runge–Kutta method has at least the consistency order 1.
For every explicit Runge–Kutta method of consistency order $p = 2$, $\alpha_1 = 0$ holds.
The condition $\sum_{\ell=1}^{m} \gamma_\ell = 1$ is necessary for the consistency of positive order of a Runge–Kutta method.
Every single-step method of consistency order p defines a quadrature formula of exactness degree p.

Application 22.1 Both attractive and repulsive forces act between particles such as atoms or molecules. The attractive forces dominate for large distances and the repulsive forces for small distances. This is often described with a so-called Lennard–Jones potential $V(r) = -c_1 r^{-2} + c_2 r^{-4}$, which defines the acting force through certain negative gradients, using the fact that the derivative $V'(r) = 2c_1 r^{-3} - 4c_2 r^{-5}$ is negative for $r^2 < r_0^2 = 2c_2/c_1$ and positive for $r > r_0$. With

Newton's law of inertia, the trajectories of N interacting particles with unit mass can be described by the system of differential equations

$$y_i'' = - \sum_{j=1,\ldots,N,\, j\neq i} \nabla_{y_i} V(\|y_i - y_j\|) = - \sum_{j=1,\ldots,N,\, j\neq i} V'(\|y_i - y_j\|) \frac{y_i - y_j}{\|y_i - y_j\|}$$

for $i = 1, 2, \ldots, N$ with suitable initial data. In this way, systems of particles such as water droplets can be simulated, which however leads to extremely large systems of differential equations. Use various MATLAB routines to simulate systems of 10–40 particles, which are distributed on a grid with grid spacing $d = 1$ and have no initial velocities, in the time interval $[0, T]$ with $T = 100$ with the parameters $c_1 = 10$ and $c_2 = 2$.

Chapter 23
Multistep Methods

23.1 General Multistep Methods

Multistep methods are like Runge–Kutta methods usually based on quadrature formulas, but they avoid function evaluations at the intermediate steps and instead use the values calculated in the previous time steps. The starting point is the integral representation of a differential equation $y' = f(t, y)$ on the interval $[t_{k+m-1}, t_{k+m}]$, that is

$$y(t_{k+m}) = y(t_{k+m-1}) + \int_{t_{k+m-1}}^{t_{k+m}} f(s, y(s)) \, ds.$$

The integral on the right-hand side is approximated using the function values at the time steps $t_k, t_{k+1}, \ldots, t_{k+m}$, that is

$$\int_{t_{k+m-1}}^{t_{k+m}} f(s, y(s)) \, ds \approx \tau \sum_{\ell=0}^{m} \beta_\ell f(t_{k+\ell}, y(t_{k+\ell})).$$

This can be interpreted as a generalised quadrature formula, where a function on the interval $[t_k, t_{k+m}]$ is interpolated and the interpolant is then integrated over the subinterval $[t_{k+m-1}, t_{k+m}]$, see Fig. 23.1. The function values at the time steps only need to be determined once and can be reused in later time steps. The above integration of the differential equation can also generally be carried out over a larger interval $[t_{k+m-n}, t_{k+m}]$ with $1 \le n \le m$.

Definition 23.1 An *m-multistep method* is a method of the form

$$\sum_{\ell=0}^{m} \alpha_\ell y_{k+\ell} = \tau \Phi(t_k, y_k, y_{k+1}, \ldots, y_{k+m}, \tau)$$

Fig. 23.1 Multistep methods can be interpreted as the application of a generalised quadrature formula

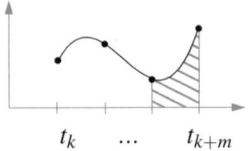

with real coefficients $(\alpha_\ell)_{\ell=0,\ldots,m}$, where $\alpha_m = 1$ applies. The method is called *explicit* if Φ does not depend on y_{k+m} and *implicit* otherwise. It is called *linear* if coefficients $(\beta_\ell)_{\ell=0,\ldots,m}$ exist, such that

$$\Phi(t_k, y_k, \ldots, y_{k+m}, \tau) = \sum_{\ell=0}^{m} \beta_\ell f(t_{k+\ell}, y_{k+\ell}).$$

Remarks 23.1

(i) A linear multistep method is explicit exactly when $\beta_m = 0$ applies.
(ii) In order to perform a step of a multistep method, the values $y_k, \ldots, y_{k+m-1}$ must already be available. At the start of the method, approximations $y_1, \ldots, y_{m-1}$ can be determined with single-step methods.

Examples 23.1

(i) The *leapfrog method* is defined by

$$y_{k+2} = y_k + 2\tau f(t_{k+1}, y_{k+1}).$$

(ii) The *Adams-Bashforth* and *Adams-Moulton methods* are explicit and implicit linear multistep methods of the form

$$y_{k+m} = y_{k+m-1} + \tau \sum_{\ell=0}^{m} \beta_\ell f(t_{k+\ell}, y_{k+\ell}),$$

(iii) So-called *backward-differentiation-formulas* or *BDF* methods use the Lagrange interpolation polynomial $q \in \mathscr{P}_m$ with $q(t_{k+i}) = y_{k+i}$, $i = 0, 1, \ldots, m$, and determine y_{k+m} as the solution of the equation

$$q'(t_{k+m}) = f(t_{k+m}, y_{k+m}).$$

23.2 Consistency

To determine the accuracy of a multistep method, a local solution is inserted into the numerical scheme.

23.2 Consistency

Definition 23.2 For $z_k \in \mathbb{R}$, let $z : [t_k, t_{k+m}] \to \mathbb{R}$ be the solution of the initial value problem $z' = f(t, z)$ in $(t_k, t_{k+m}]$ with $z(t_k) = z_k$. The *local consistency error* of a multistep method is defined by

$$\widetilde{\mathscr{C}}\big(t_k, z(t_k), \tau\big) = \frac{1}{\tau} \sum_{\ell=0}^{m} \alpha_\ell z(t_{k+\ell}) - \Phi\big(t_k, z(t_k), \ldots, z(t_{k+m}), \tau\big).$$

A multistep method is called *consistent* of order p, if for all $f \in C^p([0, T] \times \mathbb{R})$, $k = 0, 1, \ldots, K - m$ and $z \in C^{p+1}([t_k, t_{k+m}])$ we have

$$\widetilde{\mathscr{C}}\big(t_k, z(t_k), \tau\big) = \mathscr{O}(\tau^p).$$

For linear multistep methods, simple criteria for the consistency of order p emerge, as the following result, which follows the presentation in [9], shows.

Proposition 23.1 *The linear m-step method*

$$\sum_{\ell=0}^{m} \alpha_\ell y_{k+\ell} = \tau \sum_{\ell=0}^{m} \beta_\ell f(t_{k+\ell}, y_{k+\ell})$$

is consistent of order $p \geq 1$ if and only if

$$\sum_{\ell=0}^{m} \alpha_\ell = 0, \quad \sum_{\ell=0}^{m} \big(\alpha_\ell \ell^q - \beta_\ell q \ell^{q-1}\big) = 0, \quad q = 1, 2, \ldots, p.$$

Proof The Taylor approximations

$$z(t_{k+\ell}) = z(t_k + \ell\tau) = \sum_{q=0}^{p} \frac{(\ell\tau)^q}{q!} z^{(q)}(t_k) + \mathscr{O}(\tau^{p+1}),$$

$$z'(t_{k+\ell}) = z'(t_k + \ell\tau) = \sum_{q=1}^{p} \frac{(\ell\tau)^{q-1}}{(q-1)!} z^{(q)}(t_k) + \mathscr{O}(\tau^p)$$

as well as $f\big(t_{k+\ell}, z(t_{k+\ell})\big) = z'(t_{k+\ell})$ show

$$\widetilde{\mathscr{C}}\big(t_k, z(t_k), \tau\big) = \sum_{\ell=0}^{m} \Big[\frac{\alpha_\ell}{\tau} z(t_{k+m}) - \beta_\ell z'(t_{k+\ell})\Big]$$

$$= \sum_{\ell=0}^{m} \Big[\frac{\alpha_\ell}{\tau} \sum_{q=0}^{p} \frac{(\ell\tau)^q}{q!} z^{(q)}(t_k) - \beta_\ell \sum_{q=1}^{p} \frac{(\ell\tau)^{q-1}}{(q-1)!} z^{(q)}(t_k)\Big] + \mathscr{O}(\tau^p)$$

$$= \frac{1}{\tau}\sum_{\ell=0}^{m}\alpha_\ell z(t_k) + \sum_{\ell=0}^{m}\sum_{q=1}^{p}\left[\frac{\alpha_\ell}{\tau}\frac{(\ell\tau)^q}{q!} - \beta_\ell\frac{(\ell\tau)^{q-1}}{(q-1)!}\right]z^{(q)}(t_k) + \mathcal{O}(\tau^p)$$

$$= \frac{1}{\tau}\sum_{\ell=0}^{m}\alpha_\ell z(t_k) + \sum_{q=1}^{p}\sum_{\ell=0}^{m}\frac{\tau^{q-1}}{q!}\left[\alpha_\ell \ell^q - \beta_\ell q \ell^{q-1}\right]z^{(q)}(t_k) + \mathcal{O}(\tau^p).$$

Under the given conditions, both sums vanish. The reversal of the statement follows by considering the functions $z(t) = t^n$, $z(0) = 0$, $n = 1, 2, \ldots, p$, as well as $z(t) = 1$ as solutions of suitable differential equations. □

23.3 Adams Methods

Adams methods are linear multistep methods based on the approximation of local integral representations of a differential equation, that is, with a polynomial p that approximates the mapping $s \mapsto f(s, y(s))$ in the interval $I_{k,m} = [t_{k+m-1}, t_{k+m}]$, one uses

$$y(t_{k+m}) = y(t_{k+m-1}) + \int_{I_{k,m}} f(s, y(s))\,ds$$

$$\approx y_{k+m-1} + \int_{I_{k,m}} p(s)\,ds.$$

The polynomial $p \in \mathscr{P}_{\tilde{m}}$ of degree $\tilde{m} \leq m$ is chosen as the Lagrange interpolation polynomial for the nodes and values $(t_{k+\ell}, f(t_{k+\ell}, y_{k+\ell}))_{\ell=0,\ldots,\tilde{m}}$, that is

$$p(s) = \sum_{\ell=0}^{\tilde{m}} f(t_{k+\ell}, y_{k+\ell})L_\ell(s),$$

with the Lagrange basis $(L_\ell)_{\ell=0,\ldots,\tilde{m}}$ for the nodes $t_k, t_{k+1}, \ldots t_{k+\tilde{m}}$. With the coefficients

$$\beta_\ell = \frac{1}{\tau}\int_{I_{k,m}} L_\ell(s)\,ds$$

the multistep method results in

$$y_{k+m} = y_{k+m-1} + \tau\sum_{\ell=0}^{\tilde{m}} \beta_\ell f(t_{k+\ell}, y_{k+\ell}).$$

23.3 Adams Methods

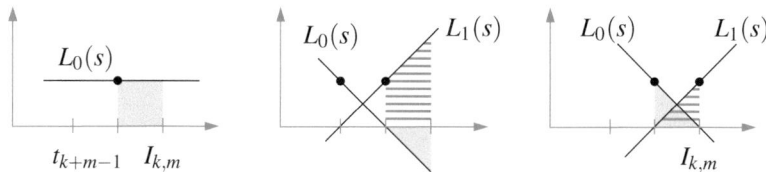

Fig. 23.2 The coefficients of the Adams methods result from the integration of the Lagrange basis polynomials in the interval $I_{k,m} = [t_{k+m-1}, t_{k+m}]$; the cases $m = 1, \widetilde{m} = 0$ (left), $m = 2, \widetilde{m} = 1$ (middle) and $m = 1, \widetilde{m} = 1$ (right) are shown

Typical choices of the parameter $\widetilde{m}$ are $\widetilde{m} = m - 1$ and $\widetilde{m} = m$, leading to explicit and implicit methods, respectively, which are referred to as Adams-Bashforth and Adams-Moulton methods. It should be noted that the Lagrange basis polynomials are only integrated in the interval $[t_{k+m-1}, t_{k+m}]$, in the case $\widetilde{m} = m - 1$ thus outside the node range $[t_k, t_{k+m-1}]$, see Fig. 23.2. The sum of the coefficients β_ℓ is always 1.

Example 23.2

(i) For the explicit Adams-Bashforth method with $m = 1$ and $\widetilde{m} = 0$, $L_0(s) = 1$ is constant and $\beta_0 = 1$ results. In the case of $m = 2$ and $\widetilde{m} = 1$, the functions L_0 and L_1 are linear and their integrals yield the coefficients $\beta_0 = -1/2$ and $\beta_1 = 3/2$.
(ii) For the implicit Adams-Moulton method with $m = 1$ and $\widetilde{m} = 1$, the functions L_0 and L_1 are linear and the coefficients $\beta_0 = 1/2$ and $\beta_1 = 1/2$ result. In the case of $m = 2$ and $\widetilde{m} = 2$, the integration of the quadratic polynomials L_0, L_1 and L_2 yields the coefficients $\beta_0 = -1/12, \beta_1 = 8/12$ and $\beta_2 = 5/12$.

In general, the coefficients shown in Tables 23.1 and 23.2 result.

Remarks 23.2

(i) The Adams-Bashforth method with m steps has the consistency order m.
(ii) In each step of the Adams-Bashforth method, only one new function evaluation is required.

Remark 23.3 The Adams-Moulton method with m steps has the consistency order $m + 1$. It is well-defined, provided $\tau \|\beta\|_1 L < 1$ with the uniform Lipschitz constant L with respect to the second argument of f.

Table 23.1 Coefficients of the Adams-Bashforth methods

m	β_0	β_1	β_2	β_3
1	1			
2	$-1/2$	$3/2$		
3	$5/12$	$-16/12$	$23/12$	
4	$-9/24$	$37/24$	$-59/24$	$55/24$

Table 23.2 Coefficients of the Adams-Moulton method for $m = 1, \ldots, 4$

m	β_0	β_1	β_2	β_3	β_4
1	1/2	1/2			
2	$-1/12$	8/12	5/12		
3	1/24	$-5/24$	19/24	9/24	
4	$-19/720$	106/720	$-264/720$	646/720	251/720

23.4 Predictor-Corrector Method

The Adams-Moulton and Adams-Bashforth methods can be combined into an explicit method, which retains the higher consistency order of the Adams-Moulton method. The idea is to perform a step of a fixed point iteration, the *corrector step*, for the Adams-Moulton method, where the initial value is determined by the Adams-Bashforth method, that is, it results from a *predictor step*.

Algorithm 23.1 (Adams-Bashforth-Moulton Method) *Let $y_0 \in \mathbb{R}$, $\tau > 0$ and $f \in C^0([0, T] \times \mathbb{R})$ and $m \in \mathbb{N}$. Furthermore, let initial values $y_1, y_2, \ldots, y_{m-1} \in \mathbb{R}$ be given. Set $k = 0$ and $K = \lfloor T/\tau \rfloor$.*

(1) Determine the auxiliary value $\widetilde{y}_{k+m} \in \mathbb{R}$ with the Adams-Bashforth method, that is, calculate

$$\widetilde{y}_{k+m} = y_{k+m-1} + \tau \sum_{\ell=0}^{m-1} \beta_\ell^{AB} f(t_{k+\ell}, y_{k+\ell}).$$

(2) Perform a step of a fixed point iteration of the Adams-Moulton method with initial value $\widetilde{y}_{k+m}$, that is, calculate

$$y_{k+m} = y_{k+m-1} + \tau \sum_{\ell=0}^{m-1} \beta_\ell^{AM} f(t_{k+\ell}, y_{k+\ell}) + \tau \beta_m^{AM} f(t_{k+m}, \widetilde{y}_{k+m}).$$

(3) Stop if $k + m > K$; otherwise increase $k \to k + 1$ and repeat step (1).

One iteration step of the Adams-Bashforth-Moulton method can be written, neglecting the arguments t_k and τ, as

$$y_{k+m} = y_{k+m-1} + \tau \Phi^{AM}\left(y_k, \ldots, y_{k+m-1}, y_{k+m-1} + \tau \Phi^{AB}(y_k, \ldots, y_{k+m-1})\right),$$

thereby defining an explicit increment function Φ^{ABM}, which is not linear.

Proposition 23.2 *Let the function f be uniformly Lipschitz continuous in the second argument. Then the Adams-Bashforth–Moulton method with m steps has the consistency order $m + 1$.*

Proof For a local solution $z : [t_k, t_{k+m}] \to \mathbb{R}$, due to the consistency order $m+1$ of the Adams-Moulton method we have

$$\tau \widetilde{\mathscr{C}}(t_k, z(t_k), \tau) = z(t_{k+m}) - z(t_{k+m-1})$$
$$- \tau \Phi^{AM}\Big(z(t_k), \ldots, z(t_{k+m-1}), z(t_{k+m-1}) + \tau \Phi^{AB}\big(z(t_k), \ldots, z(t_{k+m-1})\big)\Big)$$
$$= z(t_{k+m}) - z(t_{k+m-1}) - \tau \Phi^{AM}\big(z(t_k), \ldots, z(t_{k+m-1}), z(t_{k+m})\big)$$
$$+ \tau \Big[\Phi^{AM}\big(z(t_k), \ldots, z(t_{k+m-1}), z(t_{k+m})\big)$$
$$- \Phi^{AM}\Big(z(t_k), \ldots, z(t_{k+m-1}) + \tau \Phi^{AB}\big(z(t_k), \ldots, z(t_{k+m-1})\big)\Big)\Big]$$
$$= z(t_{k+m}) - z(t_{k+m-1}) - \tau \Phi^{AM}\big(z(t_k), \ldots, z(t_{k+m})\big) + \tau[\ldots]$$
$$= \mathscr{O}(\tau^{m+2}) + \tau[\ldots].$$

The uniform Lipschitz continuity of f or the increment function Φ^{AM} with respect to the last argument shows

$$\tau |[\ldots]| \leq \tau L \big| z(t_{k+m}) - z(t_{k+m-1}) + \tau \Phi^{AB}\big(t_k, z(t_k), \ldots, z(t_{k+m-1})\big) \big|.$$

Due to the consistency of order m of the Adams-Bashforth method, $\tau |[\ldots]| = \mathscr{O}(\tau^{m+2})$ and thus the claimed consistency order follows. □

Remark 23.4 More generally, an explicit method of consistency order p_{expl} can be used to determine a starting value and subsequently ν repetitions of a fixed point iteration with an implicit method of consistency order p_{impl} can be performed. The resulting predictor-corrector method has the consistency order $p = \min\{p_{expl} + \nu, p_{impl}\}$, see for example [9].

23.5 Learning Objectives, Quiz and Application

You should be able to construct multistep methods, clarify their advantages and disadvantages compared to single-step methods and be able to give some examples. You should be able to derive a criterion for determining the consistency order of a multistep method. You should moreover be able to explain the ideas of combining explicit and implicit multistep methods into predictor-corrector methods.

Quiz 23.1 Decide for each of the following statements whether it is true or false. You should be able to justify your answer.

Fig. 23.3 The double pendulum consists of two combined pendulums

Adams-Bashforth methods for $m \geq 1$ define linear, implicit multistep methods with consistency order $m + 1$.
Necessary for the consistency $p \geq 1$ of a multistep method is the condition $\sum_{\ell=0}^{m} \beta_\ell = 1$.
The explicit Euler method is a multistep method with two steps.
Predictor-corrector methods can be interpreted as implicit multistep methods.
The use of the *leapfrog* method and the subsequent execution of 3 fixed point iterations with the Adams-Moulton method with $m = 3$ leads to a method of consistency order $p = 5$.

Application 23.1 Interacting oscillations often lead to undesirable effects such as the spilling of a liquid being transported in a vessel, or the strong shaking of a washing machine running at certain speeds. The mathematical description can in these cases lead to differential equations for which small changes in the data can have large effects on the solutions, which is also referred to as chaotic behaviour. Numerically, the processes are therefore usually only approximable for short periods of time. An example is the double pendulum, where another pendulum is attached to the arm of a pendulum, see Fig. 23.3. Let ϕ_1 and ϕ_2 denote the deflection angles with respect to the respective rest positions. Then the pendulum movements in the case of equal masses and pendulum lengths and with a suitable scaling of the gravitational acceleration are described by the system of differential equations

$$2\phi_1'' + \phi_2'' \cos(\phi_1 - \phi_2) + \phi_2' \sin(\phi_1 - \phi_2) + 2\sin(\phi_1) = 0,$$
$$\phi_2'' + \phi_1'' \cos(\phi_1 - \phi_2) - \phi_1' \sin(\phi_1 - \phi_2) + \sin(\phi_2) = 0$$

Simulate the system with the initial data $\phi_1(0) = \pi/2$, $\phi_2(0) = 0$, $\phi_1'(0) = 0$, $\phi_2'(0) = 0$ in the time interval $[0, T]$ with $T = 100$. Perturb the initial data and use different MATLAB routines for the numerical solution. Visualise your results.

Chapter 24
Convergence of Multistep Methods

24.1 Difference Equations

For single-step methods, the consistency of a method already implies its convergence. This is generally false for multistep methods. In these, the equation

$$\frac{1}{\tau} \sum_{\ell=0}^{m} \alpha_\ell y_{k+\ell} = 0$$

defines approximations of the trivial problem $y'(t) = 0$. The following example shows that even a high order of consistency does not necessarily lead to meaningful approximations. In this chapter, we follow the presentations in [1, 8, 9].

Example 24.1 The multistep method defined by $m = 2$ and $\alpha_2 = 1$, $\alpha_1 = 4$, $\alpha_0 = -5$ and $\beta_2 = 0$, $\beta_1 = 4$, $\beta_0 = 2$

$$y_{k+2} + 4y_{k+1} - 5y_k = \tau\big(4f(t_{k+1}, y_{k+1}) + 2f(t_k, y_k)\big)$$

has the order of consistency $p = 3$. Thus, approximate solutions of the differential equation $y' = 0$ satisfy $y_{k+2} + 4y_{k+1} - 5y_k = 0$. Solutions of this equation are given by linear combinations $y_k = \gamma_1 v_k + \gamma_2 w_k$ of the special solutions

$$v_k = \lambda_1^k, \quad w_k = \lambda_2^k,$$

where $\lambda_1 = 1$ and $\lambda_2 = -5$ are the roots of the polynomial $q(\lambda) = \lambda^2 + 4\lambda - 5$. For the initial values $y_0 = 1$ and $y_1 = 1 + \delta$ we get $\gamma_1 = 1 + \delta/6$ and $\gamma_2 = -\delta/6$ and the solution

$$y_k = 1 + \delta/6 - (-5)^k \delta/6$$

Fig. 24.1 Unbounded solution of the difference equation $y_{k+2} + 4y_{k+1} - 5y_k = 0$

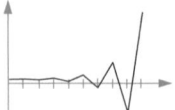

is unbounded for any $\delta \neq 0$, see Fig. 24.1. Thus, the exact solution $y(t) = 1$ of the problem $y'(t) = 0$, $y(0) = 1$, is not meaningfully approximated. The initial value $y_1 = 1 + \delta$ can be interpreted as an approximation of $y(t_1)$ of a single-step method.

Definition 24.1 Given $\alpha_0, \alpha_1, \ldots, \alpha_m \in \mathbb{R}$ with $\alpha_m = 1$, the equation

$$\sum_{\ell=0}^{m} \alpha_\ell y_{k+\ell} = 0$$

is called a *(linear homogeneous) difference equation*. A sequence $(y_k)_{k \geq 0}$ is a solution of the difference equation if it is satisfied for every $k \in \mathbb{N}_0$.

Remark 24.1 For every vector $(y_k)_{k=0,\ldots,m-1}$ of initial values, there exists a uniquely determined solution of a difference equation.

The behaviour of solutions of the difference equation can be analysed using an eigenvalue problem.

Lemma 24.1 *A sequence $(y_k)_{k \geq 0}$ is a solution of the difference equation defined by $(\alpha_\ell)_{\ell=0,\ldots,m}$ if and only if for the vectors*

$$Y_k = [y_k, y_{k+1}, \ldots, y_{k+m-1}]^\mathsf{T}$$

the relation $Y_{k+1} = A Y_k$ holds for $k = 0, 1, \ldots$, where the companion matrix $A \in \mathbb{R}^{m \times m}$ is defined by

$$A = \begin{bmatrix} 0 & 1 & & \\ & \ddots & \ddots & \\ & & 0 & 1 \\ -\alpha_0 & -\alpha_1 & \ldots & -\alpha_{m-1} \end{bmatrix}.$$

If A has the linearly independent eigenvectors $v_1, v_2, \ldots, v_m \in \mathbb{R}^m$ with associated eigenvalues $\lambda_1, \lambda_2, \ldots, \lambda_m$ and $\gamma_1, \gamma_2, \ldots \gamma_m \in \mathbb{R}$ are the coefficients of the vector Y_0 with respect to this basis, then it follows

$$Y_k = A^k Y_0 = \sum_{j=1}^{m} \lambda_j^k \gamma_j v_j.$$

The eigenvalues $\lambda_1, \lambda_2, \ldots, \lambda_m$ are exactly the roots of the characteristic polynomial $q(\lambda) = \lambda^m + \alpha_{m-1} \lambda^{m-1} + \cdots + \lambda \alpha_1 + \alpha_0$.

Proof Exercise. □

24.2 Zero-Stability

For a multistep method to lead to meaningful approximations, solutions of the associated homogeneous difference equation should be bounded. In the case of the diagonalisability of the companion matrix, this is the case, provided that $|\lambda_i| \leq 1$ for $i = 1, 2, \ldots, m$.

Definition 24.2 A linear, homogeneous difference equation is called *zero-stable*, if every solution of the difference equation is bounded.

The following definition and the subsequent proposition define a sufficient and necessary criterion for the zero-stability of a difference equation.

Definition 24.3 The polynomial $q \in \mathscr{P}_m$ satisfies *Dahlquist's root condition*, if every root $\lambda \in \mathbb{C}$ of q satisfies the estimate $|\lambda| \leq 1$, and, in the case a given root λ satisfies $|\lambda| = 1$, the root is simple.

Remark 24.2 Multiple roots $\lambda \in \mathbb{C}$ with $|\lambda| \geq 1$ always lead to unbounded solutions, because if $q(\lambda) = q'(\lambda) = 0$, then for $y_k = k\lambda^k$, it follows that

$$\sum_{\ell=0}^{m} \alpha_\ell y_{k+\ell} = k\lambda^k \sum_{\ell=0}^{m} \alpha_\ell \lambda^\ell + \lambda^{k+1} \sum_{\ell=0}^{m} \alpha_\ell \ell \lambda^{\ell-1} = k\lambda^k q(\lambda) + \lambda^{k+1} q'(\lambda) = 0.$$

If the characteristic polynomial of the companion matrix of a difference equation satisfies Dahlquist's root condition, then the equation is zero-stable and consequently its solutions are bounded.

Proposition 24.1 *Assume that the characteristic polynomial $q(z) = z^m + \alpha_{m-1} z^{m-1} + \cdots + \alpha_1 z + \alpha_0$ of the companion matrix A satisfies Dahlquist's root condition. Then there exists a regular matrix $R \in \mathbb{C}^{m \times m}$, such that, with the norm $\|\cdot\|_R : x \mapsto \|Rx\|_\infty$ and the induced operator norm $\|B\|_R = \sup_{\|x\|_R = 1} \|Bx\|_R$, we have $\|A\|_R \leq 1$.*

Proof Let $\lambda_1, \lambda_2, \ldots, \lambda_r \in \mathbb{C}$ be the complex eigenvalues of A with multiplicities $s_1, s_2, \ldots, s_r$. The main theorem about the Jordan normal form implies the existence of a regular matrix $T \in \mathbb{C}^{m \times m}$ and of matrices $J_i \in \mathbb{C}^{s_i \times s_i}$, $i = 1, 2, \ldots, r$, such that

$$T^{-1} A T = J = \begin{bmatrix} J_1 & & & \\ & J_2 & & \\ & & \ddots & \\ & & & J_r \end{bmatrix}, \quad J_i = \begin{bmatrix} \lambda_i & 1 & & \\ & \ddots & \ddots & \\ & & \ddots & 1 \\ & & & \lambda_i \end{bmatrix}.$$

If a Jordan block of size $s_i \geq 2$ and consequently an eigenvalue $|\lambda_i| < 1$ exists, define

$$\varepsilon = \min\{1 - |\lambda_i| : i = 1, 2, \ldots, r, \; |\lambda_i| < 1\},$$

and otherwise set $\varepsilon = 1$. Let $D \in \mathbb{R}^{m \times m}$ be the diagonal matrix with the entries $d_{jj} = \varepsilon^{j-1}$ for $j = 1, 2, \ldots, m$. Then we have

$$\widetilde{J} = D^{-1} T^{-1} A T D = \begin{bmatrix} \widetilde{J}_1 & & & \\ & \widetilde{J}_2 & & \\ & & \ddots & \\ & & & \widetilde{J}_r \end{bmatrix}, \quad \widetilde{J}_i = \begin{bmatrix} \lambda_i & \varepsilon & & \\ & \ddots & \ddots & \\ & & \ddots & \varepsilon \\ & & & \lambda_i \end{bmatrix}.$$

For the row sum norm of the scaled Jordan blocks $\widetilde{J}_i$ we have due to the choice of ε, that $\|\widetilde{J}_i\|_\infty \leq 1$, and it follows $\|\widetilde{J}\|_\infty \leq 1$. With $R = D^{-1} T^{-1}$ it follows for the induced operator norm with the replacement $y = D^{-1} T^{-1} x$, that

$$\|A\|_R = \sup_{\|x\|_R = 1} \|Ax\|_R = \sup_{\|D^{-1} T^{-1} x\|_\infty = 1} \|D^{-1} T^{-1} A x\|_\infty$$

$$= \sup_{\|y\|_\infty = 1} \|D^{-1} T^{-1} A T D y\|_\infty \leq \|\widetilde{J}\|_\infty.$$

This proves the claim. □

Example 24.2 For Adams methods, we have $\alpha_m = 1$, $\alpha_{m-1} = -1$ and $\alpha_\ell = 0$ otherwise, so that the characteristic polynomial $q(z) = z^m - z^{m-1}$ has the $(m-1)$-fold zero $\lambda = 0$ as well as the simple zero $\lambda = 1$. Consequently, Adams methods are zero-stable. Similarly, the general zero-stability of single-step methods is obtained.

24.3 Convergence

For a multistep method, the zero-stability of the associated difference equation is a necessary criterion for the convergence of the method. Every multistep method

$$\sum_{\ell=0}^{m} \alpha_\ell y_{k+\ell} = \tau \Phi(t_k, y_k, y_{k+1}, \ldots, y_{k+m}, \tau)$$

can be represented with the vectors $Y_k \in \mathbb{R}^m$, $k = 0, 1, \ldots, K - m + 1$, and the function $\Psi : [0, T] \times \mathbb{R}^m \times \mathbb{R}^m \times \mathbb{R} \to \mathbb{R}^m$ defined by

24.3 Convergence

$$Y_k = [y_k, y_{k+1}, \ldots, y_{k+m-1}]^T,$$

$$\Psi(t_k, Y_k, Y_{k+1}, \tau) = [0, \ldots, 0, \Phi(t_k, y_k, y_{k+1}, \ldots, y_{k+m}, \tau)]^T$$

as well as the companion matrix $A \in \mathbb{R}^{m \times m}$ in the form

$$Y_{k+1} = A Y_k + \tau \Psi(t_k, Y_k, Y_{k+1}, \tau),$$

This is the structure of a single-step method and an error analysis can be carried out similarly. The validity of Dahlquist's root condition allows the influence of the matrix A to be controlled.

Proposition 24.2 *Suppose that the multistep method*

$$\sum_{\ell=0}^{m} \alpha_\ell y_{k+\ell} = \tau \Phi(t_k, y_k, y_{k+1}, \ldots, y_{k+m}, \tau)$$

is consistent of order p and the polynomial $q(z) = z^m + \alpha_{m-1} z^{m-1} + \cdots + \alpha_1 z + \alpha_0$ satisfies Dahlquist's root condition. Furthermore, assume that Φ is uniformly Lipschitz continuous in the arguments $y_k, \ldots, y_{k+m}$, that is, there exists a constant $L \geq 0$ such that for all $t \in [0, T]$, $v, w \in \mathbb{R}^{m+1}$ and $\tau > 0$ we have

$$|\Phi(t, v_0, \ldots, v_m, \tau) - \Phi(t, w_0, \ldots, w_m, \tau)| \leq L\big(|v_0 - w_0| + \cdots + |v_m - w_m|\big).$$

If the initial values $y_0, y_1, \ldots, y_{m-1}$ are chosen such that

$$\max_{k=0,\ldots,m-1} |y_k - y(t_k)| \leq C_0 \tau^p$$

with a constant $C_0 \geq 0$ independent of τ, then there exist constants C_1, C_2, C_3 such that

$$\max_{k=0,1,\ldots,K} |y_k - y(t_k)| \leq C_1 T \tau^p \exp(C_2 L T)$$

for all $0 < \tau \leq C_3$.

Proof In this proof, c stands for a constant that can increase from step to step, but does not depend on τ and K. For $k = 0, 1, \ldots, K$ let $e_k = y_k - y(t_k)$. By definition of the consistency term $\widetilde{\mathscr{C}}_k = \widetilde{\mathscr{C}}(t_k, y(t_k), \tau)$ we have

$$\sum_{\ell=0}^{m} \alpha_\ell e_{k+\ell} = \tau \big[\Phi(t_k, y_k, \ldots, y_{k+m}, \tau) - \Phi(t_k, y(t_k), \ldots, y(t_{k+m}), \tau) - \widetilde{\mathscr{C}}_k\big]$$

for $k = 0, 1, \ldots, K - m$ and let τr_k be the right-hand side. The Lipschitz continuity of Φ and the consistency of the method imply that

$$\tau |r_k| \leq \tau L \sum_{\ell=0}^{m} |e_{k+\ell}| + c\tau^{p+1}.$$

With the vectors

$$E_k = [e_k, e_{k+1}, \ldots, e_{k+m-1}]^\mathsf{T}, \quad G_k = [0, \ldots, 0, r_k]^\mathsf{T}$$

and the companion matrix A it follows

$$E_{k+1} = A E_k + \tau G_k.$$

Let $\|\cdot\|_R$ be a norm on $\mathbb{R}^m$ such that $\|A\|_R \leq 1$ with the induced operator norm. The equivalence of norms on the vector space $\mathbb{R}^m$ shows that

$$\|\tau G_k\|_R \leq c\tau |r_k| \leq c\tau^{p+1} + c\tau L \sum_{\ell=0}^{m} |e_{k+\ell}| \leq c\tau L \big(\|E_{k+1}\|_R + \|E_k\|_R\big) + c\tau^{p+1}.$$

With the scheme for the vectors E_k, it follows that

$$\|E_{k+1}\|_R \leq \|A E_k\|_R + \|\tau G_k\|_R \leq \|A\|_R \|E_k\|_R + c\tau |r_k|$$
$$\leq \|E_k\|_R + c\tau L \big(\|E_k\|_R + \|E_{k+1}\|_R\big) + c\tau^{p+1}$$

or

$$(1 - c'\tau L)\|E_{k+1}\|_R \leq (1 + c'\tau L)\|E_k\|_R + c''\tau^{p+1}.$$

Subtracting $(1 - c'\tau L)\|E_k\|_R$ from both sides leads to

$$\|E_{k+1}\|_R - \|E_k\|_R \leq 4c'\tau L \|E_k\|_R + 2c''\tau^{p+1},$$

where $c'\tau L \leq 1/2$ or $1 - c'\tau L \geq 1/2$ was assumed. With $c_1 = c'$, $c_2 = 4c'$ and $c_3 = 2c''$ and a summation of this equation over $k = 0, 1, \ldots, K'$ with $K' \leq K - m$ results in

$$\|E_{K'+1}\|_R \leq \|E_0\|_R + c_2 \tau L \sum_{k=0}^{K'} \|E_k\|_R + c_3 \tau^{p+1} K'.$$

The discrete Gronwall lemma and $K'\tau \leq T$ show that

$$\max_{k=0,1,\ldots,K-m+1} \|E_k\|_R \le \bigl(\|E_0\|_R + c_3 T \tau^p\bigr) \exp\bigl(c_2 L T\bigr).$$

Since $\|E_0\|_R \le c\tau^p$ and $|e_{k+\ell}| \le c\|E_k\|_R$ for $k = 0, 1, \ldots, K-m+1$ and $\ell = 0, 1, \ldots, m-1$ the claimed estimate follows. □

Remark 24.3 A two-step method with quadratic consistency can be initialized with the implicit or explicit Euler method. If, e.g., y_1 is obtained via the explicit scheme $y_1 = y_0 + \tau f(0, y_0)$ then, noting the consistency error

$$y(\tau) - y_0 - \tau f(0, y_0) = y(\tau) - y(0) - \tau y'(0) = \frac{\tau^2}{2} y''(\xi)$$

for some $\xi \in [0, \tau]$, we find that $y_1 - y(t_1) = \mathcal{O}(\tau^2)$. In the convergence analysis for the Euler schemes one factor τ is needed to control the accumulation of consistency errors over several steps $t_k = k\tau$, $k = 1, 2, \ldots, K$.

24.4 Learning Objectives, Quiz and Application

You should be able to specify stability problems of multistep methods and explain Dahlquist's root condition. You should be able to provide a proof sketch for the derivation of error estimates for multistep methods.

Quiz 24.1 Decide for each of the following statements whether it is true or false. You should be able to justify your answer.

Zero stability is a necessary criterion for the convergence of a multistep method of positive consistency order.	
Every single-step method satisfies Dahlquist's root condition.	
The validity of Dahlquist's root condition for a multistep method implies the zero stability of the associated difference equation.	
The recursion formula $y_{k+2} = y_{k+1} - (1/4) y_k$ is zero-stable.	
If a multistep method satisfies Dahlquist's root condition, then for the associated companion matrix $\varrho(A) < 1$.	

Application 24.1 The simulation of electrical circuits allows the prediction of the voltages falling on the components. As an example, we consider an RLC circuit, which consists of a resistor, an inductor and a capacitor, as shown in Fig. 24.2. According to Ohm's law, the voltage drop across the resistor U_R is proportional to the current I_R flowing through it, i.e. $U_R = R I_R$. The current I_C flowing through the capacitor is proportional to the voltage change, that is, $I_C = C U'_C$. On the other hand, the voltage drop U_L at the coil is proportional to the current change, that is, $U_L = L I'_L$. Kirchhoff's laws state that the sum of the currents flowing

Fig. 24.2 Diagram of an RLC circuit

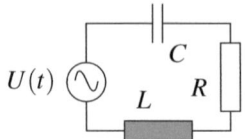

through a node of a circuit is zero and that the sum of the voltages belonging to a mesh disappears. For the RLC circuit with time-dependent voltage source $U(t)$, the equations are therefore

$$U(t) = U_R(t) + U_L(t) + U_C(t),$$
$$I(t) = I_R(t) = I_L(t) = I_C(t).$$

Derive the differential equation

$$I'' + \frac{R}{L}I' + \frac{1}{LC}I = U'$$

for the current $I(t)$ flowing through the circuit and simulate this for the initial values $I(0) = 0\,\text{A}$, $I'(0) = 0.5\,\text{A/s}$, the proportionality factors $R = 47\,\Omega$, $L = 20\,\text{mH}$, $C = 0.1\,\mu\text{F}$ and the alternating voltage $U(t) = \sin(50 \cdot 2\pi t)230\,\text{V}$. Solve the initial value problem with various MATLAB routines and test other values of the capacitance. Present the voltages U_R, U_L and U_C as functions of time in the interval $[0, T]$ with $T = 10\,\text{ms}$ comparatively in a graph.

Chapter 25
Stiff Differential Equations

25.1 Stiffness

The convergence studies of the Euler methods in the important special case $y' = \lambda y$ show that error estimates for the explicit method are valid under a condition $\tau|\lambda| \leq c$, while this condition is not necessary for the implicit variant if $\lambda < 0$. In applications, differential equations of the form $y' = Ay$ occur, where the matrix A has negative eigenvalues. Implicit methods are particularly well suited for this class. In this chapter, we follow the presentations in [1, 7, 8].

Example 25.1 For $\lambda < 0$, the solution of the initial value problem

$$y' = \lambda y, \quad y(0) = y_0,$$

is given by $y(t) = y_0 e^{\lambda t}$ and it holds that $|y(t)| \leq |y_0|$ for all $t \geq 0$

(a) With the explicit Euler method, we get $y_k = (1 + \tau\lambda)^k y_0$, $k \geq 0$, and this sequence is bounded exactly when $|1 + \tau\lambda| \leq 1$ holds, thus when $\tau \leq 2/|\lambda|$ holds.
(b) With the implicit Euler method, we get $y_k = (1 - \tau\lambda)^{-k} y_0$, $k \geq 0$, and since $1 - \tau\lambda \geq 1$ holds, the sequence is bounded for any choice of $\tau > 0$.

The difficulties of explicit methods become even more apparent in systems of differential equations.

Example 25.2 For $\lambda_1, \lambda_2 < 0$, the bounded solution of the system

$$\begin{bmatrix} y_1 \\ y_2 \end{bmatrix}' = \frac{1}{2} \begin{bmatrix} \lambda_1 + \lambda_2 & \lambda_1 - \lambda_2 \\ \lambda_1 - \lambda_2 & \lambda_1 + \lambda_2 \end{bmatrix} \begin{bmatrix} y_1 \\ y_2 \end{bmatrix}, \quad \begin{bmatrix} y_1(0) \\ y_2(0) \end{bmatrix} = \begin{bmatrix} 2 \\ 0 \end{bmatrix}$$

© The Author(s), under exclusive license to Springer-Verlag GmbH, DE, part of Springer Nature 2025
S. Bartels, *Numerical Mathematics 3x9*, La Matematica per il 3+2 160,
https://doi.org/10.1007/978-3-662-70890-3_25

is given by

$$\begin{bmatrix} y_1(t) \\ y_2(t) \end{bmatrix} = \begin{bmatrix} e^{\lambda_1 t} + e^{\lambda_2 t} \\ e^{\lambda_1 t} - e^{\lambda_2 t} \end{bmatrix}.$$

With the explicit Euler method, the approximations are

$$y_{1,k} = (1 + \tau\lambda_1)^k + (1 + \tau\lambda_2)^k,$$

$$y_{2,k} = (1 + \tau\lambda_1)^k - (1 + \tau\lambda_2)^k,$$

and the sequence $(y_{1,k}, y_{2,k})_{k \geq 0}$ is bounded exactly when $|1 + \tau\lambda_1| \leq 1$ and $|1 + \tau\lambda_2| \leq 1$ hold. For example, if $\lambda_1 = -1$ and $\lambda_2 = -10^\alpha$ with $\alpha \geq 2$ the contributions $e^{\lambda_2 t} = e^{-10^\alpha t}$ to the exact solution are negligible for $t \geq 10^{-\alpha/2}$, but the time step size is determined by λ_2 in the form $\tau \leq 2/|\lambda_2| = 2 \cdot 10^{-\alpha}$.

The occurrence of large negative eigenvalues in a differential equation leads to the concept of stiffness.

Definition 25.1 The differential equation $y' = f(t, y)$ is called *stiff*, if the Jacobian matrix $Df(t, \bar{y}) \in \mathbb{R}^{n \times n}$ with respect to the argument y has an eigenvalue $\lambda \in \mathbb{C}$ with the property $\text{Re}(\lambda) \ll 0$ for some $t \geq 0$ and $\bar{y} \in \mathbb{R}^n$.

Remark 25.1 The local behaviour of a solution y at time $t_* \geq 0$ is described by the matrix $A = Df(t_*, y(t_*))$ and its eigenvalues, particularly in terms of perturbations. The function z defined by $y(t_* + s) = y(t_*) + z(s)$ satisfies for small values $s \geq 0$ in the case of an autonomous equation the linear differential equation

$$z'(s) = y'(t_* + s) = f(y(t_*) + z(s)) \approx f(y(t_*)) + Az(s),$$

where initial values $z(0) = z_0$ are considered to assess the effects of perturbations. If A is diagonalisable and the eigenvalues have exclusively negative real parts, then the solution y is stable in the sense that small perturbations do not lead to large changes in the solution.

25.2 A-Stability

The following concept of stability is used to identify suitable numerical methods for stiff differential equations.

Definition 25.2 A numerical method is called *A-stable* or *unconditionally stable*, if for every complex diagonalisable matrix $A \in \mathbb{R}^{n \times n}$ whose eigenvalues all have non-positive real parts, the approximations $(y_k)_{k \geq 0}$ of the differential equation $y' = Ay$ defined by the method are bounded for all initial data and all time step sizes $\tau > 0$.

25.2 A-Stability

Due to the diagonalisability of the matrix A, it is sufficient to consider scalar equations in which A is replaced by a number $\lambda \in \mathbb{C}$ with $\operatorname{Re}(\lambda) \leq 0$ to prove A-stability.

Example 25.3 The implicit Euler method is A-stable, but the explicit Euler method is not.

The A-stability of single-step methods can be analysed using so-called stability functions.

Definition 25.3 A function $g : S \to \mathbb{C}$ with $S \subset \mathbb{C}$ is called *stability function* of the single-step method defined by the increment function Φ, if for all $\lambda \in \mathbb{C}$, $y_0 \in \mathbb{R}$, $\tau > 0$ with $\tau\lambda \in S$ and all $k \in \mathbb{N}_0$ for the approximations of the initial value problem $y' = \lambda y$, $y(0) = y_0$, we have

$$y_{k+1} = y_k + \tau \Phi(t_k, y_k, y_{k+1}, \tau) = g(\tau\lambda) y_k.$$

Necessary and sufficient for the A-stability of a method is that $|g(z)| \leq 1$ for all $z \in \mathbb{C}$ with $\operatorname{Re}(z) \leq 0$.

Examples 25.4

(i) For the explicit Euler method, $g(z) = 1 + z$.
(ii) For the implicit Euler method, $g(z) = 1/(1-z)$.
(iii) For the trapezoidal or midpoint method, $\Phi(t, y_k, y_{k+1}, \tau) = \lambda(y_k + y_{k+1})/2$ and thus $g(z) = (2+z)/(2-z)$.

For Runge–Kutta methods, a closed formula for the stability function can be given.

Lemma 25.1 *Let $\alpha \in \mathbb{R}^m$, $\beta \in \mathbb{R}^{m \times m}$ and $\gamma \in \mathbb{R}^m$ be the coefficients of a Runge–Kutta method. Then, with the vector $e = [1, 1, \ldots, 1]^\mathsf{T} \in \mathbb{R}^m$, the associated stability function is given by*

$$g(z) = 1 + z\gamma^\mathsf{T}(I_m - z\beta)^{-1} e.$$

For strict lower triangular matrices β, g is well-defined for all $z \in \mathbb{C}$.

Proof For Runge–Kutta methods, $y_{k+1} = y_k + \tau\gamma^\mathsf{T}\eta^k$ applies with the solution $\eta^k = [\eta^k_1, \eta^k_2, \ldots, \eta^k_m]^\mathsf{T} \in \mathbb{R}^m$ of the system of equations

$$\eta^k_\ell = f\left(t_k + \tau\alpha_\ell, y_k + \tau \sum_{j=1}^m \beta_{\ell j} \eta^k_j\right)$$

for $\ell = 1, 2, \ldots, m$. In the special case $f(t, y) = \lambda y$, we get

$$\eta^k = \lambda(y_k e + \tau\beta\eta^k)$$

or $\eta^k = (I_m - \lambda\tau\beta)^{-1}(\lambda y_k e)$, from which the claimed identity for g follows. If β is a strict lower triangular matrix, then the matrix $I_m - z\beta$ is invertible for every $z \in \mathbb{C}$. □

By forward substitution, polynomial expressions are obtained in the case of explicit Runge–Kutta methods. For implicit methods, rational functions are obtained.

Example 25.5 The classic Runge–Kutta method is defined by $m = 4$ as well as $\alpha = [0, 1/2, 1/2, 1]^T$, $\gamma = [1, 2, 2, 1]^T/6$ and $\beta \in \mathbb{R}^{4\times 4}$ with the non-vanishing entries $\beta_{21} = \beta_{32} = 1/2$ and $\beta_{43} = 1$. This results in

$$g(z) = 1 + z + \frac{z^2}{2} + \frac{z^3}{6} + \frac{z^4}{24}.$$

Various statements can be derived from the stability function of a single-step method.

Proposition 25.1 *Let* $g : S \to \mathbb{C}$ *be the stability function of a single-step method with the property that* $\{z \in \mathbb{C} : \operatorname{Re}(z) \leq 0\} \subset S$.

(i) *The method is A-stable if and only if* $|g(z)| \leq 1$ *for all* $z \in \mathbb{C}$ *with* $\operatorname{Re}(z) \leq 0$.
(ii) *In the case of an explicit Runge–Kutta method,* $\lim_{|z|\to\infty} |g(z)| = \infty$, *i.e. explicit Runge–Kutta methods are not A-stable.*
(iii) *If the method is consistent of order* $p \geq 0$, *then* $|e^z - g(z)| \leq c|z|^{p+1}$ *for* $0 < |z| \leq c$.

Proof

(i) If $|g(z)| \leq 1$, then the boundedness of the sequence $y_k = g(\tau\lambda)^k y_0$, $k \in \mathbb{N}$, immediately follows, and obviously this condition is also necessary.
(ii) For explicit Runge–Kutta methods, it follows from the representation $g(z) = 1 + z\gamma^T(I_m - z\beta)^{-1}e$ and Cramer's rule, that g is a polynomial in z, from which the claim follows.
(iii) Let $\lambda \in \mathbb{C}$ with $|\lambda| = 1$. For the solution $y(t) = e^{t\lambda}$, of the initial value problem $y' = \lambda y$, $y(0) = 1$, and the approximation $y_1 = g(\tau\lambda)y_0 = g(\tau\lambda)$, it follows with the definition of the consistency term, that

$$|e^{\tau\lambda} - g(\tau\lambda)| = |y(\tau) - y_1| \leq c\tau^{p+1} = c|\tau\lambda|^{p+1}$$

for all $0 < \tau \leq c'$ applies. With $z = \tau\lambda$, the claim follows.
□

If a polynomial $g \in \mathscr{P}_m$ satisfies $|g(z) - e^z| \leq c|z|^{p+1}$, then the first $p+1$ coefficients of g coincide with those of the exponential function.

Corollary 25.1 *An m-stage, explicit Runge–Kutta method has at most consistency order m.*

25.3 Gradient Flows

Table 25.1 Overview of various stability concepts

Stability concept	Test equation	Meaning	Example
Zero-stability	$y' = 0$	Necessary convergence criterion	Adams method
A-stability	$y' = \lambda y$	Avoidance of a step size condition	Trapezoidal method
L-stability	$y' = \lambda y$	Numerical damping property	Implicit Euler method

Proof The stability function is a polynomial of degree m and this can approximate the function e^z at most up to an error $\mathcal{O}(z^{m+1})$. □

The unconditional boundedness of approximations is a meaningful requirement for numerical methods. In some situations, however, too rapid decay of the approximation solutions can be undesirable. For the implicit Euler method, for example, we have $|g(z)| \to 0$ for $z \to -\infty$, which leads to a strong damping behaviour for large step sizes. For the trapezoidal method, on the other hand, $|g(z)| \to 1$ for $|z| \to \infty$, so no numerical damping for large time step sizes occurs, but oscillations must be expected.

Definition 25.4 A single-step method is called *L-stable*, if it is A-stable and in addition $\lim_{\mathrm{Re}(z) \to -\infty} g(z) = 0$ holds.

The A-stability property thus describes the unconditional stability of a method and the L-stability property indicates additional damping properties of the method for large step sizes. An overview of various stability concepts is shown in Table 25.1.

Remarks 25.2

(i) The implicit Euler method is L-stable.
(ii) The trapezoidal method is A-stable, but not L-stable.

25.3 Gradient Flows

An important class of stiff differential equations are *gradient flows*, in which the right-hand side is given by the negative gradient of a function, i.e. autonomous differential equations of the form

$$y'(t) = -\nabla G(y(t)), \quad y(0) = y_0,$$

with a given function $G \in C^1(\mathbb{R}^n)$. These initial value problems can be interpreted as continuous descent methods for the minimisation of the function G. The value of the function G is reduced along the path $t \mapsto y(t)$, because if one multiplies the differential equation with $-y'(t)$, it follows

$$-\|y'(t)\|^2 = -y'(t) \cdot y'(t) = \nabla G(y(t)) \cdot y'(t) = \frac{d}{dt} G(y(t)).$$

If the function G is *coercive*, i.e. $G(w) \to \infty$ for $|w| \to \infty$, it follows that solutions remain bounded and are defined for all $t \in [0, \infty)$, even if G' is not globally Lipschitz continuous. If G is μ-*convex*, i.e. there exists a number $\mu > 0$ such that $G(s) + (\mu/2)|s|^2$ is convex, then the implicit Euler method is well-defined and stable for $\tau < 1/\mu$. Note that gradient flows in general do not define linear differential equations of the form $y' = Ay$.

Proposition 25.2 *Let $G \in C^2(\mathbb{R}^n)$ be μ-convex. For $0 < \tau < \mu^{-1}$ and $y_0 \in \mathbb{R}^n$, the sequence*

$$y_{k+1} = y_k - \tau \nabla G(y_{k+1})$$

is uniquely defined and for all $\ell \geq 0$ we have

$$G(y_\ell) + \frac{1}{2\tau} \sum_{k=0}^{\ell-1} \|y_{k+1} - y_k\|^2 \leq G(y_0).$$

Proof Let $y_k \in \mathbb{R}^n$ for a $k \geq 0$ be given. The mapping

$$H_{k+1}(s) = \frac{1}{2\tau} \|s - y_k\|^2 + G(s)$$

is strictly convex for $0 < \tau < \mu^{-1}$, i.e. $D^2 H_{k+1}(s)$ is positive definite for all $s \in \mathbb{R}^n$, and its unique minimum $y_{k+1} \in \mathbb{R}^n$ is attained in a compact set $\overline{B_r(0)}$. For this the optimality condition applies

$$0 = \nabla H_{k+1}(y_{k+1}) = \frac{1}{\tau}(y_{k+1} - y_k) + \nabla G(y_{k+1}).$$

With the convexity property

$$\nabla H_{k+1}(v)(w - v) + H_{k+1}(v) \leq H_{k+1}(w),$$

which holds for all $v, w \in \mathbb{R}^n$, it follows with $v = y_{k+1}$ and $w = y_k$, that

$$G(y_{k+1}) + \frac{1}{2\tau} \|y_{k+1} - y_k\|^2 \leq G(y_k)$$

and the summation of this estimate over $k = 0, 1, \ldots, \ell - 1$ proves the claim. $\square$

Remarks 25.3

(i) From coercivity or growth properties of G it follows, that $y_{k+1} - y_k \to 0$ for $k \to \infty$, and thus the convergence $y_{k_\ell} \to y_*$ of a subsequence to a stationary point or a minimum y_* of G, i.e. $\nabla G(y_*) = 0$.

(ii) Note that no Lipschitz continuity of ∇G was assumed.

(iii) The equation $y_{k+1} = y_k - \tau \nabla G(y_{k+1})$ is solved with a fixed point iteration or the Newton method.

(iv) If G is given as the sum of a convex and a concave function, i.e. $G = G^{cx} + G^{cv}$, then the *implicit-explicit* method

$$y_{k+1} = y_k - \tau \nabla G^{cx}(y_{k+1}) - \tau \nabla G^{cv}(y_k)$$

is preferable to the implicit method due to the better solvability with the Newton method. Often, *semi-implicit* methods based on the linearisation $\nabla G(y_{k+1}) \approx \nabla G(y_k) + D^2 G(y_k)(y_{k+1} - y_k)$, i.e.

$$y_{k+1} = y_k - \tau \left[\nabla G(y_k) - D^2 G(y_k)(y_{k+1} - y_k) \right]$$

provide a good alternative to the implicit method. This corresponds exactly to the execution of one step of the Newton method for the implicit scheme.

25.4 Heat Equation

Stiff differential equations occur in the spatial discretisation of parabolic partial differential equations such as the heat equation. In a one-dimensional situation, a function $u : [0, T] \times [a, b] \to \mathbb{R}$ is sought, which solves the initial boundary value problem

$$\begin{aligned}
\partial_t u(t, x) - \kappa \partial_x^2 u(t, x) &= f(t, x) & (t, x) &\in (0, T] \times (a, b), \\
u(0, x) &= u_0(x) & x &\in [a, b], \\
u(t, a) = 0, \; u(t, b) &= 0 & t &\in (0, T]
\end{aligned}$$

where the right-hand side $f \in C^0([0, T] \times [a, b])$ and the initial values $u_0 \in C^0([a, b])$ are given. The function u describes the temperature distribution in a thin metal wire, whose ends are constantly kept at temperature 0 and at time $t = 0$ has the temperature distribution u_0. The right-hand side f describes possible heat sources and sinks in the wire. Similar to the approximation of a first derivative, a second derivative can be approximated with a difference quotient. For a step size $h > 0$ we have

$$\frac{u(x-h) - 2u(x) - u(x+h)}{h^2} = u''(x) + \mathcal{O}(h^2).$$

With the spatial step size $h = (b-a)/M$ and the grid points $x_j = a + jh$, $j = 0, 1, \ldots, M$ and a time step size $\tau > 0$ and the time steps $t_k = k\tau$, $k = 0, 1, \ldots, K$, approximations

$$U_j^k \approx u(t_k, x_j)$$

are sought. Replacing the time derivative and the second spatial derivative with difference quotients leads to the identities

$$\frac{1}{\tau}(U_j^{k+1} - U_j^k) - \frac{\kappa}{h^2}(U_{j-1}^{k+1} - 2U_j^{k+1} - U_{j+1}^{k+1}) = F_j^{k+1}$$

for $k = 0, 1, \ldots, K - 1$, $j = 1, 2, \ldots, M - 1$ and $F_j^{k+1} = f(t_{k+1}, x_j)$. At the boundary nodes, the boundary conditions $U_0^{k+1} = U_M^{k+1} = 0$ are used. The equations for $j = 1, 2, \ldots, M - 1$ can be written simultaneously as

$$\frac{1}{\tau}\begin{bmatrix} U_1^{k+1} \\ U_2^{k+1} \\ \vdots \\ U_{M-1}^{k+1} \end{bmatrix} - \frac{\kappa}{h^2}\begin{bmatrix} -2 & 1 & & \\ 1 & \ddots & \ddots & \\ & \ddots & \ddots & 1 \\ & & 1 & -2 \end{bmatrix}\begin{bmatrix} U_1^{k+1} \\ U_2^{k+1} \\ \vdots \\ U_{M-1}^{k+1} \end{bmatrix} = \frac{1}{\tau}\begin{bmatrix} U_1^k \\ U_2^k \\ \vdots \\ U_{M-1}^k \end{bmatrix} + \begin{bmatrix} F_1^{k+1} \\ F_2^{k+1} \\ \vdots \\ F_{M-1}^{k+1} \end{bmatrix}.$$

With the vectors $\widehat{U}^k = [U_1^k, U_2^k, \ldots, U_{M-1}^k]^\mathsf{T}$ and $F^k = [F_1^k, F_2^k, \ldots, F_{M-1}^k]^\mathsf{T}$ this is equivalent to

$$\widehat{U}^{k+1} = \widehat{U}^k + \tau\left(\frac{\kappa}{h^2} A \widehat{U}^{k+1} + F^{k+1}\right)$$

for $k = 0, 1, \ldots, K - 1$ with the initial data $\widehat{U}^0 = [u_0(x_1), \ldots, u_0(x_{M-1})]^\mathsf{T}$. This can be interpreted as an implicit time discretisation of a system of linear differential equations, that is of the initial value problem

$$U'(t) = \frac{\kappa}{h^2} A U(t) + F(t), \quad U(0) = \widehat{U}_0,$$

with the symmetric and negative definite matrix $A \in \mathbb{R}^{(M-1)\times(M-1)}$ for which $\operatorname{cond}_2(A) \sim h^{-2}$ applies, that is, a stiff differential equation.

25.5 Learning Objectives, Quiz and Application

You should be aware of the problems of explicit methods with stiff differential equations and you should be able to name and explain terms for categorising the stability of a numerical method. You should be able to define gradient flows and prove the basic properties of the application of the implicit Euler method.

Quiz 25.1 Decide for each of the following statements whether it is true or false. You should be able to justify your answer.

The differential equation $y' = e^{5t} \sin(y)$ is stiff.
The explicit Euler method is A-stable and the implicit Euler method L-stable.
The trapezoidal method is consistent of order $p = 2$ and A-stable.
The stability function of every explicit Runge–Kutta method is unbounded.
The Richardson method for the iterative solution of $Ax = b$ corresponds to the application of the explicit Euler method on $y' = -(Ay - b)$.

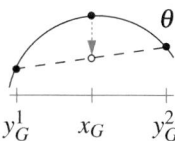

Fig. 25.1 Diffusion processes strive for a state of equilibrium

Application 25.1 Heat conduction and diffusion processes are balancing processes that aim for a stationary, i.e. a temporally invariant state. For a mathematical description, we consider a metal body and model it as a uniform particle grid. To achieve a balancing effect, we use the assumption that the change in temperature at each internal grid point is proportional to the deviation of the temperature from the average of the temperatures at the neighbouring points, i.e.

$$\partial_t \theta(t, x_G) = -\frac{\kappa}{h^2}\left(\theta(t, x_G) - \frac{1}{|\mathcal{N}(x_G)|} \sum_{y_G \in \mathcal{N}(x_G)} \theta(t, y_G)\right),$$

where $\mathcal{N}(x_G)$ denotes the set of neighbouring points of x_G with cardinality $|\mathcal{N}(x_G)|$ and h is the grid size, see Fig. 25.1. At the boundary points, the temperature is given by the value 0 and at the time $t = 0$ by values $\theta_0(x_G)$.

(i) Show that the right-hand side of the differential equation in the case of a twice continuously differentiable function θ for $h \to 0$ converges to $\kappa \theta''(t, x_G)$ or $\kappa \Delta \theta(t, x_G)$.
(ii) Show that the heat conduction process can be formulated as a system of differential equations $Y' = AY$ in $(0, T]$ with initial condition $Y(0) = Y_0$ and specify the matrix A for the case of a metal plate, which is described as a uniform grid of the domain $(0, 1)^2$.
(iii) Use the grid size $h = 1/20$ as well as the implicit and explicit Euler method with different time step sizes and randomly generated initial data. Assess for which step sizes you obtain useful approximations and whether it is a stiff differential equation.

Chapter 26
Step Size Control

26.1 A Posteriori Error Control

We next derive an error estimate that depends on the calculated approximations and the non-constant step sizes. The estimate allows for an optimal local adjustment of the step sizes on particular features of solutions and thus leads to efficient methods, see Fig. 26.1.

The sequence $(y_k)_{k=0,\dots,K}$ is defined by the implicit Euler method

$$y_{k+1} = y_k + \tau_{k+1} f(y_{k+1})$$

with possibly non-constant step sizes $\tau_{k+1} = t_{k+1} - t_k > 0$. We identify τ with the sequence $(\tau_k)_{k=1,\dots,K}$ and assume that $t_K = T$.

Definition 26.1 The *affine-linear interpolant* $\widehat{y}_\tau : [0, T] \to \mathbb{R}$ is defined for $t \in [t_k, t_{k+1}]$, $k = 0, 1, \dots, K-1$, by

$$\widehat{y}_\tau(t) = \frac{t - t_{k+1}}{t_k - t_{k+1}} y_k + \frac{t - t_k}{t_{k+1} - t_k} y_{k+1}.$$

The *piecewise constant interpolant* $\overline{y}_\tau : [0, T] \to \mathbb{R}$ is defined for $t \in (t_k, t_{k+1}]$, $k = 0, 1, \dots, K-1$, by

$$\overline{y}_\tau(t) = y_{k+1}.$$

The interpolants are exemplarily shown in Fig. 26.2.
By definition of $\widehat{y}_\tau$, for $t \in (t_k, t_{k+1})$

$$\widehat{y}_\tau'(t) = \frac{1}{\tau_{k+1}}(y_{k+1} - y_k)$$

© The Author(s), under exclusive license to Springer-Verlag GmbH,
DE, part of Springer Nature 2025
S. Bartels, *Numerical Mathematics 3x9*, La Matematica per il 3+2 160,
https://doi.org/10.1007/978-3-662-70890-3_26

Fig. 26.1 Adaptive control of local step sizes

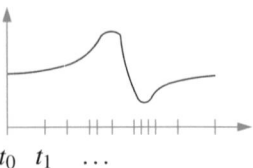

Fig. 26.2 Piecewise linear and piecewise constant interpolants of given approximation values

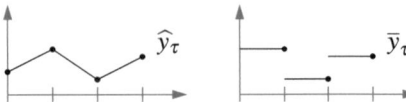

and with the definition of $\bar{y}_\tau$, it follows that the implicit Euler method can be written for all $t \in (t_k, t_{k+1})$ in the form

$$\widehat{y}'_\tau(t) = f(\bar{y}_\tau(t)) = f(\widehat{y}_\tau(t)) + (f(\bar{y}_\tau(t)) - f(\widehat{y}_\tau(t))).$$

The function $\widehat{y}_\tau$ therefore solves the autonomous equation $y' = f(y)$ outside the time steps $(t_k)_{k=1,2,\ldots,K-1}$ up to the residual

$$R_\tau = f(\widehat{y}_\tau) - f(\bar{y}_\tau).$$

This observation can be quantified and leads to an *a posteriori error estimate*.

Proposition 26.1 *Let $f : \mathbb{R} \to \mathbb{R}$ be Lipschitz continuous with constant $L \geq 0$ and $y \in C^1([0, T])$ the solution of the initial value problem $y' = f(y)$, $y(0) = y_0$. Then*

$$\sup_{t \in [0,T]} |y(t) - \widehat{y}_\tau(t)|^2 \leq \frac{L}{3} \Big(\sum_{k=0}^{K-1} \tau_{k+1} |y_{k+1} - y_k|^2 \Big) \exp(3LT).$$

Proof Subtracting the identities

$$y' = f(y), \quad \widehat{y}'_\tau = f(\widehat{y}_\tau) + (f(\bar{y}_\tau) - f(\widehat{y}_\tau))$$

shows that the error $e_\tau = y - \widehat{y}_\tau$ satisfies the equation

$$e'_\tau = (f(y) - f(\widehat{y}_\tau)) - (f(\bar{y}_\tau) - f(\widehat{y}_\tau))$$

Multiplying this equation by e and using the product rule $(|e_\tau|^2)' = 2e'_\tau e_\tau$, we find that

$$\frac{d}{dt} \frac{1}{2} |e_\tau|^2 = e'_\tau e_\tau = (f(y) - f(\widehat{y}_\tau))e_\tau - (f(\bar{y}_\tau) - f(\widehat{y}_\tau))e_\tau$$
$$\leq L|e_\tau|^2 + L|\widehat{y}_\tau - \bar{y}_\tau||e_\tau|.$$

26.1 A Posteriori Error Control

With the inequality $2ab \leq a^2 + b^2$ it follows

$$\frac{d}{dt}\frac{1}{2}|e_\tau|^2 \leq L|e_\tau|^2 + \frac{L}{2}|\widehat{y}_\tau - \overline{y}_\tau|^2 + \frac{L}{2}|e_\tau|^2$$

$$= \frac{3L}{2}|e_\tau|^2 + \frac{L}{2}|\widehat{y}_\tau - \overline{y}_\tau|^2.$$

Since e is continuous and piecewise differentiable, the integration of this inequality over $(0, t)$ shows that

$$|e_\tau(t)|^2 - |e_\tau(0)|^2 \leq 3L \int_0^t |e_\tau(s)|^2 \, ds + L \int_0^T |\widehat{y}_\tau(s) - \overline{y}_\tau(s)|^2 \, ds.$$

Using that $e_\tau(0) = 0$, an application of the Gronwall lemma implies that for all $t \in [0, T]$ we have

$$|e_\tau(t)|^2 \leq L \left(\int_0^T |\widehat{y}_\tau(s) - \overline{y}_\tau(s)|^2 \, ds \right) \exp(3LT).$$

On each interval (t_k, t_{k+1}) we have

$$\widehat{y}_\tau(s) - \overline{y}_\tau(s) = \frac{1}{\tau_{k+1}}\big((t_{k+1} - s)y_k + (s - t_k)y_{k+1}\big) - \frac{t_{k+1} - t_k}{\tau_{k+1}}y_{k+1}$$

$$= \frac{s - t_{k+1}}{\tau_{k+1}}(y_{k+1} - y_k).$$

This leads to

$$\int_0^T |\widehat{y}_\tau(s) - \overline{y}_\tau(s)|^2 \, ds = \sum_{k=0}^{K-1} \frac{(y_{k+1} - y_k)^2}{\tau_{k+1}^2} \int_{t_k}^{t_{k+1}} (t_{k+1} - s)^2 \, ds$$

$$= \sum_{k=0}^{K-1} \frac{(y_{k+1} - y_k)^2}{\tau_{k+1}^2} \frac{\tau_{k+1}^3}{3}.$$

This proves the assertion. □

Remark 26.1 The estimate of the proposition is called *a posteriori* error estimate, since it bounds the approximation error $y - \widehat{y}_\tau$ *after* the calculation of the numerical solution by computable quantities.

26.2 Adaptive Algorithm

The a posteriori error estimate allows for an adaptive adjustment of the time step sizes, that is the step size τ_{k+1} should be reduced until the error indicator defined by $\eta_{k+1} = |y_{k+1} - y_k|$ satisfies the estimate $\eta_{k+1} \leq \delta$ with a given tolerance δ satisfied. Conversely, an inequality $\eta_{k+1} \leq \delta$ motivates the enlargement of the step size in the following time step.

Algorithm 26.1 (Step Size Control) *Let $\delta > 0$, $y_0 \in \mathbb{R}$ and $\tau_1 > 0$. Set $k = 0$ and $t_0 = 0$.*

(1) Compute y_{k+1} using

$$y_{k+1} = y_k + \tau_{k+1}\Phi(t_k, y_k, y_{k+1}, \tau_{k+1}).$$

(2) If $\eta_{k+1} > \delta$, then set $\tau_{k+1} \to \tau_{k+1}/2$ and repeat (1).
(3) Stop if $t_{k+1} = t_k + \tau_{k+1} = T$; otherwise increase $k \to k+1$, set $\tau_{k+1} = \min\{2\tau_k, T - t_k\}$, and repeat step (1).

26.3 Control Procedure

If no a posteriori error estimation is available, an alternative possibility for step size control is obtained via the use of a so-called *control procedure*. This is based on an additional scheme of higher consistency order than the actually used method. If $(y_k)_{k=0,\ldots,K}$ are approximations of order $\mathcal{O}(\tau^p)$ and $(\tilde{y}_k)_{k=0,\ldots,K}$ are approximations of order $\mathcal{O}(\tau^q)$ with $q > p$, then it follows

$$|y(t_k) - y_k| \leq |y(t_k) - \tilde{y}_k| + |\tilde{y}_k - y_k| = \mathcal{O}(\tau^q) + |\tilde{y}_k - y_k|.$$

With the reverse triangle inequality, $|a| - |b| \leq |a - b|$, one also obtains

$$|\tilde{y}_k - y_k| - \mathcal{O}(\tau^q) = |\tilde{y}_k - y_k| - |\tilde{y}_k - y(t_k)| \leq |y_k - y(t_k)|.$$

Overall, up to terms of order $\mathcal{O}(\tau^q)$, we have that

$$|y(t_k) - y_k| \approx \eta_k = |\tilde{y}_k - y_k|,$$

that is, the computable quantity η_k approximates the actual error up to terms of higher order.

26.4 Extrapolation

An extrapolated scheme can serve as a control procedure for step size control. For the approximations $(y_k^\tau)_{k=0,\ldots,K}$ of the exact solution $y : [0, T] \to \mathbb{R}$ calculated with a single-step procedure of consistency order p, it can be shown that the error $y(t_k) - y_k$ can be represented by a function $\varphi(\tau)$ that $\phi(\tau) = O(\tau^p)$ for $\tau \to 0$. A Taylor expansion of the function φ leads to the representation

$$y(t_k) - y_k^\tau = c_1 \tau^p + c_2 \tau^{p+1} + o(\tau^{p+1}).$$

If the same procedure is used with the step size $\tau/2$, we obtain through $y_{2k}^{\tau/2}$ a further approximation of $y(t_k)$ and may assume the error representation

$$y(t_k) - y_{2k}^{\tau/2} = c_1 2^{-p} \tau^p + c_2 2^{-2p} \tau^{p+1} + o(\tau^{p+1})$$

The multiplication of the second equation by 2^p and subsequent subtraction from the first equation lead to

$$\left(1 - 2^p\right) y(t_k) - y_k^\tau + 2^p y_{2k}^{\tau/2} = c_2 \left(1 - 2^{-p}\right) \tau^{p+1} + o(\tau^{p+1}),$$

that is, the term $c_1 \tau^p$ is eliminated. This implies

$$\widetilde{y}_k^\tau = \frac{y_k^\tau - 2^p y_{2k}^{\tau/2}}{1 - 2^p} = y(t_k) - c_2 \frac{1 - 2^{-p}}{1 - 2^p} \tau^{p+1} + o(\tau^{p+1}),$$

so that the computable expression $\widetilde{y}_k^\tau$ approximates the function value $y(t_k)$ up to an error term of order $\mathcal{O}(\tau^{p+1})$, see Fig. 26.3. We have thus constructed a procedure where the effort is approximately doubled, but the error is reduced by the factor τ and not merely 2^{-p}. This approach can be rigorously analysed and generalised.

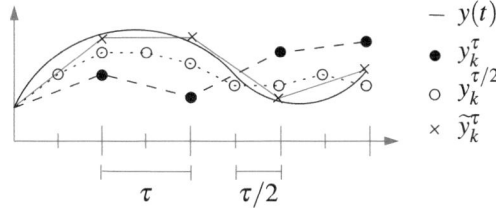

Fig. 26.3 Construction of a control procedure through extrapolation of the approximation solutions y^τ and $y^{\tau/2}$

26.5 Learning Objectives, Quiz and Application

You should be able to explain the basic concepts of step size control and derive an a posteriori error estimation. You should be able to explain the role of control procedures.

Quiz 26.1 Decide for each of the following statements whether they are true or false. You should be able to justify your answer.

The approximation error of the implicit Euler method cannot be bounded without knowledge of the exact solution.	
In the adaptive algorithm, a step size is determined with which all approximations are calculated.	
For all $a, b \in \mathbb{R}$ and $\gamma > 0$, $ab \leq \gamma a^2/2 + b^2/(2\gamma)$ holds.	
If $y \in C^0([0, T])$ and $t_k = k\tau$ for $k = 0, 1, \ldots, K$ with $\tau = T/K$, then $\max_{k=0,\ldots,K-1} |y(t_{k+1}) - y(t_k)| \to 0$ for $\tau \to 0$.	
By extrapolating a single-step method of consistency order $p \geq 1$ with step sizes τ and $\tau/2$, a method of consistency order $p + 1$ is obtained.	

Application 26.1 In simple markets, the price p of a product is determined by the supply s and the demand d. Demand decreases with increasing price, while supply increases with increasing price. Consequently, $p \mapsto d(p)$ is a monotonically decreasing and $p \mapsto s(p)$ a monotonically increasing function. A difference between supply and demand leads to a change in price, that is, we have

$$p'(t) = \alpha\big(d(p) - s(p)\big).$$

(i) Show that under suitable conditions on d and s, a state of equilibrium is reached and this is attained exponentially quickly in the case of small perturbations.
(ii) In reality, the number of products purchased may be less than the demand, as for example the price does not yet correspond to the actual value of the product, but this number is moving towards the demand. Modify and extend the model to account for this delay effect.
(iii) How can the model take into account the dependence on external factors such as the availability of required raw materials?

Chapter 27
Symplectic, Shooting and dG Methods

27.1 Hamiltonian Systems

A Hamiltonian system describes the dynamics of N bodies in three-dimensional space using a differentiable function $H : \mathbb{R}^{N \times 3} \times \mathbb{R}^{N \times 3} \to \mathbb{R}$ and the system of differential equations

$$q' = \partial_p H(q, p), \quad p' = -\partial_q H(q, p)$$

in the interval $(0, T]$ with initial data for q and p. The functions $q_i, p_i : [0, T] \to \mathbb{R}^3$, $i = 1, 2, \ldots, N$, describe the positions and impulses of the bodies and H is the sum of kinetic and potential energy, that is, for example,

$$H(q, p) = \sum_{i=1}^{N} \frac{\|p_i\|^2}{2m_i} + V(q_1, q_2, \ldots, q_N),$$

with the masses m_i, $i = 1, 2, \ldots, N$, of the bodies.

Examples 27.1

(i) The pendulum described by the equation $\phi'' = -(g/\ell)\sin(\phi)$ can be represented as a Hamiltonian system of the function

$$H(\phi, \psi) = \frac{1}{2m\ell^2}\psi^2 - mg\ell\cos(\phi)$$

because it follows that $m\ell^2 \phi' = \psi$ and $\psi' = -mg\ell\sin(\phi)$.

(ii) Multi-body problems such as solar systems can be described by Hamiltonian systems.

(iii) Through $[\partial_p H(q, p), -\partial_q H(q, p)]^T$ a tangent vector to the graph of H at the point (p, q) is defined. The associated Hamiltonian system thus follows a level line of the function H.

Hamiltonian systems fulfil conservation principles for total angular momentum and total energy.

Example 27.2 The total energy of a Hamiltonian system is constant, because we have

$$\frac{d}{dt}H(q, p) = \partial_p H(q, p)p' + \partial_q H(q, p)q' = 0.$$

If we combine the variables q and p into a vector $z \in \mathbb{R}^{2n}$ with $n = dN$ and $d \in \{1, 2, 3\}$ and identify $H(q, p) = H(z)$, then a Hamiltonian system can be written as

$$z' = J\nabla H(z), \quad z(0) = z_0$$

with the matrix

$$J = \begin{bmatrix} & I_n \\ -I_n & \end{bmatrix}.$$

The matrix J fulfils the identities $J^T = -J = J^{-1}$ and defines the skew-symmetric bilinear form

$$\omega(z_1, z_2) = z_1^T J z_2.$$

This expression corresponds to an oriented area of the parallelogram spanned by z_1 and z_2. In the case $n = 1$ for example, we have that $\omega(z_1, z_2) = \det[z_1, z_2]$ for $z_1, z_2 \in \mathbb{R}^2$.

Definition 27.1 A matrix $A \in \mathbb{R}^{2n \times 2n}$ is called *symplectic*, if

$$\omega(Az_1, Az_2) = \omega(z_1, z_2)$$

for all $z_1, z_2 \in \mathbb{R}^{2n}$ or equivalently $A^T J A = J$. A differentiable mapping $\Psi : \mathbb{R}^{2n} \to \mathbb{R}^{2n}$ is called *symplectic*, if its differential $D\Psi(z)$ for all $z \in \mathbb{R}^{2n}$ is a symplectic matrix.

Symplectic mappings preserve the oriented area of parallelograms. Symplecticity is the characteristic property of Hamiltonian systems.

Proposition 27.1 *For a Hamiltonian system $z' = J\nabla H(z)$, $z(0) = z_0$, with $H \in C^2(\mathbb{R}^{2n})$ the flow $\phi_t : \mathbb{R}^{2n} \to \mathbb{R}^{2n}$, $z_0 \mapsto z(t)$, which assigns the state $z(t)$ at time t to an initial configuration z_0, is a symplectic mapping for every $t \in \mathbb{R}$.*

Fig. 27.1 According to the second Keplerian law, the radius vector of a planet sweeps out equal segments in equal time intervals

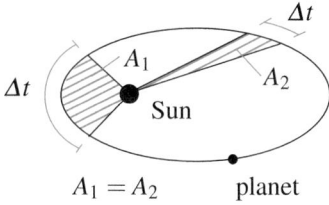

Proof For the mapping $t \mapsto \phi_t(z_0)$ we have

$$\frac{d}{dt}\phi_t(z_0) = \frac{d}{dt}z(t) = z'(t) = J\nabla H(z(t)) = J\nabla H(\phi_t(z_0)).$$

Differentiating this identity with respect to z_0 leads to

$$\frac{d}{dt}D\phi_t(z_0) = JD^2H(\phi_t(z_0))D\phi_t(z_0).$$

To prove the symplecticity, we consider $F(t) = D\phi_t(z_0)^T J D\phi_t(z_0)$ and note that from $\phi_0(z_0) = z_0$ for all $z_0 \in \mathbb{R}^{2n}$ the identity $F(0) = J$ follows. For the derivative, using the symmetry of the Hessian matrix D^2H, we have that

$$F'(t) = \left[\frac{d}{dt}D\phi_t(z_0)\right]^T J[D\phi_t(z_0)] + [D\phi_t(z_0)]^T J\left[\frac{d}{dt}D\phi_t(z_0)\right]$$
$$= [D\phi_t(z_0)]^T D^2H(\phi_t(z_0)) J^T J[D\phi_t(z_0)]$$
$$+ [D\phi_t(z_0)]^T J^2 D^2 H(\phi_t(z_0))[D\phi_t(z_0)] = 0,$$

where $J^T J = I_{2n} = -J^2$ was used. Thus, it follows $F(t) = J$ for all $t \in [0, T]$ or the symplecticity of ϕ_t. □

Remark 27.1 Kepler's laws for determining planetary orbits can be interpreted as a consequence of the symplecticity or the conservation properties of Hamiltonian systems. The second Keplerian law, for example, postulates that a radius vector drawn from the Sun to the planet sweeps out equal areas in equal times, see Fig. 27.1.

27.2 Symplectic Methods

In order to meaningfully capture the dynamics of a Hamiltonian system, that is to approximate the energy and momentum conservation properties well, the numerical methods used should also define symplectic mappings. A single-step method of the form

$$\begin{bmatrix} q_{k+1} \\ p_{k+1} \end{bmatrix} = \begin{bmatrix} q_k \\ p_k \end{bmatrix} + \tau \begin{bmatrix} \Phi_1(t_k, q_k, p_k, q_{k+1}, p_{k+1}, \tau) \\ \Phi_2(t_k, q_k, p_k, q_{k+1}, p_{k+1}, \tau) \end{bmatrix}$$

defines in the case of well-posedness for $k = 0, 1, \ldots, K - 1$ the mappings

$$\Psi^{k+1} : (q_k, p_k) \mapsto (q_{k+1}, p_{k+1}).$$

Definition 27.2 A numerical method is called *symplectic*, if the mappings defined by it, $\Psi^{k+1} : (q_k, p_k) \mapsto (q_{k+1}, p_{k+1})$, $k = 0, 1, \ldots, K - 1$, for each Hamilton function $H \in C^2(\mathbb{R}^{2n})$ are symplectic.

The symplecticity of a method can be checked with the following criterion in the case $n = 1$.

Lemma 27.1 *A mapping* $\Psi : \mathbb{R}^2 \to \mathbb{R}^2$ *is symplectic if and only if* $\det D\Psi = 1$ *holds.*

Proof We have

$$D\Psi^T J D\Psi = \begin{bmatrix} \partial_1 \Psi_1 & \partial_1 \Psi_2 \\ \partial_2 \Psi_1 & \partial_2 \Psi_2 \end{bmatrix} \begin{bmatrix} 0 & 1 \\ -1 & 0 \end{bmatrix} \begin{bmatrix} \partial_1 \Psi_1 & \partial_2 \Psi_1 \\ \partial_1 \Psi_2 & \partial_2 \Psi_2 \end{bmatrix} = \begin{bmatrix} 0 & \det D\Psi \\ -\det D\Psi & 0 \end{bmatrix},$$

from which the claim follows. □

We check the symplecticity for some standard methods.

Examples 27.3

(i) For the explicit Euler method, $\Psi^{k+1} = \Psi$ holds with

$$\begin{bmatrix} q_{k+1} \\ p_{k+1} \end{bmatrix} = \Psi(q_k, p_k) = \begin{bmatrix} \Psi_1(q_k, p_k) \\ \Psi_2(q_k, p_k) \end{bmatrix} = \begin{bmatrix} q_k \\ p_k \end{bmatrix} + \tau \begin{bmatrix} \partial_p H(q_k, p_k) \\ -\partial_q H(q_k, p_k) \end{bmatrix}$$

and thus

$$\partial_1 \Psi_1 = 1 + \tau \partial_p \partial_q H, \qquad \partial_2 \Psi_1 = \tau \partial_p \partial_p H,$$
$$\partial_1 \Psi_2 = -\tau \partial_q \partial_q H, \qquad \partial_2 \Psi_2 = 1 - \tau \partial_p \partial_q H,$$

as well as $\det D\Psi = 1 + \mathcal{O}(\tau^2)$, so that the method is not symplectic.

(ii) For the *partitioned Euler method*

$$\begin{bmatrix} q_{k+1} \\ p_{k+1} \end{bmatrix} = \begin{bmatrix} q_k \\ p_k \end{bmatrix} + \tau \begin{bmatrix} \partial_p H(q_k, p_{k+1}) \\ -\partial_q H(q_k, p_{k+1}) \end{bmatrix}$$

27.2 Symplectic Methods

the right side depends on p_{k+1}, so that

$$\partial_1 \Psi_1 = \frac{\partial q_{k+1}}{\partial q_k} = 1 + \tau \partial_q \partial_p H(q_k, p_{k+1}) + \tau \partial_p^2 H(q_k, p_{k+1}) \frac{\partial p_{k+1}}{\partial q_k},$$

$$\partial_2 \Psi_1 = \frac{\partial q_{k+1}}{\partial p_k} = \tau \partial_p^2 H(q_k, p_{k+1}) \frac{\partial p_{k+1}}{\partial p_k},$$

$$\partial_1 \Psi_2 = \frac{\partial p_{k+1}}{\partial q_k} = -\tau \partial_q^2 H(q_k, p_{k+1}) - \tau \partial_q \partial_p H(q_k, p_{k+1}) \frac{\partial p_{k+1}}{\partial q_k},$$

$$\partial_2 \Psi_2 = \frac{\partial p_{k+1}}{\partial p_k} = 1 - \tau \partial_q \partial_p H(q_k, p_{k+1}) \frac{\partial p_{k+1}}{\partial p_k}.$$

The last two equations can be solved and lead to

$$\partial_1 \Psi_2 = \frac{\partial p_{k+1}}{\partial q_k} = -\tau \big(1 + \tau \partial_q \partial_p H(q_k, p_{k+1})\big)^{-1} \partial_q^2 H(q_k, p_{k+1}),$$

$$\partial_2 \Psi_2 = \frac{\partial p_{k+1}}{\partial p_k} = \big(1 + \tau \partial_q \partial_p H(q_k, p_{k+1})\big)^{-1}.$$

Hence, $\det D\Psi = 1$, so the method is symplectic.
(iii) The midpoint method is symplectic.
(iv) The implicit Euler method is not symplectic.

Remark 27.2 In the case of a Hamiltonian function of the form

$$H(q, p) = \sum_{i=1}^{N} \frac{\|p_i\|^2}{2m_i} + \frac{1}{2} \sum_{\substack{i,j=1 \\ i \neq j}}^{N} V(\|q_i - q_j\|)$$

the systems of equations defined by the partitioned Euler method can be explicitly solved at each time step.

The advantages of symplectic methods can be illustrated using the example of the linearised pendulum.

Example 27.4 We consider the Hamiltonian function

$$H(q, p) = \frac{1}{2} p^2 + \frac{1}{2} q^2$$

for which the solutions of the Hamiltonian system

$$q' = p, \quad p' = -q$$

are given by $q(t) = a \sin(t) + b \cos(t)$ and $p(t) = a \cos(t) - b \sin(t)$. The total energy $H(q(t), p(t))$ of each solution is constant. With the difference quotient

$d_t a_{k+1} = (a_{k+1} - a_k)/\tau$ and the θ-method

$$d_t q_{k+1} = p_{k+\theta_2} = (1-\theta_2)p_k + \theta_2 p_{k+1},$$
$$d_t p_{k+1} = -q_{k+\theta_1} = (1-\theta_1)q_k + \theta_1 q_{k+1},$$

the explicit and implicit Euler, the midpoint and the partitioned Euler methods can be described by the choices

$$\theta = (0,0), \quad \theta = (1,1), \quad \theta = (1/2, 1/2), \quad \theta = (0,1)$$

respectively. We use the formula

$$(a-b)(\theta a + (1-\theta)b) = \frac{1}{2}(a^2 - b^2) - \frac{1-2\theta}{2}(a-b)^2,$$

which is obtained by adding and subtracting $a/2$, and multiply the equations of the θ-method by $q_{k+\theta_1}$ and $p_{k+\theta_2}$ respectively. The subsequent addition of the equations in the cases of the explicit and implicit Euler and the midpoint methods leads to

$$\frac{1}{2\tau}(q_{k+1}^2 - q_k^2) + \frac{1}{2\tau}(p_{k+1}^2 - p_k^2) = \frac{1-2\theta_1}{2\tau}(q_{k+1} - q_k)^2 + \frac{1-2\theta_2}{2\tau}(p_{k+1} - p_k)^2.$$

We sum over $k = 0, 1, \ldots, \ell - 1$, multiply by τ and obtain

$$H(q_\ell, p_\ell) - H(q_0, p_0) = \frac{1-2\theta_1}{2}\sum_{k=0}^{\ell-1}(q_{k+1} - q_k)^2 + \frac{1-2\theta_2}{2}\sum_{k=0}^{\ell-1}(p_{k+1} - p_k)^2.$$

In the case of the explicit Euler method, the right-hand side is generally positive and there is an increase in the total energy, while in the case of the implicit Euler method there is a decrease. For the midpoint method, the right-hand side vanishes and the energy is exactly conserved. For the partitioned Euler method, multiplying the equations by $q_{k+1/2}$ and $p_{k+1/2}$ yields

$$H(q_\ell, p_\ell) - H(q_0, p_0) = (-\tau p_\ell q_\ell + \tau p_0 q_0)/2.$$

With $\tau|pq| \leq \tau(p^2 + q^2)/2$ it follows

$$\frac{1-\tau/2}{1+\tau/2}H(q_0, p_0) \leq H(q_\ell, p_\ell) \leq \frac{1+\tau/2}{1-\tau/2}H(q_0, p_0).$$

The results of corresponding numerical experiments are shown in Fig. 27.2.

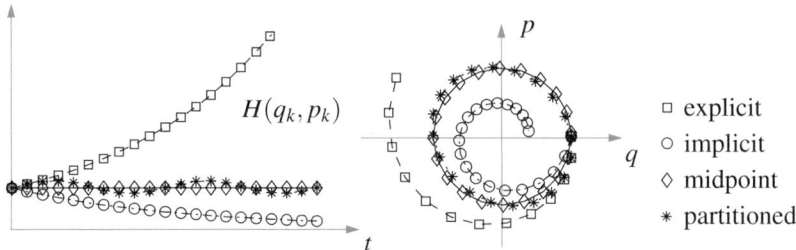

Fig. 27.2 Application of various methods to a Hamiltonian system; symplectic methods such as the midpoint and the partitioned Euler method preserve physically relevant quantities

27.3 Shooting Method

In one-dimensional *boundary value problems*, a function $u : [a, b] \to \mathbb{R}$ is sought, which satisfies a differential equation within the interval and boundary conditions at both interval ends. A one-dimensional boundary value problem of second order is, for example,

$$u''(x) = f\big(x, u(x), u'(x)\big), \quad x \in (a, b),$$
$$u(a) = \alpha, \quad u(b) = \beta.$$

This can describe the trajectory of a ball that is thrown at location a at height α so that it reaches height β at location b. One-dimensional boundary value problems can be solved iteratively with the numerical methods constructed for initial value problems. In the above model problem, we are looking for a parameter $s \in \mathbb{R}$, such that the solution $y : [a, b] \to \mathbb{R}$ of the initial value problem

$$y''(x) = f\big(x, y(x), y'(x)\big), \quad x \in (a, b),$$
$$y(a) = \alpha, \quad y'(a) = s$$

has the property $y(b) = \beta$ and thus fulfils the boundary value problem. Since y depends on the parameter s, we write y_s for the solution of the initial value problem in the following. Intuitively, the sought number $s \in \mathbb{R}$ is the launch angle necessary to achieve the height β at location b. We define the mapping

$$F : \mathbb{R} \to \mathbb{R}, \quad s \mapsto y_s(b) - \beta$$

and try to determine a root s^* of F. With the Newton method, this is done approximately for a starting value s_0 through the iteration

$$s_{i+1} = s_i - \frac{F(s_i)}{F'(s_i)},$$

Fig. 27.3 Different initial velocities s_i lead to different values at the final time point

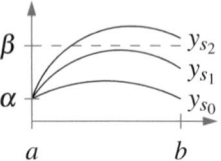

where $F(s) = y_s(b) - \beta$ and $F'(s) = \partial_s y_s(b)$ applies. The so-called *shooting method* is illustrated in Fig. 27.3.

The function $v(x) = \partial_s y_s(x)$ is for given $s \in \mathbb{R}$ and y_s the solution of the initial value problem differentiated with respect to s, that is

$$v''(x) = \partial_2 f\bigl(x, y_s(x), y_s'(x)\bigr)v(x) + \partial_3 f\bigl(x, y_s(x), y_s'(x)\bigr)v'(x), \quad x \in (a, b),$$
$$v(a) = 0, \; v'(a) = 1.$$

The function v is thus the solution of a linear initial value problem, which can be solved with little effort. To achieve convergence of the Newton method, the starting value s_0 must generally be close enough to s^*.

27.4 Discontinuous Galerkin Methods

We multiply the autonomous differential equation $y' = f(y)$ with a function ϕ, integrate the product over the interval $[t_k, t_{k+1}]$ and perform a partial integration, so that we obtain the identity under utilisation of the continuity property $y(t_k^+) = y(t_k^-)$

$$-\int_{t_k}^{t_{k+1}} y(t)\phi'(t)\,\mathrm{d}t + y(t_{k+1}^-)\phi(t_{k+1}^-) - y(t_k^-)\phi(t_k^+) = \int_{t_k}^{t_{k+1}} f(y(t))\phi(t)\,\mathrm{d}t$$

where $g(t_m^\pm)$ denotes the right and left-hand limits $\lim_{\varepsilon \to 0} g(t_m \pm \varepsilon)$ for $\varepsilon > 0$. The idea of the *discontinuous Galerkin method* is to consider discontinuous approximations $y_\tau : [0, T] \to \mathbb{R}$, and to partially reverse the above reformulation to derive a defining equation for y_τ. We replace y with y_τ in the above equation and use

$$-\int_{t_k}^{t_{k+1}} y_\tau(t)\phi'(t)\,\mathrm{d}t = \int_{t_k}^{t_{k+1}} y_\tau'(t)\phi(t)\,\mathrm{d}t + y_\tau(t_k^+)\phi(t_k^+) - y_\tau(t_{k+1}^-)\phi(t_{k+1}^-).$$

This leads to the integral equation

$$\int_{t_k}^{t_{k+1}} y_\tau'(t)\phi(t)\,\mathrm{d}t + \bigl[y_\tau(t_k^+) - y_\tau(t_k^-)\bigr]\phi(t_k^+) = \int_{t_k}^{t_{k+1}} f(y_\tau(t))\phi(t)\,\mathrm{d}t,$$

Fig. 27.4 Discontinuous Galerkin methods approximate the solution by discontinuous, piecewise polynomial functions

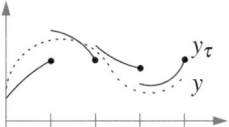

where $y_\tau(t_0^-) = y_0$. The numerical solution is now sought as a piecewise polynomial $y_\tau|_{(t_k,t_{k+1}]} \in \mathscr{P}_\ell|_{(t_k,t_{k+1}]}$, so that the integral equation holds for all $k = 0, 1, \ldots, K - 1$ and all $\phi \in \mathscr{P}_\ell|_{(t_k,t_{k+1}]}$, see Fig. 27.4.

Example 27.5 For $\ell = 0$ we obtain the implicit Euler method, because if we set $y_{k+1} = y_\tau|_{(t_k,t_{k+1}]}$, it follows using $y_\tau'|_{(t_k,t_{k+1}]} = 0$ and with $\phi = 1$, that $y_{k+1} - y_k = (t_{k+1} - t_k) f(y_{k+1})$.

27.5 Learning Objectives, Quiz and Application

You should be able to define Hamiltonian systems and explain the significance of symplectic methods. You should be able to motivate shooting methods and explain their algorithmic implementation. Moreover, you should be able to demonstrate characteristic properties of discontinuous Galerkin methods.

Quiz 27.1 Decide for each of the following statements whether it is true or false. You should be able to justify your answer.

Hamiltonian systems are special gradient flows.	
Orthogonal matrices are symplectic.	
The partitioned Euler method has consistency order $p = 2$.	
Every boundary value problem can be formulated in a unique way as an initial value problem.	
The approximate solution of the discontinuous Galerkin method is a discontinuous function.	

Application 27.1 To simulate the outer solar system, we use the Hamilton function

$$H(q, p) = \sum_{i=1}^{N} \frac{\|p_i\|^2}{2m_i} - \frac{\gamma}{2} \sum_{\substack{i,j=1 \\ i \neq j}}^{N} \frac{m_i m_j}{\|q_i - q_j\|}$$

with the momenta $p_i \in \mathbb{R}^3$ and positions $q_i \in \mathbb{R}^3$, $i = 1, 2, \ldots, N$, of the considered planets. Use the resulting Hamiltonian system and the approximations given in Table 27.1 in solar masses SM $\approx 2 \cdot 10^{30}$ kg and astronomical units AU $\approx 150 \cdot 10^9$ m or AU/day to describe three planets.

Table 27.1 Data for simulating a simple solar system

Planet	Mass	Initial position	Initial velocity
Sun	1	$(0, 0, 0)$	$(0, 0, 0) \cdot 10^{-3}$
Jupiter	1/1000	$(-3, -4, -1)$	$(5, -4, -2) \cdot 10^{-3}$
Saturn	3/10000	$(10, -3, -2)$	$(2, 5, 2) \cdot 10^{-3}$

Use as the gravitational constant the simple approximation of the heliocentric gravitational constant $\gamma = 3 \cdot 10^{-4} \mathrm{AU}^3/(\mathrm{SM} \cdot \mathrm{day}^2)$. Simulate the system numerically using the explicit and the partitioned Euler methods with different step sizes. Plot the orbits of the planets and consider the total energy of the system as a function of time. Experimentally determine the length of a Jupiter year.

Part IV
Problems and Projects

Chapter 28
Problems on Numerical Linear Algebra

28.1 Basic Concepts

Problem 28.1.1 Let $\widetilde{\phi} = f \circ g$ be a method for the mathematical operation ϕ and let the operation defined by g be ill-conditioned. Show that the method $\widetilde{\phi}$ is generally unstable.

Problem 28.1.2 Show that the addition of two non-negative or non-positive numbers is well conditioned.

Problem 28.1.3 For $p > 0$, $\beta > 1$ and $j = 1, 2, 3, 4$ let the sequences $(a_n^{(j)})_{n \in \mathbb{N}}$ be defined by

$$a_n^{(1)} = n^p, \quad a_n^{(2)} = \beta^n, \quad a_n^{(3)} = n!, \quad a_n^{(4)} = \log_2 n.$$

For which pairs $1 \leq i, j \leq 4$ does $a_n^{(i)} = \mathcal{O}(a_n^{(j)})$ hold?

Problem 28.1.4 Under what conditions on $a, b, c, d \in \mathbb{R}$ is the calculation of an intersection point of the two lines $x \mapsto ax + b$ and $x \mapsto cx + d$ a well conditioned problem?

Problem 28.1.5 How can cancellation effects be avoided in the practical calculation of the expressions

$$\frac{1 - 2x}{1 + 2x} - \frac{1}{1 + x}, \quad \frac{e^x - 1}{x}$$

for $x \neq 0$ with $|x| \ll 1$?

Problem 28.1.6 Discuss the conditioning of determining the roots of a quadratic equation $x^2 + px + q = 0$ as well as the stability of their calculation with the pq-

formula $x_{1,2} = -p/2 \pm (p^2/4 - q)^{1/2}$. Consider especially the cases $p^2 \approx 4q$ and $p^2 \gg 4|q|$.

Problem 28.1.7 For which $x \in \mathbb{R}$ must cancellation effects be expected in the approximate calculation $e^x \approx \sum_{k=0}^{n} x^k/k!$? How can these be avoided?

Problem 28.1.8 Determine the order of magnitude of the effort for matrix-vector multiplication, matrix-matrix multiplication and the computation of the determinant of a matrix using Laplace's expansion theorem.

Problem 28.1.9 Assume that a computer operates with 10^9 floating point operations per second *(flops)* and there are three algorithms with effort $\mathcal{O}(n)$, $\mathcal{O}(n^3)$ and $\mathcal{O}(n!)$ to solve the same task. How many seconds, hours, days or years do the algorithms approximately need for the problem sizes $n = 10^k$ with $k = 1, 2, \ldots, 6$?

Problem 28.1.10 Let $\phi(0)$ be a well conditioned operation with $\phi'(0) = 3$. Examine for given $x_1, x_2 \in \mathbb{R}$ the conditioning of the operation $\phi(x_1 + x_2)$.

Project 28.1.1 The functions $f, g : \mathbb{R}_{>0} \to \mathbb{R}$ defined by

$$f(x) = \frac{1}{x} - \frac{1}{x+1}, \quad g(x) = \frac{1}{x(x+1)}$$

agree, but motivate two different methods for numerical computation. Determine for $x_k = 10^k, k = 1, 2, \ldots, 15$, the expression

$$\delta_k = \frac{|f(x_k) - g(x_k)|}{|g(x_k)|}$$

in MATLAB and arrange the results in a table. What do you observe and how do you explain the observations?

Project 28.1.2 Implement the recursive calculation of the determinant of a square matrix using Laplace's expansion theorem in MATLAB, C++ or Python. Measure manually, or with the help of the commands tic ... toc or clock(), the runtimes for the calculation of det A with the matrix $A \in \mathbb{R}^{n \times n}$ defined by $a_{ii} = 2$ and $a_{ij} = (-1)^j/(n-1)$ for $n = 10, 20, 40, 80$.

28.2 Operator Norm and Condition Number

Problem 28.2.1 For fixed norms $\|\cdot\|$ on $\mathbb{R}^n$ and on $\mathbb{R}^m$, let $\|\cdot\|_{op}$ denote the induced operator norm on $\mathbb{R}^{m \times n}$. Prove the following statements:

(i) The operator norm $\|\cdot\|_{op}$ defines a norm on $\mathbb{R}^{m \times n}$.
(ii) We have

28.2 Operator Norm and Condition Number

$$\|A\|_{op} = \sup_{\|x\|=1} \|Ax\| = \inf\{c \geq 0 : \forall x \in \mathbb{R}^n \|Ax\| \leq c\|x\|\}$$

and the supremum and the infimum are attained, provided $A \neq 0$.

(iii) In the case $A \neq 0$, for $x \in \mathbb{R}^m$ with $\|x\| \leq 1$ and $\|Ax\| = \|A\|_{op}$ it holds that $\|x\| = 1$.

(iv) Show that $\|A\|_{\mathcal{G}} = \max_{1 \leq i, j \leq n} |a_{ij}|$ defines a norm but not an operator norm on $\mathbb{R}^{n \times n}$.

Problem 28.2.2

(i) Show that the spectral radius is bounded by any operator norm $\|\cdot\|_{op}$ on the space of square matrices, i.e. we have $|\lambda| \leq \|A\|_{op}$ for every eigenvalue $\lambda \in \mathbb{R}$ of A.

(ii) Let $\|\cdot\|$ be a norm on $\mathbb{R}^n$ and let $\|\cdot\|_{op}$ be the induced operator norm. Show that for every regular matrix $D \in \mathbb{R}^{n \times n}$ a norm $\|\cdot\|_D : x \mapsto \|Dx\|$ is defined on $\mathbb{R}^n$ and construct for every matrix $A \in \mathbb{R}^{n \times n}$ a matrix $M_{D,A}$ such that, for the operator norm induced by $\|\cdot\|_D$, denoted $\|\cdot\|_{op,D}$, we have $\|A\|_{op,D} = \|M_{D,A}\|_{op}$.

(iii) Show that the inequality $\|A\|_2 \leq \|A\|_{op,D}$ is generally false.

Problem 28.2.3 For $1 \leq p < \infty$, a norm is defined on $\mathbb{R}^\ell$ by $\|x\|_p = \left(\sum_{j=1}^\ell |x_j|^p\right)^{1/p}$. The induced operator norm is also denoted by $\|\cdot\|_p$.

(i) Show that $\|A\|_1 = \max_{k=1,\ldots,n} \sum_{j=1}^m |a_{jk}|$ holds for all $A \in \mathbb{R}^{m \times n}$.

(ii) For the symmetric matrix $B \in \mathbb{R}^{n \times n}$, let

$$\rho(B) = \max\{|\lambda| : \lambda \text{ is an eigenvalue of } B\}.$$

Show, that $\|A\|_2 = \sqrt{\rho(A^\top A)}$ holds for all $A \in \mathbb{R}^{m \times n}$.

Problem 28.2.4 Let $A = \begin{bmatrix} a & b \\ b & c \end{bmatrix}$ with $a, b, c \in \mathbb{R}$, such that $\det A \neq 0$. Determine $\text{cond}_1(A)$, $\text{cond}_2(A)$ and $\text{cond}_\infty(A)$ and discuss for which ratios of a, b and c the corresponding linear equation systems are ill conditioned.

Problem 28.2.5 Let $A \in \mathbb{R}^{n \times n}$ be invertible and let $\|\cdot\|$ be an induced operator norm on $\mathbb{R}^{n \times n}$. Show that

$$\|A^{-1}\| = \left(\inf_{\|x\|=1} \|Ax\|\right)^{-1}$$

and $\|A^{-1}\| \geq \|A\|^{-1}$ hold.

Problem 28.2.6

(i) Let $A \in \mathbb{R}^{n \times n}$. Show that

$$\|A\|_2^2 \leq \|A\|_1 \|A\|_\infty$$

holds and verify the statement explicitly for $A = \begin{bmatrix} a & b \\ b & c \end{bmatrix}$.

(ii) Show that for every matrix $A \in \mathbb{R}^{n \times n}$ the estimates

$$n^{-1/2}\|A\|_2 \leq \|A\|_1 \leq n^{1/2}\|A\|_2,$$

$$n^{-1}\|A\|_\infty \leq \|A\|_1 \leq n\|A\|_\infty$$

hold and provide matrices $A \in \mathbb{R}^{n \times n}$ that show that the estimates cannot be improved.

Problem 28.2.7 For $A \in \mathbb{R}^{n \times n}$ the Frobenius norm is defined by $\|A\|_{\mathscr{F}}^2 = \sum_{1 \leq i,j \leq n} a_{ij}^2$. Show that

$$\|A\|_{\mathscr{F}} = \sqrt{\mathrm{tr}(A^T A)}.$$

Conclude that the Frobenius norm is compatible with the operator norm induced by the Euclidean norm in the sense that

$$\|A\|_2 \leq \|A\|_{\mathscr{F}} \leq \sqrt{n}\|A\|_2.$$

Use the identity $\mathrm{tr}(A^T A) = \lambda_1 + \cdots + \lambda_n$ with the non-negative eigenvalues $\lambda_1, \ldots, \lambda_n$ of $A^T A$. Can the estimates also be proven without using the eigenvalues?

Problem 28.2.8 Let $A \in \mathbb{R}^{m \times n}$.

(i) Show that $(\mathrm{Im}\, A^T)^\perp = \ker A$ with

$$V^\perp = \{v \in \mathbb{R}^n : v \cdot w = 0 \text{ for all } w \in V\}$$

for $V \subset \mathbb{R}^n$ and conclude $\mathbb{R}^n = \mathrm{Im}\, A^T + \ker A$.

(ii) Prove the dimension formula $n = \dim(\mathrm{Im}\, A) + \dim(\ker A)$ and conclude that $\mathrm{rank}\, A = \mathrm{rank}\, A^T$, where for a matrix M the column rank of M is defined by $\mathrm{rank}\, M = \dim \mathrm{Im}\, M$.

(iii) Show that

$$\ker A^T A = \ker A.$$

28.2 Operator Norm and Condition Number

Problem 28.2.9

(i) Let $A \in \mathbb{R}^{n \times m}$ and $B \in \mathbb{R}^{m \times p}$. With natural numbers $n_1, n_2, m_1, m_2, p_1, p_2$ let $A_{ij} \in \mathbb{R}^{n_i \times m_j}$, $B_{jk} \in \mathbb{R}^{m_j \times p_k}$, such that

$$A = \begin{bmatrix} A_{11} & A_{12} \\ A_{21} & A_{22} \end{bmatrix}, \quad B = \begin{bmatrix} B_{11} & B_{12} \\ B_{21} & B_{22} \end{bmatrix}$$

holds. Determine matrices $C_{ik} \in \mathbb{R}^{n_i \times p_k}$, so that a corresponding partitioning also holds for $C = AB$.

(ii) Show that for every regular matrix $A \in \mathbb{R}^{n \times n}$ the identity $(A^\mathsf{T})^{-1} = (A^{-1})^\mathsf{T}$ holds, which justifies the notation $A^{-\mathsf{T}}$.

Problem 28.2.10 Let $A \in \mathbb{R}^{n \times n}$ be regular and $1 \leq m \leq n$, such that the upper left $m \times m$ submatrix $A_{11} = (a_{ij})_{1 \leq i,j \leq m}$ is also regular. Let A be decomposed according to

$$A = \begin{bmatrix} A_{11} & A_{12} \\ A_{21} & A_{22} \end{bmatrix}.$$

Show that $S = A_{22} - A_{21} A_{11}^{-1} A_{12}$ is regular and that A^{-1} is given by

$$A^{-1} = \begin{bmatrix} A_{11}^{-1} + A_{11}^{-1} A_{12} S^{-1} A_{21} A_{11}^{-1} & -A_{11}^{-1} A_{12} S^{-1} \\ -S^{-1} A_{21} A_{11}^{-1} & S^{-1} \end{bmatrix}.$$

Project 28.2.1 Write programs in C++ and MATLAB that calculate the operator norm $\|\cdot\|_\infty$ of a matrix $A \in \mathbb{R}^{m \times n}$. Measure manually, or with the help of the commands `clock()` or `tic \ldots toc`, for the Hilbert matrix $H \in \mathbb{R}^{n \times n}$ with entries $h_{ij} = 1/(i+j-1)$, $1 \leq i, j \leq n$, the runtimes of the programs for $n = 10^k$, $k = 1, 2, \ldots, 4$. Also compare your programs with the runtime of the MATLAB routine `norm(H,inf)`.

Project 28.2.2 The set $N_2(1) = \{x \in \mathbb{R}^2 : \|x\|_2 = 1\}$ can be approximated in MATLAB using `plot(X,Y,'-b')`, with `Phi=(0:dphi:2*pi)` and `X=cos(Phi)`, `Y=sin(Phi)` for example using `dphi=0.01`. Plot the deformed set $A(N_2(1))$ for matrices

$$\begin{bmatrix} k & 0 \\ 0 & k \end{bmatrix}, \quad \begin{bmatrix} k_1 & 0 \\ 0 & k_2 \end{bmatrix}, \quad \begin{bmatrix} c & s \\ -s & c \end{bmatrix}, \quad \begin{bmatrix} 1 & k \\ 0 & 1 \end{bmatrix}, \quad \begin{bmatrix} c' & s' \\ s & c' \end{bmatrix}$$

with suitable numbers $k, k_1, k_2 \in \mathbb{R}$, $c = \cos(\theta)$, $s = \sin(\theta)$ for $\theta \in [0, 2\pi]$. Using the commands `hold on/off`, the sets can be displayed in one graphics window, and by changing the argument `'-b'` with different colours. Finally, replace $N_2(1)$ with $N_1(1)$ and $N_\infty(1)$.

28.3 Matrix Factorisations

Problem 28.3.1 Let $A \in \mathbb{R}^{n \times n}$ be a positive definite matrix, i.e. $x^T A x > 0$ for all $x \in \mathbb{R}^n \setminus \{0\}$.

(i) Show that A is regular.
(ii) Show that for all $1 \leq k \leq n$ the $k \times k$ submatrix $A_k = (a_{ij})_{1 \leq i,j \leq k}$ is also positive definite.
(iii) Show that all real eigenvalues of A are positive.

Problem 28.3.2 Let $A \in \mathbb{R}^{n \times n}$ be a strictly diagonally dominant matrix, i.e.

$$\sum_{j=1,\ldots,n, j \neq i} |a_{ij}| < |a_{ii}|, \quad i = 1, 2, \ldots, n.$$

(i) Show that the submatrices $A_k = (a_{ij})_{1 \leq i,j \leq k}$ for $k = 1, 2, \ldots, n$ are also strictly diagonally dominant.
(ii) Show that the matrix A is regular.

Hint: To prove (ii), show that for a suitable norm $\|\cdot\|$ on $\mathbb{R}^n$ the estimate $\|Ax\| > 0$ holds for all $x \in \mathbb{R}^n \setminus \{0\}$, and deduce from this that A is injective.

Problem 28.3.3 Show that the invertible (normalised) lower triangular matrices form a group, i.e. if $L, L_1, L_2 \in \mathbb{R}^{n \times n}$ are (normalised) lower triangular matrices and $\det L \neq 0$, then L^{-1} and $L_1 L_2$ are also (normalised) lower triangular matrices.

Problem 28.3.4 Let $A \in \mathbb{R}^{n \times n}$, a lower triangular matrix L and an upper triangular matrix U with $A = LU$ be given. Show that, for $k = 1, 2, \ldots, n$ and the left, upper $k \times k$ submatrices A_k, L_k and U_k of A, L and U respectively, the decomposition $A_k = L_k U_k$ also holds.

Problem 28.3.5

(i) Show that $A_1 = \begin{bmatrix} 0 & 0 \\ 1 & 0 \end{bmatrix}$ does not have a normalised LU decomposition and $A_2 = \begin{bmatrix} 0 & 1 \\ 1 & 0 \end{bmatrix}$ does not have a Cholesky decomposition.
(ii) Calculate the normalised LU decomposition of A_3 and the Cholesky decomposition of A_4 with

$$A_3 = \begin{bmatrix} 5 & 3 & 1 \\ 10 & 8 & 8 \\ 15 & 11 & 10 \end{bmatrix}, \quad A_4 = \begin{bmatrix} 9 & 12 & 9 \\ 12 & 41 & 22 \\ 9 & 22 & 38 \end{bmatrix},$$

if they exist.

Problem 28.3.6 Let $A \in \mathbb{R}^{n \times n}$ be symmetric and positive definite.

28.3 Matrix Factorisations

(i) Show that there exists a uniquely determined normalised lower triangular matrix $L \in \mathbb{R}^{n \times n}$ and a diagonal matrix $D \in \mathbb{R}^{n \times n}$ with positive diagonal entries, such that $A = LDL^T$ holds.

(ii) Develop a method for determining L and D that avoids the use of the square root function, and determine the matrices L and D for

$$A = \begin{bmatrix} 9 & 12 & 9 \\ 12 & 41 & 22 \\ 9 & 22 & 38 \end{bmatrix}.$$

Problem 28.3.7 Let $(v_1, v_2, \ldots, v_n)$ be a basis of $\mathbb{R}^n$.

(i) Show that the matrix $G \in \mathbb{R}^{n \times n}$ defined by $g_{ij} = v_i \cdot v_j$ is symmetric and positive definite.
(ii) Show that G is invertible and G^{-1} is also symmetric and positive definite.
(iii) Construct a lower triangular matrix $L \in \mathbb{R}^{n \times n}$, such that for $W = LV$ the identity $W^T W = I_n$ holds, where $V = [v_1, v_2, \ldots, v_n] \in \mathbb{R}^{n \times n}$.

Problem 28.3.8 Let $A \in \mathbb{R}^{n \times n}$ be symmetric with non-negative eigenvalues. Construct a symmetric matrix $B \in \mathbb{R}^{n \times n}$ with $A = B^2 = BB$ and show that $\text{cond}_2(B) = \text{cond}_2(A)^{1/2}$, provided A is regular.

Problem 28.3.9 Let $A \in \mathbb{R}^{n \times n}$ be a symmetric and positive definite matrix. Show that $\lambda_{max}(A^{-1}) = 1/\lambda_{min}(A)$ holds.

Problem 28.3.10

(i) How can the LU decomposition be simplified in the case of symmetric matrices and what effort does this entail?
(ii) Let $A \in \mathbb{R}^{n \times n}$ be a band matrix with bandwidth m, i.e. $a_{ij} = 0$ if $|i - j| > m$. How large is the effort of calculating the LU decomposition, provided it exists?

Project 28.3.1 Write a C++ or Python program with functions `solve_upper` and `solve_lower` for solving systems of linear equations with regular upper or lower triangular matrix. The solutions of $Ux = b$ and $Lx = b$ are given by backward or forward running loops through

$$x_j = \left(b_j - \sum_{k=j+1}^{n} u_{jk} x_k\right)/u_{jj}, \qquad x_j = \left(b_j - \sum_{k=1}^{j-1} \ell_{jk} x_k\right)/\ell_{jj},$$

where the empty sum has the value zero. Test the routines for the systems of equations $A_\ell x = b_\ell$, $\ell = 1, 2$, with

$$A_1 = \begin{bmatrix} 1 & 2 & 3 \\ & 4 & 5 \\ & & 6 \end{bmatrix}, \quad b_1 = \begin{bmatrix} 6 \\ 9 \\ 6 \end{bmatrix}, \quad A_2 = \begin{bmatrix} 1 & & \\ 2 & 3 & \\ 4 & 5 & 6 \end{bmatrix}, \quad b_2 = \begin{bmatrix} 3 \\ 12 \\ 28 \end{bmatrix}.$$

Project 28.3.2 Write a C++ or Python program that determines the LU decomposition for a matrix $A \in \mathbb{R}^{n \times n}$ that can be LU decomposed. Justify why the entries of the matrix A can be overwritten with the calculated entries of L, so that no new fields need to be initialised. Under what circumstances should the calculation of L be aborted? Test the implementation with the matrices

$$A_1 = \begin{bmatrix} 4 & 2 & 3 \\ 2 & 4 & 2 \\ 3 & 2 & 4 \end{bmatrix}, \quad A_2 = \begin{bmatrix} 2 & -1 & & \\ -1 & \ddots & \ddots & \\ & \ddots & \ddots & -1 \\ & & -1 & 2 \end{bmatrix},$$

to solve the systems of equations $A_i x = b_i$, $i = 1, 2$ for $b_1 = [1, 1, 1]^T$ and $b_2 = [1, \ldots, 1]^T$, where $A_2 \in \mathbb{R}^{n \times n}$ and $b_2 \in \mathbb{R}^n$ with $n = 10, 20, 40, 80$. Check your results using the MATLAB commands lu(A) and x=A\beta . What can be said about the runtime for the solution of the system of equations $A_2 x = b_2$ depending on n?

Project 28.3.3 Write a C++ or Python program that calculates the Cholesky decomposition $A = LL^T$ for a given symmetric, positive definite matrix $A = (a_{ij})_{i,j=1,\ldots,n} \in \mathbb{R}^{n \times n}$. Justify why the entries of the matrix A can be overwritten with the calculated entries of L so that no new fields need to be initialised. Under what circumstances should the calculation of L be aborted? Test the implementation with the matrices

$$A_1 = \begin{bmatrix} 4 & 2 & 3 \\ 2 & 4 & 2 \\ 3 & 2 & 4 \end{bmatrix}, \quad A_2 = \begin{bmatrix} 2 & -1 & & \\ -1 & \ddots & \ddots & \\ & \ddots & \ddots & -1 \\ & & -1 & 2 \end{bmatrix},$$

to solve the system of equations $A_i x = b_i$, $i = 1, 2$, for $b_1 = [1, 1, 1]^T$ and $b_2 = [1, \ldots, 1]^T$, where $A_2 \in \mathbb{R}^{n \times n}$ and $b_2 \in \mathbb{R}^n$ with $n = 10, 20, 40, 80$ apply. Check your results using the MATLAB commands chol(A) and x=A\beta . What can be said about the runtime for the solution of the system of equations $A_2 x = b_2$ depending on n?

Project 28.3.4 For $m \in \mathbb{N}$ and $n = m^2$, let $B_m \in \mathbb{R}^{m \times m}$ and $A_n \in \mathbb{R}^{n \times n}$ be defined by

$$A_n = \begin{bmatrix} B_m & -I_m & & \\ -I_m & \ddots & \ddots & \\ & \ddots & \ddots & -I_m \\ & & -I_m & B_m \end{bmatrix}, \quad B_m = \begin{bmatrix} 4 & -1 & & \\ -1 & \ddots & \ddots & \\ & \ddots & \ddots & -1 \\ & & -1 & 4 \end{bmatrix}.$$

28.4 Elimination Methods

Use the MATLAB routines chol and lu, to determine Cholesky and LU decompositions $L_n L_n^T = A_n$ and $M_n U_n = A_n$ and consider the errors

$$\|A_n - L_n L_n^T\|_\infty, \qquad \|A_n - M_n U_n\|_\infty,$$

which you can determine with norm(B,inf), for $n = 10^k$, $k = 1, 2, \ldots, 6$.

28.4 Elimination Methods

Problem 28.4.1 Construct a permutation matrix $P \in \mathbb{R}^{4\times 4}$, such that the matrix PA has a normalised LU decomposition, where

$$A = \begin{bmatrix} -1 & 2 & 3 & 3 \\ 1 & -4 & -2 & -5 \\ 0 & -4 & 0 & -3 \\ -1 & 10 & -5 & 17 \end{bmatrix}.$$

Solve the linear system $Ax = b$ with $b = [17, -23, -13, 51]^T$.

Problem 28.4.2 Use the Gaussian elimination method without pivot search to solve the linear system $Ax = b$ with

$$A = \begin{bmatrix} -1 & 16 & -4 & 3 \\ -3 & 20 & -22 & 0 \\ 1 & -16 & 1 & -2 \\ 3 & -6 & 4 & 2 \end{bmatrix}, \quad b = \begin{bmatrix} -24 \\ -45 \\ 20 \\ 11 \end{bmatrix}.$$

Determine the LU decomposition of A and calculate $\det A$.

Problem 28.4.3 Let

$$A = \begin{bmatrix} 1 & 0 & 1 \\ 2 & -1 & 1 \\ 2 & 2 & 3 \end{bmatrix}, \quad b = \begin{bmatrix} 5 \\ 7 \\ 14 \end{bmatrix}, \quad \tilde{b} = \begin{bmatrix} 5.5 \\ 6.5 \\ 14.5 \end{bmatrix}.$$

Calculate A^{-1}, $\mathrm{cond}_\infty(A)$ and the solutions of $Ax = b$ as well as $A\tilde{x} = \tilde{b}$.

Problem 28.4.4 Let $A \in \mathbb{R}^{n\times n}$ be a symmetric and positive definite matrix and let $b^{(1)}, b^{(2)}, \ldots, b^{(m)} \in \mathbb{R}^n$ different right sides. Let $A = LL^T$ be the Cholesky decomposition of A with the lower triangular matrix $L \in \mathbb{R}^{n\times n}$. Compare the effort of the following two approaches to solve the m linear systems $Ax^{(i)} = b^{(i)}$, $i = 1, 2, \ldots, m$:

(i) By solving the n linear systems $Az^{(j)} = e_j$ with the Cholesky decomposition of A for the canonical basis vectors $e_j \in \mathbb{R}^n$, the inverse $A^{-1} =$

$[z^{(1)}, z^{(2)}, \ldots, z^{(n)}]$ is determined and subsequently $x^{(i)} = A^{-1}b^{(i)}$ for $i = 1, 2, \ldots, m$ is determined by matrix-vector multiplication.

(ii) With the Cholesky decomposition of A, the solutions of $Ax^{(i)} = b^{(i)}$ for $i = 1, 2, \ldots, m$ are determined.

Problem 28.4.5 Let $P \in \mathbb{R}^{n \times n}$ be the permutation matrix corresponding to the bijection $\pi : \{1, 2, \ldots, n\} \to \{1, 2, \ldots, n\}$. Show that $P^\mathsf{T} = P^{-1}$ and

$$P^{-1} = [e_{\pi^{-1}(1)}, e_{\pi^{-1}(2)}, \ldots, e_{\pi^{-1}(n)}].$$

Problem 28.4.6 How does the effort of the Gaussian elimination method with column pivot search differ from that with total pivot search?

Problem 28.4.7 Show that with the canonical basis vectors $e_1, e_2, \ldots, e_m \in \mathbb{R}^m$ and $f_1, f_2, \ldots, f_n \in \mathbb{R}^n$ for $A \in \mathbb{R}^{m \times n}$, we have that

$$A = \sum_{i=1}^{m} \sum_{j=1}^{n} a_{ij} e_i f_j^\mathsf{T}.$$

Problem 28.4.8 Let $P \in \mathbb{R}^{n \times n}$ be a permutation matrix that swaps the k-th and ℓ-th entry of a vector, where $\ell > k$.

(i) Let $A \in \mathbb{R}^{n \times n}$. Determine PA and AP.

(ii) Let $L = I_n - \ell_k e_k^\mathsf{T}$ with the canonical basis vector $e_k \in \mathbb{R}^n$ and a vector $\ell_k = [0, \ldots, 0, \ell_{k+1,k}, \ldots, \ell_{n,k}]^\mathsf{T}$. Show that a vector

$$\widehat{\ell}_k = [0, \ldots, 0, \widehat{\ell}_{k+1,k}, \ldots, \widehat{\ell}_{n,k}]^\mathsf{T}$$

exists, such that with $\widehat{\ell} = I_n - \widehat{\ell}_k e_k^\mathsf{T}$ the identity $\widehat{\ell} = PLP$ holds.

Problem 28.4.9 Let $A \in \mathbb{R}^{m \times n}$. Construct a method for determining all solutions of $Ax = 0$.

Problem 28.4.10 For $k = 1, 2, \ldots, n-1$ let $L^{(k)} = I_n - \ell_k e_k^\mathsf{T}$ be defined with vectors $\ell_k = [0, \ldots, 0, \ell_{k+1,k}, \ldots, \ell_{n,k}]^\mathsf{T}$ and let $\widetilde{L} = L^{(n-1)} L^{(n-2)} \ldots L^{(1)}$. Show that

$$\widetilde{L}^{-1} = I_n + \sum_{k=1}^{n-1} \ell_k e_k^\mathsf{T}.$$

Project 28.4.1 Write a C++ or Python program, that, for an LU-decomposable matrix $A \in \mathbb{R}^{n \times n}$ and a vector $b \in \mathbb{R}^n$, solves the linear system $Ax = b$ using Gaussian elimination and determines the LU decomposition of A. Test the program with the system defined by

$$A = \begin{bmatrix} 1 & 7 & -2 & 3 \\ 5 & -1 & -4 & 0 \\ 8 & 1 & 3 & 5 \\ 4 & -4 & 4 & -4 \end{bmatrix}, \quad b = \begin{bmatrix} 21 \\ -9 \\ 39 \\ -8 \end{bmatrix}.$$

The matrix A can be overwritten by the calculated values $a_{ij}^{(k+1)}$ and ℓ_{ik}. Use your program to solve systems of linear equations with upper triangular matrix to solve the resulting system $A^{(n)}x = b^{(n)}$.

Project 28.4.2 Perturb the right side of the linear system $Ax = b$ defined by

$$a_{ij} = (i+j-1)^{-1}, \quad b_i = \sum_{k=1}^{n}(-1)^{k-1}/(i+k-1), \quad x_i = (-1)^{i-1}, \quad i, j = 1, 2, \ldots, n,$$

with the vector $d \in \mathbb{R}^n$, $d_i = 10^{-5}\cos(i\pi/n)$ for $i = 1, 2, \ldots, n$ and $n = 10$. Consider the relative error $\|x - x_d\|_2/\|x\|_2$ and compare this with the condition number of the matrix, which you can determine with the MATLAB command cond(A,2). Comment on the results.

Project 28.4.3 Implement the Gaussian elimination method with pivot search. To do this, introduce a vector $\pi \in \mathbb{N}^n$, which takes into account the row swaps. Also implement a termination criterion that ends the procedure if for the pivot element the estimate $|a_{\pi(k),k}^{(k)}| \leq 10^{-10}$ applies. When solving the resulting system of linear equations the row swaps need to be considered in the backward substitution. Test the procedure for the system $Ax = b$ where $A \in \mathbb{R}^{3\times 3}$ and $b \in \mathbb{R}^3$ are defined by

$$A = \begin{bmatrix} 0 & 1 & 0 \\ 0 & 0 & 1 \\ 1 & 0 & 0 \end{bmatrix}, \quad b = \begin{bmatrix} 1 \\ 2 \\ 3 \end{bmatrix}.$$

28.5 Least Squares Problems

Problem 28.5.1 Let $A \in \mathbb{R}^{m\times n}$, $b \in \mathbb{R}^m$ and $x, y \in \mathbb{R}^n$. Calculate the derivative of the mapping

$$t \mapsto \|A(x+ty) - b\|_2^2, \quad t \in \mathbb{R},$$

and deduce the Gaussian normal equation, if x is a solution of the associated least squares problem.

Problem 28.5.2 Let $A \in \mathbb{R}^{2\times 1}$ and $b \in \mathbb{R}^2$ be defined by

$$A = \begin{bmatrix} 2 \\ 1 \end{bmatrix}, \quad b = \begin{bmatrix} 1 \\ 3 \end{bmatrix}.$$

Determine graphically using a set square the solution of the least squares problem by orthogonally decomposing b into vectors v, w with $v \in \operatorname{Im} A$ and $w \in \ker A^\mathsf{T}$.

Problem 28.5.3 A Householder matrix $P \in \mathbb{R}^{m\times m}$ is defined for $v \in \mathbb{R}^m$ with $\|v\|_2 = 1$ by $P = I_m - 2vv^\mathsf{T}$.

(i) Show that $P = P^\mathsf{T}$ and $P^{-1} = P$ hold.
(ii) Show that a real $m \times m$ Householder matrix has $m - 1$ eigenvalues with the value 1 and one eigenvalue -1.
(iii) Construct using geometric considerations for $m = 2, 3$ a Householder matrix that maps a given vector $x \in \mathbb{R}^m$ to a multiple of $e_1 \in \mathbb{R}^m$.

Problem 28.5.4 Let $D \in \mathbb{R}^{m\times m}$ be a diagonal matrix with positive diagonal entries. The minimisation of $x \mapsto \|D(Ax - b)\|_2^2$ realises for example a different weighting of various measurement results. Determine the associated normal equation.

Problem 28.5.5 Let $A \in \mathbb{R}^{m\times n}$, $b \in \mathbb{R}^m$ and $1 < p < \infty$. Calculate the partial derivatives of the mapping

$$x \mapsto \|Ax - b\|_p^p, \quad x \in \mathbb{R}^n.$$

Determine all numbers p, for which the derivative is given by a linear mapping.

Problem 28.5.6 Calculate using the Householder method a QR decomposition for

$$A = \begin{bmatrix} 1 & 1 & 1 \\ 0 & -\sqrt{2} & \sqrt{2}/2 \\ 0 & \sqrt{2} & 5/\sqrt{2} \end{bmatrix}$$

and solve the equation $Ax = b$ for $b = [3\sqrt{2}, -1, 7]^\mathsf{T}$.

Problem 28.5.7 Let $A \in \mathbb{R}^{n\times n}$ be a regular matrix with columns $a_1, a_2, \ldots, a_n \in \mathbb{R}^n$ and let $(q_1, q_2, \ldots, q_n)$ be the resulting orthonormal basis obtained through the Gram–Schmidt process, that is

$$\tilde{q}_j = a_j - \sum_{k=1}^{j-1}(a_j \cdot q_k)q_k, \quad q_j = \frac{\tilde{q}_j}{\|\tilde{q}_j\|_2}$$

for $j = 1, 2, \ldots, n$.

28.5 Least Squares Problems

(i) Show that for $R \in \mathbb{R}^{n \times n}$ defined by $r_{kj} = a_j \cdot q_k$ for $k < j$, $r_{kj} = 0$ for $k > j$, $r_{jj} = \|\tilde{q}_j\|_2$ for $j = 1, ..., n$, it follows $A = QR$.

(ii) Calculate Q and R for

$$A = \begin{bmatrix} 1 & 2 & 0 \\ 0 & 1 & 2 \\ 1 & 0 & 2 \end{bmatrix}.$$

Problem 28.5.8 Let $A \in \mathbb{R}^{m \times n}$ and $A = QR$ be a QR decomposition. Show that R defines a Cholesky decomposition of $A^T A$.

Problem 28.5.9 Let $A \in \mathbb{R}^{n \times n}$ be regular and $A = QR$ be a QR decomposition. Show that $\operatorname{cond}_2(A) = \operatorname{cond}_2(R)$ holds.

Problem 28.5.10 Let $i < j$, $\theta \in \mathbb{R}$ and define $B = B(i, j, \theta) \in \mathbb{K}^{m \times m}$ by $b_{k\ell} = \delta_{k\ell}$ for $k \neq i, j$, $b_{ii} = b_{jj} = c$ and $b_{ij} = -b_{ji} = s$, with $c = \cos(\theta)$ and $s = \sin(\theta)$, that is

$$B(i, j, \theta) = \begin{bmatrix} 1 & & & & & & \\ & \ddots & & & & & \\ & & c & & s & & \\ & & & 1 & & & \\ & & & & \ddots & & \\ & & -s & & c & & \\ & & & & & 1 & \\ & & & & & & \ddots \end{bmatrix}.$$

(i) Show that the matrix $B(i, j, \theta)$ defines a rotation of the (i, j) plane by the angle θ.
(ii) Show that the successive multiplication of $A \in \mathbb{R}^{m \times n}$ with suitable $B(i, j, \theta)$ leads to a QR decomposition.
(iii) Is this procedure more expensive than the Householder procedure? If yes, are there classes of matrices for which it is less complex?

Project 28.5.1 Implement the Householder procedure for calculating a QR decomposition in C++ or Python. Use your program to solve the linear system $Ax = b$ with the $n \times n$ Hilbert matrix A defined by $a_{ij} = (i + j - 1)^{-1}$, $1 \leq i, j \leq n$, and the right-hand side $b = [1, 2, \ldots, n]^T$ for $n = 3$ and $n = 10$.

Project 28.5.2 From physics, it is known that bodies exposed only to gravity fly in parabolas. A body has the initial velocity $v = (v_x, v_y)$ and is at point 0 at time $t = 0$. At time t it is then at the location $x = v_x t$, $y = v_y t - \frac{1}{2}gt^2$, where g is the acceleration due to gravity. In a series of experiments, the values given in

Table 28.1 Measurement values of an experimental series

i	1	2	3	4	5	6	7
$t_i [s]$	0.1	0.2	0.6	0.9	1.1	1.2	2.0
$x_i [m]$	0.73	1.28	4.24	6.11	7.69	8.21	13.83
$y_i [m]$	0.96	1.81	4.23	5.05	5.15	4.81	0.55

Table 28.1 were measured. Formulate a suitable least squares problem and solve it in MATLAB using the QR decomposition provided by [Q,R] = qr(A), to determine the velocity v_y and the acceleration due to gravity g as accurately as possible. Create a graph using the plot command, in which the measured values and the calculated parabola are listed. To what accuracy is it meaningful to specify the results? What model errors, data errors and measurement errors occur in this experiment?

28.6 Singular Value Decomposition and Pseudoinverse

Problem 28.6.1 Let $A, B \in \mathbb{R}^{m \times n}$ and $(v_1, v_2, \ldots, v_n) \subset \mathbb{R}^n$ be a basis of $\mathbb{R}^n$. Show that from $Av_i = Bv_i$ for $i = 1, 2, \ldots, n$ the equality $A = B$ follows.

Problem 28.6.2 Determine a singular value decomposition of the matrix

$$A = \frac{1}{4}\begin{bmatrix} 3 & 1 & -1 & -3 \\ -1 & -3 & 3 & 1 \end{bmatrix}^\mathsf{T}.$$

Calculate A^+ using the singular value decomposition as well as the identity $A^+ = (A^\mathsf{T} A)^{-1} A^\mathsf{T}$. Use A^+ to solve the least squares problem defined by A and $b = [4, 1, 2, 3]^\mathsf{T}$.

Problem 28.6.3 Let $A \in \mathbb{R}^{m \times n}$. Show that the pseudoinverse A^+ is the unique solution $X \in \mathbb{R}^{n \times m}$ of the equations

$$AXA = A, \quad XAX = X, \quad (AX)^\mathsf{T} = AX, \quad (XA)^\mathsf{T} = XA$$

To prove uniqueness, assume the existence of a second solution Y, derive the identities $X = XA(YAY)(AXA)X$ and $Y = (YA)^\mathsf{T} Y(AY)^\mathsf{T}$ and show that the right-hand sides match.

Problem 28.6.4 Show that $\operatorname{rank} A^\mathsf{T} A = \operatorname{rank} AA^\mathsf{T} = \operatorname{rank} A$ holds.

Problem 28.6.5

(i) Let $V \subset \mathbb{R}^n$ be a subspace and $V^\perp$ its orthogonal complement. Show that there exists a uniquely determined matrix $P_V \in \mathbb{R}^{n \times n}$ with $P_V v = v$ for all $v \in V$ and $P_V w = 0$ for all $w \in V^\perp$.
(ii) Let $A \in \mathbb{R}^{m \times n}$. Show that $A^+ A = P_{(\ker A)^\perp}$ and $AA^+ = P_{\operatorname{Im} A}$.

28.6 Singular Value Decomposition and Pseudoinverse

Problem 28.6.6 Let $(\lambda_i, v_i) \in \mathbb{R} \times \mathbb{R}^n$, $i = 1, \ldots, n$, be eigenvalues and corresponding linearly independent eigenvectors of the matrix $A \in \mathbb{R}^{n \times n}$. Show that A can be represented as $A = VDV^{-1}$ with $V = [v_1, \ldots, v_n]$ and $D = \operatorname{diag}(\lambda_1, \ldots, \lambda_n)$.

Problem 28.6.7 Let $A \in \mathbb{R}^{n \times n}$ with eigenvalues $\lambda_1, \ldots, \lambda_n \in \mathbb{C}$ and $\|\cdot\|_{op}$ an operator norm.

(i) Show that $\|A\|_2 \leq \|A\|_{op}$ holds.
(ii) Show that $\max_{i=1,\ldots,n} |\lambda_i| \leq \|A\|_2$ holds.

Problem 28.6.8

(i) Let $A \in \mathbb{R}^{n \times n}$ and $\lambda \in \mathbb{C}$ be an eigenvalue of A. Prove the following statements:
(a) The number $\bar{\lambda}$ is an eigenvalue of A.
(b) If A is symmetric, then the eigenvalues of A are real.
(c) If A is regular, then λ^{-1} is an eigenvalue of A^{-1}.
(d) The matrix A^T has the eigenvalue λ.
(ii) Let $A, B \in \mathbb{R}^{n \times n}$ be matrices with eigenvalues λ and μ. Under what conditions is $\lambda \mu$ an eigenvalue of AB?

Problem 28.6.9

(i) Let $n \in \mathbb{N}$ be odd and $Q \in SO(n)$, i.e. we have $Q \in \mathbb{R}^{n \times n}$ with $Q^T Q = I_n$ and $\det Q = 1$. Show that Q has the eigenvalue 1.
(ii) Conclude that during a football match there are at least two points on the surface of the football that are in the same place in the surrounding space at least twice.

Problem 28.6.10 Show that the tridiagonal matrix defined by $a, b, c \in \mathbb{R}$ with $bc > 0$

$$A = \begin{bmatrix} a & b & & \\ c & a & \ddots & \\ & \ddots & \ddots & b \\ & & c & a \end{bmatrix} \in \mathbb{R}^{n \times n}$$

has the eigenvalues $\lambda_k = a + 2 \operatorname{sign}(c) \sqrt{bc} \cos(k\pi/(n+1))$, $k = 1, 2, \ldots, n$. First consider the case $a = 0$ and the vectors

$$v_k = \left((c/b)^{\ell/2} \sin(k\pi \ell/(n+1))\right)_{\ell=1,\ldots,n}.$$

Project 28.6.1 In MATLAB the singular value decomposition of a matrix A can be calculated with the command svd. For an image defined by the file img.jpg, a compression of the grayscale representation can be defined with the lines shown in Fig. 28.1. Choose as an image, for example, the section from Albrecht Dürer's picture *Melancolia I*, which shows the magic square. Explain the individual lines of the program and extend it by a calculation of the approximation error $\|X -$

```
1  RGB = imread('img.jpg');
2  G = rgb2gray(RGB);
3  D = double(G);
4  X = D/max(max(D));
5  figure(1);
6  subplot(1,2,1); imshow(X); title('original');
7  [U,S,V] = svd(X);
8  for k = 5:5:size(U,1)
9      X_comp = U(:,1:k)*S(1:k,1:k)*V(:,1:k)';
10     subplot(1,2,2); imshow(X_comp);
11     title('compressed'); pause
12 end
```

Fig. 28.1 Image compression using singular value decomposition

$X_{comp} \|_{\mathscr{F}}$. How do you assess the ratio of quality loss to reduction of storage requirements for different values of k? Test the program for another image.

Project 28.6.2 The unit square $Q = [0,1]^2 \subset \mathbb{R}^2$ can be represented in MATLAB by `fill(X,Y,0)` with `X = [0,1,1,0,0]` and `Y = [0,0,1,1,0]`. Visualise the image $A(Q)$ with the linear transformations, which are defined by the matrices

$$\begin{bmatrix} k & 0 \\ 0 & k \end{bmatrix}, \quad \begin{bmatrix} k_1 & 0 \\ 0 & k_2 \end{bmatrix}, \quad \begin{bmatrix} c & s \\ -s & c \end{bmatrix}, \quad \begin{bmatrix} 1 & k \\ 0 & 1 \end{bmatrix}, \quad \begin{bmatrix} c' & s' \\ s & c' \end{bmatrix}$$

with suitable numbers $k, k_1, k_2 \in \mathbb{R}$, $c = \cos(\theta)$, $s = \sin(\theta)$ for $\theta \in [0, 2\pi]$. Determine the eigenvalues and eigenvectors of the transformations with the MATLAB command `[V,D] = eig(A)` and interpret them geometrically.

28.7 The Simplex Method

Problem 28.7.1 Let $A \in \mathbb{R}^{m \times n}$ and $b \in \mathbb{R}^m$. Show that the set $C = \{x \in \mathbb{R}^n : x \geq 0, Ax = b\}$ has at most a finite number of corners.
Hint: Consider the zero entries of elements in C.

Problem 28.7.2 Let $f(x) = a^{\mathsf{T}} x + b$ and let $C \subset \mathbb{R}^n$ be a convex, closed and bounded set. Show that the function f takes its extreme values at the corners of C, i.e. there exist corner points $x_m, x_M \in C$ with $f(x_m) = \min_{x \in C} f(x)$ and $f(x_M) = \max_{x \in C} f(x)$.

Problem 28.7.3 Transform the minimisation of $g(y) = p^{\mathsf{T}} y$ under the constraint $Uy \leq d$ into a linear program in normal form by decomposing $y = v - w$ into non-negative vectors v and w, introducing a non-negative vector z with $Uy + z = d$ and defining suitable vectors c, d and a suitable matrix A.

28.7 The Simplex Method

Problem 28.7.4 Determine all corners of the convex sets $B_1^2(0) = \{x \in \mathbb{R}^n : \|x\|_2 \leq 1\}$ and $B_1^\infty(0) = \{x \in \mathbb{R}^n : \|x\|_\infty \leq 1\}$.

Problem 28.7.5

(i) Determine all minima of the function $f(x) = \sum_{i=1}^n |x - z_i|$ for the cases

$$z = [-1, 1]^T, \quad z = [-1, 0, 1]^T, \quad z = [0, 10]^T, \quad z = [0, 1, 10]^T.$$

How can minima be characterised?

(ii) Let $z_1, z_2, \ldots, z_n \in \mathbb{R}$ with $z_1 \leq z_2 \leq \cdots \leq z_n$. Determine a minimum of the function

$$f(x) = \sum_{i=1}^n |x - z_i|.$$

Use the formal necessary optimality condition $f'(x^*) = 0$ with the derivative $|\cdot|' = \text{sign}(\cdot)$.

Problem 28.7.6 Let $z_1, z_2, \ldots, z_n \in \mathbb{R}$. Formulate the minimisation of the function $f(y) = \sum_{i=1}^n |y - z_i|$ as a linear program.

Problem 28.7.7 Let $a \in \mathbb{R}^2$ and $c \in \mathbb{R}$. Construct through geometric considerations the minimisers of the mappings $x \mapsto \|x\|_p$ under the constraint $a^T x = \alpha$ for $p = 1, 2, \infty$.

Problem 28.7.8 Let $a \in \mathbb{R}^n \setminus \{0\}$ and let $C \subset \mathbb{R}^n$ be non-empty, bounded and strictly convex, i.e. if $\theta x_1 + (1 - \theta) x_2 \in \partial C$ for $x_1, x_2 \in C$, then $\theta = 1$ or $\theta = 0$. Show that the minimisation of the function $f(x) = a \cdot x$ under the constraint $x \in C$ has a unique solution.

Problem 28.7.9 Let $A = [4, 2, 1]$, $b = 4$ and $c = [1, 1, 1]^T$.

(i) Determine the corners of the set $\{x \in \mathbb{R}^3 : x \geq 0, Ax = b\}$ and investigate whether these are degenerate.
(ii) Carry out the simplex method for the minimisation of $f(x) = c^T x$ under the constraint $Ax = b$ and $x \geq 0$ with the starting corner $x^0 = [0, 0, 4]^T$.

Problem 28.7.10 Construct matrices $A \in \mathbb{R}^{3 \times 2}$ and vectors $b \in \mathbb{R}^2$, so that the resulting sets $M = \{x \in \mathbb{R}^3 : Ax = b, x \geq 0\}$ are (i) empty and unbounded and (ii) bounded and non-empty.

Project 28.7.1 A company produces m different products, for the manufacture of which n machines are required. The j-th machine has a maximum monthly running time of ℓ_j hours. The k-th product generates a profit of e_k euros per unit and occupies the j-th machine with t_{jk} hours per unit. The total monthly profit should be optimised without exceeding the maximum running times.

```
1 [x_1,x_2,x_3] = sphere;
2 surf([m_1+r*x_1,m_2+r*x_2,m_3+r*x_3]); hold on;
3 tetramesh([1,2,3,4],Z); hold off;
```

Fig. 28.2 Visualisation of a sphere

(i) Formulate the described situation as a maximisation problem with constraints in the form

$$\text{Maximise } f(x) = c \cdot x \text{ under the conditions } Ax \leq b, \ x \geq 0$$

where $x = (x_1, x_2, \ldots, x_m)$ are the monthly units of the different products and the inequalities are to be understood component-wise.

(ii) Use the MATLAB routine linprog to solve the problem for the data $m = 2$, $n = 3$, $e_1 = 200$, $e_2 = 600$, and $t_{11} = 1$, $t_{21} = 1$, $t_{31} = 0$, $t_{12} = 3$, $t_{22} = 1$, $t_{32} = 2$ and $\ell_1 = 150$, $\ell_2 = 180$, $\ell_3 = 140$. What is the optimal monthly profit?

Project 28.7.2 If $a \in \mathbb{R}^3$ is a vector with positive components and $\alpha \in \mathbb{R}$ is a positive number, then $\{x \in \mathbb{R}^3 : x \geq 0, \ a^T x \leq \alpha\}$ defines a tetrahedron. The centre m and the radius $r > 0$ of a sphere of maximum volume contained in the tetrahedron are to be determined. Formulate the problem as a linear program and solve it with the MATLAB routine linprog. Then determine the solution for the case $a = [1, 2, 3]^T$ and $\alpha = 4$. You can visualise your solution with the MATLAB commands shown in Fig. 28.2, where $Z \in \mathbb{R}^{4 \times 3}$ is a matrix containing the coordinates of the corners of the tetrahedron.

Hint: The distance of a point $m \in \mathbb{R}^3$ to the plane defined by a vector $v \in \mathbb{R}^3$ with $\|v\|_2 = 1$ and a number $\gamma \in \mathbb{R}$ is given by $|v^T m - \gamma|$.

28.8 Eigenvalue Problems

Problem 28.8.1

(i) Determine the Gershgorin circles of the matrix

$$A = \begin{bmatrix} 1 & 4 & 7 \\ 2 & 5 & 8 \\ 3 & 6 & 9 \end{bmatrix}.$$

(ii) Let $A \in \mathbb{R}^{n \times n}$ be strictly diagonally dominant and symmetric. Provide an explicit upper bound for the condition number $\text{cond}_2(A)$.

Problem 28.8.2 Show that the characteristic polynomial $p(\lambda) = \det(A - \lambda I_n)$ of the $n \times n$ matrix

28.8 Eigenvalue Problems

$$A = \begin{bmatrix} 0 & & & & -a_0 \\ 1 & 0 & & & -a_1 \\ & \ddots & \ddots & & \vdots \\ & & 1 & 0 & -a_{n-2} \\ & & & 1 & -a_{n-1} \end{bmatrix}$$

is given by $p(\lambda) = (-1)^n (\lambda^n + a_{n-1}\lambda^{n-1} + \cdots + a_1\lambda + a_0)$.

Problem 28.8.3

(i) Let $A \in \mathbb{R}^{n \times n}$ be symmetric with eigenvalues $\lambda_1 \geq \lambda_2 \geq \cdots \geq \lambda_n$ and let $v_1 \in \mathbb{R}^n \setminus \{0\}$ be an eigenvector corresponding to the eigenvalue λ_1. Show that

$$\lambda_2 = \max_{\substack{x \in \mathbb{R}^n \setminus \{0\} \\ x \cdot v_1 = 0}} \frac{x^\mathsf{T} A x}{\|x\|_2^2}.$$

(ii) Show that the vector $x^* \in \mathbb{R}^n \setminus \{0\}$ is an eigenvector of the symmetric matrix $A \in \mathbb{R}^{n \times n}$ if and only if $\nabla r(x^*) = 0$ holds with the function

$$r : \mathbb{R}^n \setminus \{0\} \to \mathbb{R}, \quad x \mapsto \frac{x^\mathsf{T} A x}{\|x\|_2^2}.$$

Problem 28.8.4

(i) Show that the power method also converges when the iterates are normalised with respect to a different norm.

(ii) Perform five steps of the power method for the matrices

$$A = \frac{1}{2}\begin{bmatrix} 2 & 0 & 0 \\ 0 & 5 & -1 \\ 0 & -1 & 5 \end{bmatrix}, \quad B = \begin{bmatrix} -6 & -22 & 59 \\ -4 & -6 & 22 \\ -2 & -4 & 13 \end{bmatrix}$$

with the initial vector $x_0 = [1, 1, 1]^\mathsf{T}/2$ and observe the sizes $\|\tilde{x}_k\|_2$ and $x_k^\mathsf{T} A x_k$.

Problem 28.8.5 Determine the k-th iterate of the power method for the matrix

$$A = \begin{bmatrix} 0 & 2 & & \\ & \ddots & \ddots & \\ & & \ddots & 2 \\ 2 & & & 0 \end{bmatrix}$$

with the starting vectors $x_0 = [1, 0, \ldots, 0]^\mathsf{T}$ and $x_0 = [1, 1, \ldots, 1]^\mathsf{T}$ and discuss the validity of the assumptions of the convergence result.

Problem 28.8.6 Let $A \in \mathbb{R}^{n \times n}$.

(i) Show that a Householder matrix $\tilde{H} \in \mathbb{R}^{(n-1) \times (n-1)}$ exists, such that for $B = HAH^T$ with $H = \begin{bmatrix} 1 & 00 & \tilde{H} \end{bmatrix}$ the property $b_{i1} = 0$ holds for $i > 2$.

(ii) Conclude that A can be transformed into a matrix $\hat{A} \in \mathbb{R}^{n \times n}$ with $\hat{a}_{ij} = 0$ for $i > j + 1$ using $n - 2$ similarity transformations. Discuss the required numerical effort.

(iii) Show that the property $a_{ij} = 0$ for $i > j + 1$ is preserved in the QR method.

Problem 28.8.7 Perform a step of the QR method for the matrix

$$A = \begin{bmatrix} 1 & -2 & 3 \\ 0 & 3 & 5 \\ 0 & 1 & 2 \end{bmatrix},$$

determine the eigenvalues of A using the characteristic polynomial and compare the results.

Problem 28.8.8

(i) Let $A \in \mathbb{R}^{n \times n}$ be a symmetric matrix and $G_{pq} \in \mathbb{R}^{n \times n}$ a Givens rotation. Show for the entries of the matrix $B = G_{pq}^{-1} A G_{pq}$ that

$$b_{pp} = c^2 a_{pp} + 2cs a_{pq} + s^2 a_{qq},$$
$$b_{qq} = s^2 a_{pp} - 2cs a_{pq} + c^2 a_{qq},$$
$$b_{pq} = b_{qp} = cs(a_{qq} - a_{pp}) + (c^2 - s^2) a_{pq},$$
$$b_{ip} = c a_{ip} + s a_{iq}, \quad i \in \{1, 2, \ldots, n\} \setminus \{p, q\},$$
$$b_{iq} = -s a_{ip} + c a_{iq}, \quad i \in \{1, 2, \ldots, n\} \setminus \{p, q\},$$
$$b_{ij} = a_{ij}, \quad i, j \notin \{p, q\}.$$

(ii) Infer $b_{pq} = 0$, provided $a_{pq} \neq 0$ and G_{pq} is defined by $c = \sqrt{(1 + D)/2}$ and $s = \text{sign}(a_{pq}) \sqrt{(1 - D)/2}$ with

$$D = \frac{a_{pp} - a_{qq}}{\left((a_{pp} - a_{qq})^2 + 4 a_{pq}^2\right)^{1/2}}.$$

Problem 28.8.9 Let $\|A\|_{\mathscr{F}} = \left(\sum_{i,j=1}^n a_{ij}^2\right)^{1/2}$ be the Frobenius norm.

(i) Show that $\|A\|_{\mathscr{F}}^2 = \text{tr}(A^T A)$ as well as $\text{tr}(AB) = \text{tr}(BA)$ for all $A, B \in \mathbb{R}^{n \times n}$ and infer $\|Q^{-1} B Q\|_{\mathscr{F}} = \|B\|_{\mathscr{F}}$ for $B \in \mathbb{R}^{n \times n}$, $Q \in O(n)$.

(ii) Show that $\|A\|_2 \leq \|A\|_{\mathscr{F}}$ for all $A \in \mathbb{R}^{n \times n}$.

28.8 Eigenvalue Problems

Problem 28.8.10 Construct a symmetric matrix $A_k \in \mathbb{R}^{3\times 3}$ with an entry $(A_k)_{ij} = 0$, so that for the next iterate A_{k+1} in the Jacobi method $(A_{k+1})_{ij} \neq 0$ holds.

Project 28.8.1 Implement the von Mises power method to approximate the smallest and largest eigenvalue as well as the corresponding eigenvectors of the $n \times n$ matrix

$$A = \begin{bmatrix} 2 & -1 & & \\ -1 & \ddots & \ddots & \\ & \ddots & \ddots & -1 \\ & & -1 & 2 \end{bmatrix}$$

for $n = 4, 16, 64, 256$. Use the MATLAB command $\text{x = B\textbackslash c}$ to solve linear systems $Bx = c$. Use different starting vectors and a suitable termination criterion for the iteration and determine the errors of the approximations using the exact values $\lambda_{min} = 2 - 2\cos(\pi/(n+1))$ and $\lambda_{max} = 2 - 2\cos(n\pi/(n+1))$.

Project 28.8.2

(i) Use the MATLAB routine [Q,R] = qr(A), to implement the QR method and terminate the iteration if $\|A_k - A_{k+1}\|_2 / \|A_k\|_2 \leq 10^{-5}$ holds. What would be another meaningful termination criterion? Approximate with your program the eigenvalues of the matrices $A \in \mathbb{R}^{n\times n}$, $n = 4, 10, 20$ and $B, B^\mathsf{T} \in \mathbb{R}^{3\times 3}$ defined by

$$A = \begin{bmatrix} 2 & -1 & & \\ -1 & \ddots & \ddots & \\ & \ddots & \ddots & -1 \\ & & -1 & 2 \end{bmatrix}, \quad B = \begin{bmatrix} -1 & -10 & 29 \\ -2 & -4 & 18 \\ -1 & -3 & 11 \end{bmatrix}$$

and discuss the prerequisites of the proposition about the convergence of the method based on these examples.

(ii) Implement the Jacobi method with the termination criterion $\mathcal{N}(A_k) \leq 10^{-4}$ in MATLAB and test it for the matrix $A \in \mathbb{R}^{n\times n}$ defined by

$$a_{ij} = \sin(|i - j|\pi/n) - 2\delta_{ij}$$

for $i, j = 1, 2, \ldots, n$ with $n = 2, 4, 8, 16$. Modify the program to obtain an implementation of the cyclic Jacobi method, that is, the search for the largest entry is omitted and all entries are successively treated. Observe graphically the size of the entries of the iterates using the MATLAB commands $\text{[X,Y] = meshgrid(1:n,1:n)}$, surf(X,Y,A) and view(-270,90). Consider the number of iterations required depending on n.

28.9 Iterative Solution Methods

Problem 28.9.1 Construct a matrix $M \in \mathbb{R}^{2\times 2}$, which is a contraction with respect to an operator norm and not with respect to another.

Problem 28.9.2 Show that for the iterates of the fixed point iteration $x^{k+1} = \Phi(x^k)$ with the contraction $\Phi : \mathbb{R}^n \to \mathbb{R}^n$ the error estimate

$$\|x^k - x^*\| \leq \frac{q}{1-q} \|x^k - x^{k-1}\|$$

holds. How is this estimation relevant for practical purposes?

Problem 28.9.3

(i) Let $T \in \mathbb{R}^{n \times n}$ be a regular matrix and $\|\cdot\|$ a norm on $\mathbb{R}^n$. Show that $\|x\|_T = \|Tx\|$ for $x \in \mathbb{R}^n$ defines another norm on $\mathbb{R}^n$.

(ii) Let $R \in \mathbb{R}^{n \times n}$ and $D \in \mathbb{R}^{n \times n}$ be an invertible diagonal matrix. Show that for $T = D^{-1}RD$ and $i, j = 1, 2, \ldots, n$ we have

$$t_{ij} = \frac{d_{jj}}{d_{ii}} r_{ij}.$$

Problem 28.9.4 For a given matrix $A \in \mathbb{R}^{n \times n}$ we define a directed graph G with node set $\{1, 2, \ldots, n\}$ and edges e_{ij}, $1 \leq i, j \leq n$, from node i to node j, if $a_{ij} \neq 0$. Show that A is irreducible if and only if one can reach every node j from any node i along the edges with $j \neq i$.

Problem 28.9.5

(i) Let $A_1, A_2 \in \mathbb{R}^{n \times n}$ be defined by

$$A_1 = \begin{bmatrix} 2 & -1 & & \\ -1 & \ddots & \ddots & \\ & \ddots & \ddots & -1 \\ & & -1 & 2 \end{bmatrix}, \quad A_2 = \begin{bmatrix} 4 & -1 & & \\ -1 & \ddots & \ddots & \\ & \ddots & \ddots & -1 \\ & & -1 & 4 \end{bmatrix}.$$

Investigate these matrices with respect to diagonal dominance and irreducibility.

(ii) Show that in the case of the matrix A_2 for the iteration matrix M^J of the Jacobi method the estimate $\varrho(M^J) \leq 1/2$ holds.

Problem 28.9.6 Perform 5 steps of the Richardson, Jacobi and Gauss-Seidel methods for

$$A = \begin{bmatrix} 2 & -1 & 0 \\ -1 & 2 & -1 \\ 0 & -1 & 2 \end{bmatrix}, \quad b = \begin{bmatrix} 1 \\ 1 \\ 1 \end{bmatrix}$$

28.9 Iterative Solution Methods

with $\omega = 1$ and $\omega = 1/10$ and $x^0 = [1, 1, 1]^T$ respectively. Compare the iterates with the exact solution of the system of equations.

Problem 28.9.7 Show that $A \in \mathbb{R}^{n \times n}$ is irreducible if and only if for all $i, j \in \{1, 2, \ldots, n\}$ there exists a sequence $i_1, i_2, \ldots, i_\ell \in \{1, 2, \ldots, n\}$ with $i_1 = i$ and $i_\ell = j$ and $a_{i_k i_{k+1}} \neq 0$ for $k = 1, 2, \ldots, \ell - 1$.

Problem 28.9.8 Show that $A \in \mathbb{R}^{n \times n}$ is reducible if and only if there exists a permutation matrix $P \in \{0, 1\}^{n \times n}$, such that

$$PAP^T = \begin{bmatrix} B_{11} & B_{12} \\ 0 & B_{22} \end{bmatrix}$$

with suitable matrices B_{11}, B_{12} and B_{22} and a null matrix $0 \in \mathbb{R}^{k \times \ell}$ where $k + \ell = n$.

Problem 28.9.9 For the matrix $A \in \mathbb{R}^{n \times n}$ let $\varrho(I_n - A) < 1$. Show that A is invertible and that the inverse A^{-1} is given by the convergent series

$$A^{-1} = \sum_{i=0}^{\infty} (I_n - A)^i.$$

Hint: Consider the matrix $B = I_n - A$ and argue as when determining the value of the geometric series.

Problem 28.9.10 Show that the iteration process defined by $(D + U)x^{k+1} = -Lx^k + b$ converges in the case of an irreducible and diagonally dominant matrix $A = U + D + L \in \mathbb{R}^{n \times n}$ for any initial value $x^0 \in \mathbb{R}^n$.

Project 28.9.1 Use the equivalent representations

$$x_i^{k+1} = a_{ii}^{-1} \left(b_i - \sum_{j \neq i} a_{ij} x_j^k \right),$$

$$x_i^{k+1} = a_{ii}^{-1} \left(b_i - \sum_{j < i} a_{ij} x_j^{k+1} - \sum_{j > i} a_{ij} x_j^k \right)$$

of the Jacobi and Gauss-Seidel methods to implement these in C++. Test your programs for the system of linear equations $Ax = b$ with

$$A = \begin{bmatrix} 2 & -1 & & \\ -1 & \ddots & \ddots & \\ & \ddots & \ddots & -1 \\ & & -1 & 2 \end{bmatrix}, \quad b = \begin{bmatrix} 1 \\ 1 \\ \vdots \\ 1 \end{bmatrix}$$

and the starting vector $x^0 = [1, 1, \ldots, 1]^T \in \mathbb{R}^n$ for $n = 10, 20, 40$. Terminate the iteration when $\|x^k - x^{k+1}\|_2 \leq \delta$ with $\delta = 10^{-5}$. Comment on the dependence of the number of iterations on the dimension n of the system of equations.

Project 28.9.2 Implement the Richardson method in MATLAB and test it for the system of linear equations $Ax = b$ with

$$A = \begin{bmatrix} 2 & -1 & & \\ -1 & \ddots & \ddots & \\ & \ddots & \ddots & -1 \\ & & -1 & 2 \end{bmatrix}, \quad b = \frac{1}{n^2} \begin{bmatrix} 1 \\ 1 \\ \vdots \\ 1 \end{bmatrix}$$

and a starting vector $x^0 \in \mathbb{R}^n$ randomly generated with the MATLAB command randn(n,1) for $n = 10, 20, 40, 80$ and the parameters $\omega = 1, 1/10, 1/n$. Visualise the iterations using plot([0:1/n:1],[0,x']) and observe the behaviour of these curves for several different initial values. Try to identify and characterise different phases of the iteration.

Chapter 29
Problems on Numerical Analysis

29.1 General Condition Number and Machine Numbers

Problem 29.1.1 Show that given a basis $b \geq 2$ every number $x \in \mathbb{R} \setminus \{0\}$ can be represented in the form

$$x = \pm b^e \sum_{k=1}^{\infty} d_k b^{-k}$$

with $d_1, d_2, \cdots \in \{0, 1, \ldots, b-1\}$ and $e \in \mathbb{Z}$ where $d_1 \neq 0$ can be chosen.

Problem 29.1.2

(i) Calculate the number of floating point numbers as well as the positive extrema g_{min} and g_{max} for the IEEE formats *single* and *double precision*.
(ii) Determine $\mathrm{rd}(\pi)$ for $b = 2$, $p = 5$ and $b = 10$, $p = 4$.
(iii) How can the occurrence of *overflow* be avoided when calculating $(a^2 + b^2)^{1/2}$ if $\max\{|a|, |b|\} > g_{max}^{1/2}$ and $|a|, |b| \leq g_{max}/2$?

Problem 29.1.3

(i) Represent the numbers 142, 237 and 1111 for the bases $b = 2, 4$ and 10 with the precision $p = 10$ and the exponent limits $e_{min} = -10$ and $e_{max} = 10$ as normalised floating point numbers.
(ii) Determine the 25th decimal place of $1/7$.
(iii) Why is the number $1/10$ in the binary system only representable by an infinite series?

Problem 29.1.4

(i) Let $\phi = \phi_1 \circ \cdots \circ \phi_J$, where the suboperations $\phi_1, \ldots, \phi_J$ are well conditioned. Show that ϕ is well conditioned.
(ii) Let $g \in C^1(\mathbb{R})$. Discuss the conditioning of the determination of the roots of g and illustrate the results graphically.

Problem 29.1.5 Let $\phi : \mathbb{R}^2 \to \mathbb{R}^2$, $\phi(p, q) = (x_1, x_2)$, be the operation of determining the roots x_1, x_2 of the quadratic polynomial $x^2 + px + q$. Determine a subset $W \subset \mathbb{R}^2$ on which ϕ is well-defined. Calculate for $(p, q) \in W$ the relative condition number $\kappa_\phi(p, q)$ and discuss for which pairs (p, q) the operation is well conditioned.

Problem 29.1.6 Identify possible problems when evaluating the pq-formula $x_{1,2} = -p/2 \pm (p^2/4 - q)^{1/2}$ for determining the roots of the quadratic polynomial $x^2 + px + q$. Construct a stable algorithm by utilising the relationship $x_1 x_2 = q$.

Problem 29.1.7

(i) Show that the set of regular $n \times n$ matrices defines an open subset of $\mathbb{R}^{n \times n}$.
(ii) Show that for $E \in \mathbb{R}^{n \times n}$ and sufficiently small numbers $h \in \mathbb{R}$ the matrix $I_n + hE$ is regular with

$$(I_n + hE)^{-1} = \sum_{k=0}^{\infty} (-1)^k h^k E^k.$$

Problem 29.1.8 Assume that the floating point addition is given by $x +_G y = \mathrm{rd}(x + y)$.

(i) Prove that the harmonic series $\sum_{k=1}^{\infty} 1/k$ converges in floating point arithmetic.
(ii) Show that the floating point addition $+_G$ is not associative.

Problem 29.1.9

(i) Show that the operation $\phi(x) = (1/x) - (1/(x+1))$ is well conditioned for large numbers $x \in \mathbb{R}$.
(ii) Show that the method $\tilde{\phi}(x) = (1/x) - (1/(x+1))$ is unstable.
(iii) Show that the method $\tilde{\phi}(x) = 1/(x(x+1))$ is stable.
  Hint: Identify the dominant terms of the expression

$$\tilde{\phi}(\tilde{x}) = \left(\frac{1+\varepsilon_2}{x(1+\varepsilon_1)} - \frac{1+\varepsilon_4}{(x(1+\varepsilon_1)+1)(1+\varepsilon_3)} \right)(1+\varepsilon_5)$$

and consider the quotient $|\tilde{\phi}(\tilde{x}) - \phi(x)|/|\phi(x)|$. Use approximations $1/(1+\varepsilon) \approx 1 - \varepsilon$ and $1/(1+\varepsilon+1/x) \approx 1 - \varepsilon - 1/x$.

Problem 29.1.10 Use the Gaussian elimination method without or with pivot search to solve the system of equations

29.2 Polynomial Interpolation

$$\begin{bmatrix} 0.1 \cdot 10^{-3} & 1 \\ 1 & 1 \end{bmatrix} \begin{bmatrix} x_1 \\ x_2 \end{bmatrix} = \begin{bmatrix} 1 \\ 2 \end{bmatrix}.$$

Use decimal numbers with precision $p = 3, 4, 5$, i.e., work using suitable rounding with numbers of the form $\pm 0.d_1 d_2 \ldots d_p \cdot 10^e$ with $e \in \mathbb{Z}$ and $d_1, d_2, \ldots, d_p \in \{0, 1, \ldots, 9\}$.

Project 29.1.1 To determine the rounding accuracy of a computer, let $x = 1$ and replace x with $x/2$ as long as the expression $1 + x > 1$ is evaluated as true by the computer. Determine experimentally in C++ the value of x for which this procedure stops. Define for this purpose x as a variable of type float or double.

Project 29.1.2 We consider the numerical determination of the Euler number e, which is characterised by the limit values

$$e = \lim_{n \to \infty} (1 + 1/n)^n, \quad e = \lim_{n \to \infty} \sum_{k=0}^{n} \frac{1}{k!}$$

Use only basic arithmetic operations and finite approximations of the above limit values with $n = 10^j$, $j = 1, 2, \ldots, 15$, to approximate e. Determine the approximation errors using the reference approximation $e \approx 2.718281828459045$ and display these with 15 decimal places in a table. Evaluate your results.

Project 29.1.3 The solution of a system of linear equations $Ax = b$ is given by Cramer's rule as $x_i = \det A_i / \det A$, $i = 1, 2, \ldots, n$, where $A_i \in \mathbb{R}^{n \times n}$ is obtained from A by replacing the i-th column of A with the vector b. In MATLAB A_i can be generated with the commands A_i=A and A_i(:,i)=b;. Implement Cramer's rule in MATLAB and test your program for the system of equations $Ax = b$ with

$$A = \begin{bmatrix} 0.2161 & 0.1441 \\ 1.2969 & 0.8648 \end{bmatrix}, \quad b = \begin{bmatrix} 0.1440 \\ 0.8642 \end{bmatrix}.$$

The exact solution is given by $x = [2, -2]^T$. Determine for the numerical solution $\tilde{x}$ the forward error $\|x - \tilde{x}\|_\infty / \|x\|_\infty$ as well as the backward error $\|A\tilde{x} - b\|_\infty / \|b\|_\infty$. Consider the condition number of A and compare the errors with those of the numerical solution $\hat{x}$ calculated by the Gaussian elimination method with pivot search, which you can determine in MATLAB with x=A\beta .

29.2 Polynomial Interpolation

Problem 29.2.1

(i) Let $f \in C^2([a, b])$ with the property $f(a) = f(b)$ and $f'(a) = f'(b) = 0$. Provide an optimal lower bound for the number of roots of f''.

(ii) For nodes $x_0 < x_1 < \cdots < x_n$, let $w(x) = \prod_{j=0}^{n}(x - x_j)$ be the node polynomial and L_i, $i = 0, 1, \ldots, n$, the i-th Lagrange basis polynomial. Show that

$$L_i(x) = \frac{w(x)}{(x - x_i)w'(x_i)}.$$

Problem 29.2.2 Let $a \leq x_0 < x_1 < \cdots < x_n \leq b$ be given nodes and $(v_0, v_1, \ldots, v_n)$ polynomials of maximum degree n.

(i) Show that the matrix $V \in \mathbb{R}^{(n+1)\times(n+1)}$ defined by $V_{ij} = v_i(x_j)$, $i, j = 0, 1, \ldots, n$, is regular if and only if from $\sum_{i=0}^{n} \alpha_i v_i(x) = 0$ for all $x \in [a, b]$ it follows that $\alpha_i = 0$ for $i = 0, 1, \ldots, n$.
(ii) Show that in the case of the monomials $v_i(x) = x^i$, $i = 0, 1, \ldots, n$, we have

$$\det V = \prod_{0 \leq i < j \leq n} (x_j - x_i).$$

Problem 29.2.3 Let $f(x) = \sin(\pi x)$ for $x \in [0, 1]$, $x_0 = 0$ and $x_i = i/n$, $i = 0, 1, \ldots, n$ if $n > 0$. Calculate and sketch the interpolation polynomial of f for $n = 0, 1, \ldots, 4$.

Problem 29.2.4 Let $f(x) = \sin(\pi x)$ for $x \in [0, 1]$, $x_0 = 0$ and $x_i = i/n$ for $i = 1, 2, \ldots, n$ if $n > 0$. Calculate and sketch the Hermite interpolation polynomials for $n = 0, 1, 2$ and $\ell_i = \ell$, $i = 0, 1, \ldots, n$, with $\ell = 0, 1, 2$.

Problem 29.2.5

(i) For $x \in [-5, 5]$ let $f(x) = (1 + x^2)^{-1} = \arctan'(x)$. Use the identities

$$\cos(\arctan(x)) = \frac{1}{(1 + x^2)^{1/2}}, \quad \sin(\arctan(x)) = \frac{x}{(1 + x^2)^{1/2}},$$

$$\sin(x)\sin(y) - \cos(x)\cos(y) = -\cos(x + y),$$

$$\sin(x)\cos(y) + \cos(x)\sin(y) = \sin(x + y),$$

to inductively prove or verify for $n = 0, 1, 2, 3$ that

$$f^{(n)}(x) = \frac{n!}{(1 + x^2)^{(n+1)/2}} \times \begin{cases} (-1)^{n/2} \cos((n + 1)\arctan(x)), & n \text{ even,} \\ (-1)^{(n+1)/2} \sin((n + 1)\arctan(x)), & n \text{ odd.} \end{cases}$$

(ii) Conclude that $\|f^{(2n)}\|_\infty = (2n)!$ and that the Lagrange interpolation polynomials of $\tilde{f}(x) = f(5x)$ in the interval $[-1, 1]$ do not necessarily converge uniformly towards $\tilde{f}$ as $n \to \infty$.

29.2 Polynomial Interpolation

Problem 29.2.6 Construct nodes $a \leq x_0 < x_1 < \cdots < x_n \leq b$ in the interval $[a, b]$, such that for the Lagrange interpolation of any function $f \in C^{n+1}([a, b])$ we have

$$\|f - p\|_{C^0([a,b])} \leq 2^{-n} \left(\frac{b-a}{2}\right)^{n+1} \frac{\|f^{(n+1)}\|_{C^0([a,b])}}{(n+1)!}.$$

Problem 29.2.7 Prove the following properties of the functions defined by $T_n(t) = \cos(n \arccos t)$ for $t \in [-1, 1]$:

(i) We have that $|T_n(t)| \leq 1$ for all $t \in [-1, 1]$.
(ii) With $T_0(t) = 1$ and $T_1(t) = t$, we have that

$$T_{n+1}(t) = 2t\, T_n(t) - T_{n-1}(t)$$

for all $t \in [-1, 1]$. In particular, $T_n \in \mathscr{P}_n|_{[-1,1]}$ and for $n \geq 1$, $T_n(t) = 2^{n-1} t^n + q_{n-1}$ with $q_{n-1} \in \mathscr{P}_{n-1}|_{[-1,1]}$.
(iii) For $n \geq 1$, T_n has the roots $t_j = \cos((j + 1/2)\pi/n)$, $j = 0, 1, \ldots, n-1$, and the $n + 1$ extreme points $s_j = \cos(j\pi/n)$, $j = 0, 1, \ldots, n$.

Problem 29.2.8

(i) Provide a method using on as few as possible basic arithmetic operations for evaluating the polynomial $(x + 3)^{16}$.
(ii) Compare the effort of the direct evaluation of the polynomial $p(x) = a_0 + a_1 x_1 + \cdots + a_n x^n$ with that of using the equivalent representation

$$p(x) = a_0 + x\Big(a_1 + x\big(a_2 + \ldots x\big(a_{n-2} + x(a_{n-1} + x a_n)\big)\ldots\big)\Big).$$

Problem 29.2.9 For $n + 1$ nodes and values $(x_0, y_0), (x_1, y_1), \ldots, (x_n, y_n)$ and $0 \leq j \leq n$ as well as $0 \leq i \leq n - j$, let $p_{i,j} \in \mathscr{P}_j$ be defined by $p_{i,j}(x_k) = y_k$, $k = i, i+1, \ldots, i+j$. The numbers $y_{i,j}$ are defined by $y_{i,0} = y_i$, $i = 0, 1, \ldots, n$, and

$$y_{i,j} = \frac{y_{i+1,j-1} - y_{i,j-1}}{x_{i+j} - x_i}$$

for $1 \leq j \leq n$ and $0 \leq i \leq n - j$.

(i) Show that $p_{i,j}(x) = y_{i,j} x^j + r_{i,j}(x)$ with a polynomial $r_{i,j} \in \mathscr{P}_{j-1}$ for $j \geq 1$ and $i = 0, 1, \ldots, n - j$.
(ii) Show that for $q_j(x) = p_{0,j}(x) - p_{0,j-1}(x)$, where $p_{0,-1} = 0$, the representation $q_j(x) = y_{0,j} \prod_{i=0}^{j-1}(x - x_i)$ holds.
(iii) Conclude that $p_{0,n}(x) = \sum_{j=0}^{n} y_{0,j} \prod_{i=0}^{j-1}(x - x_i)$.

Problem 29.2.10 Let $x_0 < x_1 < \cdots < x_n$ and $\ell \in \mathbb{N}$. For $x \in \mathbb{R}$ and $0 \leq j \leq n$ define

$$H_{j,\ell}(x) = \frac{(x-x_j)^\ell}{\ell!} \prod_{\substack{i=0 \\ i \neq j}}^{n} \left(\frac{x-x_i}{x_j-x_i}\right)^{\ell+1}.$$

Show that for the derivatives of $H_{j,\ell}$ the identities $\frac{d^k}{dx^k} H_{j,\ell}(x_m) = \delta_{k\ell}\delta_{jm}$ for $0 \leq k \leq \ell$ and $0 \leq m \leq n$ apply.

Project 29.2.1 Implement the Neville scheme in non-recursive form and use it to interpolate the function $f(x) = (1 + 25x^2)^{-1}$ using equidistant nodes $-1 = x_0 < x_1 < \cdots < x_n = 1$ as well as Chebyshev nodes $-1 \leq t_0 < t_1 < \cdots < t_n \leq 1$ to evaluate the polynomial at the points $x_a = \pi/8$ and $x_b = \pi/4$ for $n = 1, 2, 4, 8, 16, 32$. Comment on your observations.

Project 29.2.2

(i) Write a MATLAB program to determine the coefficients of an interpolation polynomial with respect to the Newton basis for given nodes $x_0 < x_1 < \cdots < x_n$ and corresponding values $y_0, \ldots, y_n$.

(ii) Test your program for the functions $f(x) = \sin(\pi x)$, $g(x) = (1 + 25x^2)^{-1}$ and $h(x) = |x|$ in the interval $[-1, 1]$ using equidistant nodes and Chebyshev nodes. Evaluate the interpolation polynomials at the points $z_j = -1 + 2j/100$, $j = 0, 1, \ldots, 100$ using the Horner scheme and plot the interpolation polynomials for $n = 1, 2, 4, 8$.

29.3 Interpolation with Splines

Problem 29.3.1

(i) Let $0 \leq a < b$ and $x \mapsto g(x)$ be the linear function that interpolates the function $f(x) = x^{1/2}$ at the nodes a and b. Show that for the error $e_m = \max_{x \in [a,b]} |g(x) - f(x)|$ the estimates $e_m \leq (b-a)^2 a^{-3/2}/8$ in the case $a > 0$ and $e_m \leq b^{1/2}/4$ in the case $a = 0$ apply.

(ii) For $n \in \mathbb{N}$ and $x_i = i/n$, $i = 0, 1, \ldots, n$, let $f_n \in \mathscr{S}^{1,0}(\mathscr{T}_n)$ be the interpolating spline function of $f(x) = x^{1/2}$ in the interval $[0, 1]$. Show that $\max_{x \in [0,1]} |f_n(x) - f(x)| \leq n^{-1/2}/4$.

(iii) In which regions is the error estimate from (ii) suboptimal?

Problem 29.3.2 For the partitioning defined by the points $x_i = (i/n)^4$, $i = 0, 1, \ldots, n$, of $[0, 1]$, let $f_n \in \mathscr{S}^{1,0}(\mathscr{T}_n)$ be the interpolating spline function of $f(x) = x^{1/2}$. Show that $\max_{x \in [0,1]} |f_n(x) - f(x)| \leq cn^{-2}$ with a constant $c > 0$ independent of n applies. Sketch f_n for $n = 2, 4, 8$.

29.3 Interpolation with Splines

Problem 29.3.3

(i) Show that for every interval $[a_0, a_1] \subset \mathbb{R}$ there are uniquely determined polynomials $q_{0,0}, q_{0,1}, q_{1,0}, q_{1,1} \in \mathcal{P}_3$ such that $q_{j,k}^{(\ell)}(a_m) = \delta_{jm}\delta_{k\ell}$ for $j, k, \ell, m = 0, 1$. Draw the polynomials for the interval $[0, 1]$.

(ii) Conclude that on each partitioning $\mathcal{T}_n$ with grid points $x_0 < x_1 < \cdots < x_n$ for given values $y_0, y_1, \ldots, y_n$ and $r_0, r_1, \ldots, r_n$ there exists a uniquely defined spline $s \in \mathcal{S}^{3,1}(\mathcal{T}_n)$ with $s(x_i) = y_i$ and $s'(x_i) = r_i$, $i = 0, 1, \ldots, n$, and provide a representation formula.

Problem 29.3.4 Show that the cubic spline interpolation task with natural boundary conditions is uniquely solvable by considering the linear subspace $\mathcal{S}_{nat}^{3,2}(\mathcal{T}_n) = \{s \in \mathcal{S}^{3,2}(\mathcal{T}_n) : s''(a) = s''(b) = 0\}$.

Problem 29.3.5 Let $\mathcal{T}_n$ be the partitioning $a = x_0 < x_1 < \cdots < x_n = b$ and let $s \in \mathcal{S}^{3,2}(\mathcal{T}_n)$ be the interpolating cubic spline of the function values $y_0 = 1$ and $y_i = 0$, $i = 1, 2, \ldots, n$ with natural boundary conditions. Show that s has only finitely many zeros on each interval $[x_{i-1}, x_i]$, $i = 1, 2, \ldots, n$, and provide a possibly accurate upper estimate for the number of zeros on each of these intervals. Sketch the function s.

Problem 29.3.6 Explicitly determine the interpolating cubic splines with natural as well as Hermite boundary conditions $s'(-1) = 0$, $s'(1) = 3$, for the nodes $x_i = -1 + i/2$ and values $y_i = (-1)^i$, $i = 0, 1, 2, \ldots, 4$, and draw these.

Problem 29.3.7 Let $\mathcal{T}_n$ be a partitioning of the interval $[a, b]$ and let $s \in \mathcal{S}^{1,0}(\mathcal{T}_n)$ and $g \in C^1([a, b])$ satisfy $s(x_i) = g(x_i)$ for $i = 0, 1, \ldots, n$. Prove the inequality

$$\sum_{i=1}^n \int_{x_{i-1}}^{x_i} |s'|^2 \, dx \le \int_a^b |g'|^2 \, dx.$$

Problem 29.3.8 The functions $B_m : \mathbb{R} \to \mathbb{R}$, $m \in \mathbb{N}$, are defined by the recursion

$$B_{m+1}(x) = \int_{x-1/2}^{x+1/2} B_m(t) \, dt$$

with the initialisation $B_0(x) = 1$ for $|x| \le 1/2$ and $B_0(x) = 0$ for $|x| > 1/2$.

(i) Show that B_m is non-negative and $B_m(x) = 0$ for $|x| > (m+1)/2$.
(ii) Show that, with the partitioning $\mathcal{T}_{m+1}$ of the interval $[-(m+1)/2, (m+1)/2]$ defined by the points $x_i = i - (m+1)/2$, $i = 0, \ldots, m+1 = n$, for each $m \in \mathbb{N}$ a spline function $B_m \in \mathcal{S}^{m,m-1}(\mathcal{T}_{m+1})$ is defined.
(iii) Determine the functions B_1, B_2 and B_3 explicitly and sketch them.

Problem 29.3.9 For $n \in \mathbb{N}$ and $i = 0, 1, \ldots, n$ the function $B_{i,n} : \mathbb{R} \to \mathbb{R}$ is defined by

$$B_{i,n}(x) = \binom{n}{i} x^i (1-x)^{n-i}.$$

(i) Show that the functions $(B_{0,n}, B_{1,n}, \ldots, B_{n,n})$ define a basis of the polynomial space $\mathscr{P}_n$.
(ii) Prove the formula $B_{i,n}(x) = (1-x) B_{i,n-1}(x) + x B_{i-1,n-1}(x)$.

Problem 29.3.10

(i) Let $P_0, P_1, \ldots, P_n \in \mathbb{R}^m$. Show that the mapping $z : [0, 1] \to \mathbb{R}^m$,

$$z(t) = \sum_{i=0}^{n} \binom{n}{i} t^i (1-t)^{n-i} P_i,$$

has the properties $z(0) = P_0$, $z(1) = P_n$ as well as $z'(0) = n(P_1 - P_0)$, $z'(1) = n(P_n - P_{n-1})$.
(ii) Construct points $P_0, P_1, P_2, P_3 \in \mathbb{R}^2$ so that the graph of the mapping z approximates the quarter circle $\{(x, y) \in \mathbb{R}^2 : y = (1-x^2)^{1/2}, 0 \leq x \leq 1\}$ as well as possible.

Project 29.3.1 The MATLAB command plot(X,Y,'r-*') graphically represents a polygonal chain defined by the vectors X and Y. If $X = [x_0, x_1, \ldots, x_n]^\mathsf{T}$ and $Y = [f(x_0), f(x_1), \ldots, f(x_n)]^\mathsf{T}$, a continuous, piecewise linear interpolation of the function f is represented. The representation of the graph can be changed in colour, line representation and marking with the optional argument r-*. Other useful commands are:

hold on, hold off, clf, axis, xlabel, ylabel, legend

(i) Graphically illustrate the piecewise linear approximation of the function $f(x) = x^{1/2}$ on the interval $[0, 1]$ with the grid points

(a) $x_i = i/n$, (b) $x_i = (i/n)^4$

for $i = 0, 1, \ldots, n$ and $n = 2, 4, 8, 16$, by comparing these with the representation of f on a very fine grid.
(ii) Write a routine for calculating an interpolating cubic spline with natural boundary conditions. Test the routine with the partitions from (i) for the function $f(x) = \sin(2\pi x)$.
(iii) Generate meaningful graphics in each case and save them in a jpg or png file. Comment on the results.

Project 29.3.2

(i) Research the term *Bézier curve* and explain it in 5 to 10 lines.
(ii) Use a drawing program (e.g., `xfig` under Unix or the online program Mathcha (https://www.mathcha.io)) to draw two identical black ellipses in rectangles with side lengths $\ell_1 = 5.0\,\text{cm}$ and $\ell_2 = 10.0\,\text{cm}$.
(iii) Each quarter of the two ellipses should be approximated using different methods such as a polygonal chain, a spline function, a Bézier curve, or a composite variant. For each segment, a maximum of 3 or 5 interpolation or control points should be used. In the case of composite curves, a position used twice counts as one point. Use different colors for different approximations.
(iv) Which function achieves the best approximation? Define a suitable distance concept for the curves and manually measure the corresponding errors.
(v) Export your graphic as a pdf file.

29.4 Discrete Fourier Transform

Problem 29.4.1

(i) Let $n \in \mathbb{N}$ and $\ell \in \mathbb{Z}$. Show that $\sum_{k=0}^{n-1} e^{i\ell k 2\pi/n} = n$ holds, if n is a divisor of ℓ, and $\sum_{k=0}^{n-1} e^{i\ell k 2\pi/n} = 0$ otherwise.
(ii) Conclude that the Fourier basis $(\omega^0, \omega^1, \ldots, \omega^{n-1}) \subset \mathbb{C}^n$ defined by $\omega^k = [\omega_n^{0k}, \omega_n^{1k}, \ldots, \omega_n^{(n-1)k}]^T$, $k = 0, 1, \ldots, n-1$, with the n-th root of unity $\omega_n = e^{i2\pi/n}$ has the property $\omega^k \cdot \omega^\ell = n\delta_{k\ell}$.

Problem 29.4.2

(i) Let the nodes $z_0, z_1, \ldots, z_{n-1} \in \mathbb{C}$ be pairwise different and the values $y_0, y_1, \ldots, y_{n-1} \in \mathbb{C}$ arbitrary. Show that there exists a uniquely determined polynomial $p(z) = \beta_0 + \beta_1 z + \cdots + \beta_{n-1} z^{n-1}$ with complex coefficients β_i, $i = 0, 1, \ldots, n-1$, such that $p(z_j) = y_j$ for $j = 0, 1, \ldots, n-1$.
(ii) Conclude the unique solvability of the complex trigonometric interpolation problem.

Problem 29.4.3 Let $w_0, w_1, \ldots, w_{n-1} \in \mathbb{C}$ and $n = 2m$. Construct $y_0, y_1, \ldots, y_{n-1} \in \mathbb{C}$, so that, with the coefficients $\beta_0, \beta_1, \ldots, \beta_{n-1} \in \mathbb{C}$ of the solution of the corresponding complex trigonometric interpolation problem and the function

$$q(x) = \sum_{k=-m}^{m-1} \beta_{k+m} e^{ikx},$$

the interpolation property $q(x_j) = w_j$ for $j = 0, 1, \ldots, n-1$ and $x_j = 2\pi j/n$ is fulfilled.

Problem 29.4.4 Calculate without using matrix-vector multiplications the Fourier synthesis $y = T_8 \beta$ of the vector

$$\beta = [0, \sqrt{2}, 1, \sqrt{2}, 0, -\sqrt{2}, -1, -\sqrt{2}]^T.$$

Problem 29.4.5

(i) Show that on the space of continuous, complex-valued functions $C^0([0, 2\pi]; \mathbb{C})$ through

$$\langle v, w \rangle = \int_0^{2\pi} v(x)\overline{w(x)}\, dx$$

a scalar product is defined.

(ii) Show that the functions $(\varphi_k : k \in \mathbb{Z})$ defined by $\varphi_k(x) = e^{ikx}$, $k \in \mathbb{Z}$, $x \in [0, 2\pi]$, define an orthogonal system, that is, we have $\langle \varphi_k, \varphi_\ell \rangle = \delta_{k\ell}$ for all $k, \ell \in \mathbb{Z}$ with $k \neq \ell$.

(iii) Show that the orthogonality of the system $(\varphi_k : k \in \mathbb{Z})$ is preserved when the integral is approximated by a Riemann sum, that is, with respect to

$$\langle v, w \rangle_n = \frac{2\pi}{n} \sum_{j=0}^{n-1} v(x_j)\overline{w(x_j)}$$

with $x_j = 2\pi j/n$, $j = 0, 1, \ldots, n-1$.

Problem 29.4.6 Let $\langle \cdot, \cdot \rangle : V \times V \to \mathbb{R}$ be a scalar product on the real, n-dimensional vector space V and let $(v_0, v_1, \ldots, v_{n-1})$ be an orthonormal basis of V. Show that for every vector $w \in V$ we have

$$w = \sum_{j=0}^{n-1} \langle w, v_j \rangle v_j.$$

Problem 29.4.7 Given $y_0, y_1, \ldots, y_{n-1} \in \mathbb{R}$, let T and p be the solutions of the real and complex trigonometric interpolation problems respectively. Show that $T(x_j) = p(x_j)$ for $x_j = 2\pi j/n$, $j = 0, 1, \ldots, n-1$, but in general $T \neq p$ holds.

Problem 29.4.8

(i) Show that the solution of the real trigonometric interpolation problem is given by the coefficients

$$a_k = \frac{2}{n} \sum_{j=0}^{n-1} y_j \cos(kx_j), \quad b_\ell = \frac{2}{n} \sum_{j=0}^{n-1} y_j \sin(\ell x_j)$$

29.4 Discrete Fourier Transform

for $k = 0, 1, \ldots, m$ and $\ell = 1, 2, \ldots, m - 1$ with $x_j = 2\pi j/n$, $j = 0, 1, \ldots, n - 1$, and $n = 2m$.

(ii) Conclude that the vectors

$$f^k = \big(\cos(kx_j)\big)_{j=0,\ldots,n-1}, \quad g^\ell = \big(\sin(\ell x_j)\big)_{j=0,\ldots,n-1}$$

for $k = 0, 1, \ldots, m$ and $\ell = 1, 2, \ldots, m - 1$ define an orthogonal basis of $\mathbb{R}^n$.

Problem 29.4.9 Let $n, m \in \mathbb{N}$ with $n = 2m$, $A, B \in \mathbb{R}^{n \times n}$ and $C = AB$. For $i, j \in \{1, 2\}$ let $A_{ij}, B_{ij}, C_{ij} \in \mathbb{R}^{m \times m}$, be the sub-blocks of A, B and C, such that

$$A = \begin{bmatrix} A_{11} & A_{12} \\ A_{21} & A_{22} \end{bmatrix}, \quad B = \begin{bmatrix} B_{11} & B_{12} \\ B_{21} & B_{22} \end{bmatrix}, \quad C = \begin{bmatrix} C_{11} & C_{12} \\ C_{21} & C_{22} \end{bmatrix}.$$

(i) Show that the computation of C with the standard method for computing the product of matrices leads to $\mathcal{O}(n^{\log_2 8})$ multiplications.

(ii) Show that with

$$M_1 = (A_{11} + A_{22})(B_{11} + B_{22}), \qquad M_2 = (A_{21} + A_{22})B_{11},$$
$$M_3 = A_{11}(B_{12} - B_{22}), \qquad M_4 = A_{22}(B_{21} - B_{11}),$$
$$M_5 = (A_{11} + A_{12})B_{22}, \qquad M_6 = (A_{21} - A_{11})(B_{11} + B_{12}),$$
$$M_7 = (A_{12} - A_{22})(B_{21} + B_{22})$$

we have

$$C_{11} = M_1 + M_4 - M_5 + M_7, \qquad C_{12} = M_3 + M_5,$$
$$C_{21} = M_2 + M_4, \qquad C_{22} = M_1 - M_2 + M_3 + M_6.$$

(iii) Let $n = 2^k$ for a $k \in \mathbb{N}$. Construct a recursive method for computing AB that uses $\mathcal{O}(7^k) = \mathcal{O}(n^{\log_2 7})$ multiplications.

Remark This approach uses the fact that the computation of an expression $(a + b)(c + d)$ is less expensive than the equivalent expression $ac + bc + ad + bd$.

Problem 29.4.10

(i) For $a_\ell, b_\ell \in \mathbb{R}$, $\ell = 0, 1, \ldots, m$, let

$$T(x) = \frac{a_0}{2} + \sum_{\ell=1}^{m} \big(a_\ell \sin(\ell x) + b_\ell \cos(\ell x)\big).$$

Construct $\delta_k \in \mathbb{C}$, $k = 0, 1, \ldots, 2m$, such that, with

$$q(x) = \sum_{k=-m}^{m} \delta_{k+m} e^{ikx},$$

there holds $T(x) = q(x)$ for all $x \in [0, 2\pi]$.

(ii) Show that the function q is real-valued if and only if $\delta_{m-k} = \overline{\delta}_{m+k}$ holds for $k = 0, 1, \ldots, m$.

Project 29.4.1 Implement the complex Fourier synthesis as a recursive function and use your routine to compute the Fourier transform of the vectors $y \in \mathbb{C}^n$ defined by $y_j = f_r(2\pi j/n)$, $j = 0, 1, \ldots, n-1$, $r = 1, 2, 3$, with $f_1(x) = \sin(5x) + (1/2)\cos(x)$ and

$$f_2(x) = \begin{cases} 1, & x \in [\pi - 1/4, \pi + 1/4], \\ 0, & x \notin [\pi - 1/4, \pi + 1/4], \end{cases} \quad f_3(x) = \begin{cases} 1, & x \in [0, \pi), \\ -1, & x \notin [\pi, 2\pi), \end{cases}$$

with $n = 2^s$, $s = 1, 2, \ldots, 5$, to compute. Graphically represent the associated complex trigonometric polynomials. Use the MATLAB implementation of complex numbers in the creation of your program.

Project 29.4.2 Let the function $f : [0, 2\pi] \to \mathbb{R}$ be defined by

$$f(x) = \begin{cases} x, & x \in [0, \pi], \\ 2\pi - x, & x \in [\pi, 2\pi]. \end{cases}$$

Use the MATLAB routine fft to compute for $n = 2^s$, $s = 1, 2, \ldots, 5$, complex coefficients $(\beta_k)_{k=0,1,\ldots,n-1}$ and $(\delta_k)_{k=0,1,\ldots,n-1}$ such that for the functions

$$p(x) = \sum_{k=0}^{n-1} \beta_k e^{ikx}, \quad q(x) = \sum_{k=-n/2}^{n/2-1} \delta_{k+n/2} e^{ikx}$$

the interpolation property $p(x_j) = f(x_j)$ or $q(x_j) = f(x_j)$ for $j = 0, 1, \ldots, n-1$ and with $x_j = 2\pi j/n$ is fulfilled. Plot the real and imaginary parts of the functions p and q and discuss your results.

29.5 Numerical Integration

Problem 29.5.1 Use the representation of the Lagrange interpolation error

29.5 Numerical Integration

$$f(x) - p(x) = \frac{f^{(n+1)}(\xi)}{(n+1)!} \prod_{j=0}^{n}(x - x_j),$$

to prove for the trapezoidal and Simpson's rule that

$$|I(f) - Q_{Trap}(f)| \leq \frac{(b-a)^3}{12} \|f''\|_{C^0([a,b])},$$

$$|I(f) - Q_{Sim}(f)| \leq \frac{(b-a)^5}{2880} \|f^{(4)}\|_{C^0([a,b])}.$$

Problem 29.5.2 Let $Q : C^0([a, b]) \to \mathbb{R}$ be a quadrature formula with $n+1$ weights and quadrature points $(x_i, w_i)_{i=0,\ldots,n}$, which is exact of degree n.

(i) Show that

$$w_i = \int_a^b L_i(x)\,dx$$

for $i = 0, 1, \ldots, n$ with the Lagrange basis polynomials $(L_i)_{i=0,\ldots,n}$ defined by the nodes $(x_i)_{i=0,\ldots,n}$.

(ii) Show that in the case of exactness of degree $2n$, we have that $w_i > 0$ for $i = 0, 1, \ldots, n$.

Problem 29.5.3 Assume that the quadrature formula $Q : C^0([a, b]) \to \mathbb{R}$ is exact of degree $2q$ and the associated weights $(w_i)_{i=0,\ldots,n}$ and nodes $(x_i)_{i=0,\ldots,n}$ are symmetrically arranged with respect to the interval midpoint $(a+b)/2$. Show that Q is exact of degree $2q + 1$.

Problem 29.5.4

(i) Let $\omega \in C^0(a, b)$ be a function that is improperly Riemann-integrable and positive outside a finite number of points. Show that by

$$\langle f, g \rangle_\omega = \int_a^b f(x)g(x)\omega(x)\,dx$$

a scalar product on $C^0([a, b])$ is defined.

(ii) Show that the polynomials $(P_n)_{n \in \mathbb{N}}$ defined by the derivatives

$$P_n(x) = \frac{1}{2^n n!} \frac{d^n}{dx^n}[(x^2 - 1)^n]$$

are orthogonal with respect to the scalar product defined by the weight function $\omega(x) = 1$ for $x \in [-1, 1]$, that is for $j \neq k$ we have $\langle P_j, P_k \rangle_\omega = 0$.

Problem 29.5.5 Let $(f, g) \mapsto \langle f, g \rangle$ be a scalar product on the space $C^0([a, b])$. Show that with the initialisations $p_0(x) = 1$ and $p_1(x) = x - \beta_0$ and the recursion rule

$$p_{j+1}(x) = (x - \beta_j) p_j(x) - \gamma_j p_{j-1}(x)$$

with the coefficients $\beta_j = \langle xp_j, p_j \rangle / \langle p_j, p_j \rangle$ and $\gamma_j = \langle p_j, p_j \rangle / \langle p_{j-1}, p_{j-1} \rangle$ a sequence of pairwise orthogonal polynomials $p_j \in \mathscr{P}_j$ is defined.

Problem 29.5.6

(i) Show that the function $\omega(x) = (1 - x^2)^{-1/2}$ on the interval $(-1, 1)$ is improperly Riemann-integrable.
(ii) Show that the Chebyshev polynomials $T_n(t) = \cos(n \arccos(t))$, $n \in \mathbb{N}_0$, are orthogonal with respect to the scalar product defined by the weight function $\omega(x) = (1 - x^2)^{-1/2}$.

Problem 29.5.7 Determine $n + 1$ quadrature points and weights in the interval $[-1, 1]$, such that the resulting quadrature formula is exact of degree $2n + 1$ for $n = 0, 1, 2$. Use the formulas to approximate the integral of the function $x \mapsto x^5$ in the interval $[-1, 1]$.

Problem 29.5.8 Let $\omega : (a, b) \to \mathbb{R}$ be a weight function. Construct polynomials $(\pi_j)_{j=0,\ldots,n}$ using the Gram–Schmidt process such that $\pi_j \in \mathscr{P}_j$ for $j = 0, 1, \ldots, n$, $\langle \pi_j, \pi_k \rangle_\omega = \delta_{jk}$ for all $0 \leq j, k \leq n$ with $j \neq k$, $\langle \pi_j, p \rangle_\omega = 0$ for all $p \in \mathscr{P}_{j-1}$ and $j = 1, 2, \ldots, n$ and the polynomials form a basis of $\mathscr{P}_n$.

Problem 29.5.9

(i) Let $f \in C^0([a, b])$ and for a partition fineness $h = (b - a)/N$ let $T(h)$ be the value of the composite trapezoidal rule, that is

$$T(h) = \frac{h}{2}\left[f(a) + 2 \sum_{i=1}^{N-1} f(a + ih) + f(b) \right].$$

Show that the extrapolation $T^*(h) = (T(h) - 2^\gamma T(h/2))/(1 - 2^\gamma)$ of the values $T(h)$ and $T(h/2)$ with a suitable parameter γ leads to the composite Simpson rule.

(ii) For $f \in C^\infty([a, b])$ and $h > 0$ let $T(h) \in \mathbb{R}$ be the value of a composite quadrature formula for the partition fineness $h > 0$ with error order $\mathcal{O}(h^\gamma)$. Construct a number $T^*(h)$ using the values $T(h)$, $T(h/2)$ and $T(h/4)$ that approximates the integral of f with an error of the order $\mathcal{O}(h^{\gamma+2})$.

Problem 29.5.10 We identify periodic functions $f \in C([0, 2\pi])$ with functions on the unit circle $\partial B_1(0) \subset \mathbb{C}$ via $f(e^{i\theta}) \equiv f(\theta)$ and set

$$I(f) = \int_0^{2\pi} f(e^{i\theta})\, d\theta.$$

29.5 Numerical Integration

Integrals over sets $\partial B_r(0)$ are defined as line integrals with $\gamma(\theta) = re^{i\theta}$ via

$$\int_{\partial B_r(0)} f(w)\,dw = \int_0^{2\pi} f(\gamma(\theta))\gamma'(\theta)\,d\theta.$$

Assume that f can be extended to a holomorphic function in $B_{\tilde{r}}(0), \tilde{r} > 1$.

(i) Show that the composite trapezoidal rule is given by

$$Q^N(f) = \frac{2\pi}{N}\sum_{k=1}^N f(e^{ik2\pi/N}),$$

and that it can also be interpreted as a composite midpoint rule.

(ii) By Cauchy's integral formula we have for $z \in B_{\tilde{r}}(0)$ and $0 < r < \tilde{r}$ that

$$f(z) = \sum_{n=0}^\infty c_n z^n, \quad c_n = \frac{f^{(n)}(0)}{n!} = \frac{1}{2\pi i}\int_{\partial B_r(0)} \frac{f(w)}{(w-0)^{n+1}}\,dw.$$

Show that $I(f) = 2\pi c_0$ and $|c_n| \leq M_r r^{-n}$ with $M_r = \max_{w \in \partial B_r(0)} |f(w)|$.

(iii) Show that

$$Q^N(f) = 2\pi \sum_{\ell=0}^\infty c_{\ell N}$$

and deduce that $|Q^N(f) - I(f)| \leq 2\pi M_r r^{-N}(1 - r^{-N})^{-1} = \mathcal{O}(r^{-N})$ for $1 < r < \tilde{r}$.

Project 29.5.1 Use the composite trapezoidal and Simpson rules, as well as a composite Gaussian 3-point quadrature formula, to approximate the integrals in the interval $[0, 1]$ of the functions

$$f(x) = \sin(\pi x)e^x, \quad g(x) = x^{1/3}$$

with step sizes $h = 2^{-\ell}, \ell = 1, 2, \ldots, 10$. Calculate the error e_h in each case and determine an experimental convergence rate γ from the approach $e_h \approx c_1 h^\gamma$ and the resulting formula

$$\gamma \approx \frac{\log(e_h/e_H)}{\log(h/H)}$$

for two successive step sizes $h, H > 0$. Compare the experimental convergence rates with the theoretical convergence rates of the methods and comment on your results. Display the pairs (h, e_h) for the different quadrature formulas comparatively as polygonal chains graphically in logarithmic axis scaling using the MATLAB command `loglog`.

Project 29.5.2

(i) Taylor's formula implies that the quotients

$$d_h^+ f(x) = \frac{f(x+h) - f(x)}{h}, \quad \widehat{d}_h f(x) = \frac{f(x+h) - f(x-h)}{2h}$$

for a given step size $h > 0$ define approximations of $f'(x)$ with the error order $\mathcal{O}(h)$ and $\mathcal{O}(h^2)$ respectively. Check this property experimentally for the example $f(x) = \tan(x)$ for $x = 1/2$ with the step sizes $h = 2^{-\ell}$, $\ell = 1, 2, \ldots, 15$.

(ii) Construct by extrapolation a quotient $\widehat{d}_h^* f(x)$, that approximates the derivative $f'(x)$ up to an error of the order $\mathcal{O}(h^4)$ and repeat the calculations. What are the advantages and disadvantages of approximating the derivative using extrapolation?

29.6 Nonlinear Problems

Problem 29.6.1

(i) Calculate three steps of the Newton method for the function $f(x) = \arctan(x)$ with the initial values $x_0 = 1, 3/2, 2$.
(ii) Repeat the calculations for the damped Newton method $x_{k+1} = x_k - \omega f(x_k)/f'(x_k)$ with the damping parameters $\omega = 1/2, 3/4$.

Problem 29.6.2 Let $f \in C^1(\mathbb{R})$ be convex, that is for all $x, y \in \mathbb{R}$ and $t \in [0, 1]$ we have that $f(tx + (1-t)y) \leq tf(x) + (1-t)f(y)$, as well as strictly monotonically increasing and let $x^* \in \mathbb{R}$ satisfy $f(x^*) = 0$. Show that the Newton method converges for every initial value $x_0 \in \mathbb{R}$.

Problem 29.6.3 Formulate sufficient conditions for the global convergence of the damped Newton method, by considering it as a fixed point iteration with the mapping $\Phi(x) = x - \omega Df(x)^{-1} f(x)$.

Problem 29.6.4

(i) Let $a, b \in \mathbb{R}$ with $a < b$. Construct points $c, d \in (a, b)$ with $c < d$, such that for the interval lengths $\ell_1 = c - a$, $\ell_2 = b - c$ and $\ell_3 = d - c$ the relations

$$\frac{\ell_3}{\ell_1} = \frac{\ell_1}{\ell_2}, \quad \frac{\ell_3}{\ell_2 - \ell_3} = \frac{\ell_1}{\ell_2}$$

hold. Consider the size $g = \ell_2/\ell_1$.

29.6 Nonlinear Problems

(ii) Based on the previous construction, formulate an interval reduction method for the approximation of minimum points, in which only one function evaluation is necessary per iteration step and the interval length is always reduced by the same factor.

Problem 29.6.5

(i) Let $g \in C^1(\mathbb{R}^n)$, $x \in \mathbb{R}^n$ and $\sigma \in (0, 1)$. Show that a number $\alpha > 0$ exists, so that with $d = -\nabla g(x) \neq 0$ we have

$$g(x + \alpha d) < g(x) - \sigma \alpha \|d\|^2.$$

(ii) Let $g \in C^2(N_g(x_0))$. Show that the search directions d_k of the descent method converge to zero when a fixed step size $0 < \alpha < 1/\gamma$ with $\gamma = \max_{x \in N_g(x_0)} \|D^2 g(x)\|$ is used instead of the Armijo condition.

Problem 29.6.6 The Heron method approximates the square root $a^{1/2}$ of a number $a \geq 0$ through the iteration $x_{k+1} = \Phi(x_k)$ with the function $\Phi(x) = (x + a/x)/2$.

(i) Show that Φ is a contraction in the interval $((a/2)^{1/2}, \infty)$.
(ii) Show that the Heron method coincides with the Newton method for the function $x \mapsto x^2 - a$ and investigate sufficient conditions for the local, quadratic convergence of the method.
(iii) Show that the Heron method can be interpreted as a descent method for the function $g(x) = x + a/x$.

Problem 29.6.7

(i) With $f_0 = f_1 = 1$ the sequence of Fibonacci numbers is defined by $f_k = f_{k-1} + f_{k-2}$ for all $k \geq 2$. Let α be the positive solution of the equation $x^2 = 1 + x$. Show that $\alpha^{k-1} \leq f_k \leq \alpha^k$ for all $k \geq 0$.
(ii) Let $(e_k)_{k \in \mathbb{N}_0}$ be a sequence of positive real numbers such that $e_0, e_1 < 1$ and $e_{k+2} \leq e_{k+1} e_k$ for all $k \geq 0$. Show that the sequence $(e_k)_{k \in \mathbb{N}_0}$ is dominated by a sequence $(\delta_k)_{k \geq 0}$ that converges to zero of order α, i.e. we have $e_k \leq \delta_k$ for all $k \in \mathbb{N}$ and there exists a $q \in \mathbb{R}$ with

$$\limsup_{k \to \infty} \delta_{k+1}/\delta_k^\alpha = q.$$

Problem 29.6.8

(i) Discuss the well-posedness of the secant method.
(ii) Show that for the approximation errors $e_k = x^* - x_k$ of the iterates of the secant method the relation

$$\frac{e_{k+1}}{e_k e_{k-1}} = \frac{g(x_k) - g(x_{k-1})}{f(x_k) - f(x_{k-1})}$$

holds with the function $g(x) = -f(x)/(x-x^*)$, provided both sides are well-defined.

(iii) Under what conditions is the right-hand side in the identity for $e_{k+1}/(e_k e_{k-1})$ bounded and what can be inferred about the convergence of the method?

Problem 29.6.9 Show that the polynomial $p(x) = x^3 - 2x^2 - 1$ has exactly one root $x^* \geq 2$ and justify the fixed point equation $\Phi(x^*) = x^*$ with $\Phi(x) = 2 + 1/x^2$. Prove that Φ is a contraction on $[2, \infty) \subset \mathbb{R}$ and calculate three steps of the fixed point iteration. What accuracy is achieved after 3 steps? How many steps are needed to achieve an accuracy of 10^{-6}?

Problem 29.6.10 For given $g \in C^2(\mathbb{R}^n)$, $x_k \in \mathbb{R}^n$ and $H_k \in \mathbb{R}^{n \times n}$ define $q_k : \mathbb{R}^n \to \mathbb{R}, d \mapsto g(x_k) + \nabla g(x_k) \cdot d + (1/2) d^\mathsf{T} H_k d$.

(i) Provide sufficient conditions for the existence of a unique minimum point $d_k \in \mathbb{R}^n$ of q_k.
(ii) Show that the iteration $x_{k+1} = x_k + d_k$ corresponds to the Newton method and the descent method with fixed step size $\alpha_k = \alpha$, provided $H_k = D^2 g(x_k)$ or $H_k = \alpha I$ is used.
(iii) Interpret the iteration geometrically.

Project 29.6.1

(i) Experimentally investigate the convergence of Heron's method for calculating a square root, i.e. the iteration rule $x_{k+1} = (x_k + a/x_k)/2$, for $a = 3/2$ and various initial values $x_0 \in \mathbb{R}$.
(ii) Repeat the execution of the commands sqrt(a) and (a^0.5) or pow(a,0.5) 10^8 times with $a = 3/2$ and discuss reasons for possible differences in runtimes.
(iii) For a holomorphic function $f : \mathbb{C} \to \mathbb{C}$ with zeros $z_1, z_2, \ldots, z_n \in \mathbb{C}$, the complex plane can be partitioned into basins of attraction $E_j \subset \mathbb{C}$, which for $j = 1, 2, \ldots, n$, are defined by

$$E_j = \{z \in \mathbb{C} : \text{Newton's method with initial value } z \text{ converges to } z_j\}$$

as well as the remainder $X = \mathbb{C} \setminus \cup_{j=1}^n E_j$. Consider the function $f(z) = z^3 - 1$ and use as initial values grid points $z_\ell = x_\ell + i y_\ell$ in the range $[-1, 1]^2 \subset \mathbb{C}^2$, which are arranged at a distance $h = 1/200$. Mark the points differently according to their belonging to the basin of attraction of a zero and display them graphically. Use the MATLAB commands shown in Fig. 29.1 with a suitably defined matrix C.

```
1  [X,Y] = meshgrid(-1:h:1,-1:h:1);
2  scatter(X(:),Y(:),15,C(:));
```

Fig. 29.1 Representation of differently coloured points

Project 29.6.2

(i) Implement the Newton and secant methods for finding the roots of a function $f : \mathbb{R} \to \mathbb{R}$ in MATLAB and test it with the function $f(x) = \exp(x) + x^2 - 2$, the starting value $x_0 \in \{-1, 0, 1\}$ and the termination criterion $|x_{k+1} - x_k| \leq 10^{-12}$. Terminate the Newton method after 100 iterations if the termination criterion is not met. Compare the number of iterations and the number of correct decimal places for the iteration steps.

(ii) Realise the root finding of f by a descent method for the function $g(x) = |f(x)|^2$ and compare the convergence speed with that of the Newton method.

(iii) Use the Newton method to approximate a root of the mapping

$$f : \mathbb{R}^3 \to \mathbb{R}^3, \quad (x_1, x_2, x_3) \mapsto \left(x_1^2 + x_2^2 - e,\ 3x_2 + 4x_3 - \sqrt{5},\ x_1^2 - \pi/4\right)$$

How can the solvability be assessed and a meaningful starting value be constructed?

29.7 Conjugate Gradients Method

Problem 29.7.1 For $A \in \mathbb{R}^{n \times n}$ and $x, y \in \mathbb{R}^n$, let $\|x\|_A = (x \cdot (Ax))^{1/2}$ and $\langle x, y \rangle_A = (Ax) \cdot y$. Show that $(x, y) \mapsto \langle x, y \rangle_A$ defines a scalar product that induces the norm $\|\cdot\|_A$ if and only if A is symmetric and positive definite.

Problem 29.7.2 Let $A \in \mathbb{R}^{n \times n}$ be symmetric and positive definite with eigenvalues $0 < \lambda_1 \leq \lambda_2 \leq \cdots \leq \lambda_n$. Show that for all $x \in \mathbb{R}^n \setminus \{0\}$ the inequality

$$\frac{(x \cdot Ax)(x \cdot A^{-1}x)}{\|x\|^4} \leq \frac{(\lambda_1 + \lambda_n)^2}{4\lambda_1 \lambda_n}$$

holds. Consider the case $\lambda_1 \lambda_n = 1$ first and use the diagonalisation $A = Q^T D Q$ and the elementary inequality $ab \leq (a+b)^2/4$.

Problem 29.7.3 Let $b \in \mathbb{R}^n$, let $A \in \mathbb{R}^{n \times n}$ be symmetric and positive definite and let $\phi(x) = \left(A^{-1}(b - Ax)\right) \cdot (b - Ax)$ for all $x \in \mathbb{R}^n$. For an approximation $\tilde{x} \in \mathbb{R}^n$, the descent method uses the search direction $\tilde{d} = -\nabla \phi(\tilde{x})$.

(i) Show that $\tilde{d} = b - A\tilde{x}$ holds and determine the minimum point $\tilde{\alpha}$ of the function $t \mapsto \phi(\tilde{x} + t\tilde{d})$.
(ii) Show that with the optimal $\tilde{\alpha}$ and $\tilde{x}^{new} = \tilde{x} + \tilde{\alpha}\tilde{d}$ we have

$$\|\tilde{x}^{new} - x^*\|_A^2 = \|\tilde{x} - x^*\|_A^2 \left(1 - \frac{\|\tilde{d}\|^4}{(\tilde{d} \cdot A\tilde{d})(\tilde{d} \cdot A^{-1}\tilde{d})}\right).$$

(iii) Let $\kappa = \mathrm{cond}_2(A) = \lambda_{min}^{-1}\lambda_{max}$ be the condition number of A. Use without proof the estimate valid for all $x \in \mathbb{R}^n \setminus \{0\}$

$$\frac{(x \cdot Ax)(x \cdot A^{-1}x)}{\|x\|^4} \leq \frac{(\lambda_{min} + \lambda_{max})^2}{4\lambda_{min}\lambda_{max}},$$

to prove that

$$\|\tilde{x}^{new} - x^*\|_A \leq \left(\frac{\kappa - 1}{\kappa + 1}\right)\|\tilde{x} - x^*\|_A.$$

Problem 29.7.4

(i) Show that the function $T_k(t) = \cos(k \arccos t)$, $t \in [-1, 1]$, can be uniquely extended as a polynomial on $\mathbb{R}$ and for $|t| \geq 1$ we have

$$T_k(t) = \frac{1}{2}\left(t + (t^2 - 1)^{1/2}\right)^k + \frac{1}{2}\left(t - (t^2 - 1)^{1/2}\right)^k.$$

(ii) Show that for all $s > 1$ we have

$$\frac{1}{2}\left(\frac{s^{1/2} + 1}{s^{1/2} - 1}\right)^k \leq T_k\left(\frac{s + 1}{s - 1}\right) \leq \left(\frac{s^{1/2} + 1}{s^{1/2} - 1}\right)^k.$$

Problem 29.7.5 Let $0 < a < b$ and $k \geq 0$. Show that the problem

$$\min\left\{\max_{t \in [a,b]} |p(t)| : p \in \mathscr{P}_k, \; p(0) = 1\right\}$$

has the unique solution

$$q(t) = T_k\left(\frac{a + b - 2t}{b - a}\right) / T_k\left(\frac{a + b}{b - a}\right)$$

where T_k is the k-th Chebyshev polynomial.
Hint: Assume that the statement is false and consider the zeros and extreme values of the difference $r = q - p$ for a suitable polynomial $p \in \mathscr{P}_k$.

Problem 29.7.6 Use the CG method to determine a solution to the system of linear equations $Ax = b$ defined by

$$A = \begin{bmatrix} 2 & -1 & 0 & 0 \\ -1 & 2 & -1 & 0 \\ 0 & -1 & 2 & -1 \\ 0 & 0 & -1 & 2 \end{bmatrix}, \quad b = \begin{bmatrix} 1 \\ 1 \\ 1 \\ 1 \end{bmatrix}$$

29.7 Conjugate Gradients Method

Start with $x_0 = [1, 0, 1, 0]^T$, calculate the Krylov space $\mathcal{K}_2(A, r_0) = \text{span}\{r_0, Ar_0\}$ and compare it with the space $\text{span}\{d_0, d_1\}$. Also verify that A is positive definite.

Problem 29.7.7

(i) Deduce from known statements that for the approximate solutions $(x_k)_{k=0,1,\ldots}$ calculated with the descent and the CG method, an estimate

$$\|x^* - x_k\|_A \leq cq^k \|x^* - x_0\|_A$$

with $c = 1, q = 1 - 2\,\text{cond}_2(A)^{-1} + 2\xi$ and $c = 2, q = 1 - 2\,\text{cond}_2(A)^{-1/2} + 2\zeta$ with numbers $0 \leq \xi \leq \text{cond}_2(A)^{-2}$ and $0 \leq \zeta \leq \text{cond}_2(A)^{-1}$ applies.

(ii) Show that $\log(1 + s) \approx s$ for $|s| \ll 1$.

(iii) For $\varepsilon_{stop} > 0$, let $M_\varepsilon = |\log(\varepsilon_{stop})|$. Conclude that with the descent and the CG method, about $M_\varepsilon \,\text{cond}(A)$ or $M_\varepsilon \,\text{cond}(A)^{1/2}$ many iterations are needed to meet the termination criterion $\|x^* - x_k\|_A \leq \varepsilon_{stop}$, if $\text{cond}_2(A) \gg 1$ applies.

Problem 29.7.8 Let $A \in \mathbb{R}^{n \times n}$ be symmetric and positive definite. For A-conjugate vectors $d_0, d_1, \ldots, d_{k-1} \in \mathbb{R}^n \setminus \{0\}$ and $b \in \mathbb{R}^n$, let $f : \mathbb{R}^k \to \mathbb{R}$ be defined by

$$f(\alpha_0, \alpha_1, \ldots, \alpha_{k-1}) = \frac{1}{2}\left\|b - A\left(x_0 + \sum_{i=0}^{k-1} \alpha_i d_i\right)\right\|_{A^{-1}}^2.$$

Calculate $\nabla f(\alpha_0, \alpha_1, \ldots, \alpha_{k-1})$.

Problem 29.7.9 Modify the Gram-Schmidt orthogonalisation process to determine for a given symmetric and positive definite matrix $A \in \mathbb{R}^{n \times n}$ a family $(d_i : i = 0, 1, \ldots, n-1)$ of non-vanishing A-conjugate vectors.

Problem 29.7.10 Let $A \in \mathbb{R}^{n \times n}$ be symmetric and positive definite. For $x \in \mathbb{R}^n$, let $\phi(x) = \|b - Ax\|_{A^{-1}}^2 / 2$ and $x^* \in \mathbb{R}^n$ satisfy $Ax^* = b$.

(i) Prove $\phi(x) - \phi(x^*) = \|x - x^*\|_A^2 / 2$ and $\nabla \phi(x) = -(b - Ax)$.

(ii) Show that $d = -\nabla \phi(x)$ is orthogonal to the level set $N_a \phi = \{y \in \mathbb{R}^n : \phi(y) = a\}$ for $a = \phi(x)$ at the point x, i.e. for every C^1-curve $c : (-\varepsilon, \varepsilon) \to \mathbb{R}^n$ with $c(t) \in N_a \phi$ for all $t \in (-\varepsilon, \varepsilon)$ and $c(0) = x$, we have $c'(0) \cdot d = 0$.

Project 29.7.1 Implement the CG and the descent method for the approximate solution of the system $Ax = b$. Compare the number of iterations required by the two methods using the example

$$A = \begin{bmatrix} 2 & -1 & & \\ -1 & \ddots & \ddots & \\ & \ddots & \ddots & -1 \\ & & -1 & 2 \end{bmatrix} \in \mathbb{R}^{n \times n}, \quad b = h^2 \begin{bmatrix} 1 \\ 1 \\ \vdots \\ 1 \end{bmatrix} \in \mathbb{R}^n$$

with $n = 10^s$, $s = 1, 2, \ldots, 5$, and $h = 2/(n+1)$. Choose the termination criterion $\|b - Ax_k\| \leq h$ and as initial value $x_0 = [0, 0, \ldots, 0]^T \in \mathbb{R}^n$. Calculate the quotient of the norms for two consecutive residuals, display these in a table and comment on your results. Visualise the numerical solution graphically using the command `plot([-1:h:1],[0,x',0])`. The curve should approximate a function $u : [-1, 1] \to \mathbb{R}$ for which $-u'' = 1$ and $u(-1) = u(1) = 0$ hold.

Project 29.7.2 For $n \geq 1$, the Hilbert matrix $H \in \mathbb{R}^{n \times n}$ is defined by the entries $h_{ij} = 1/(i + j - 1)$. The matrix H is symmetric and positive definite but ill conditioned. In MATLAB it can be generated with the command `hilb(n)`.

(i) Use the MATLAB routine cond to approximately determine the condition number of the Hilbert matrix for $n = 10^s$, $s = 1, 2, \ldots, 3$, and experimentally verify that $\text{cond}(H) = \mathcal{O}((1 + \sqrt{2})^{4n}/\sqrt{n})$ holds.

(ii) Implement the CG method and use it to solve the systems of equations $Hx = b$ with $b_i = \sum_{j=1}^n h_{ij}$, $i = 1, 2, \ldots, n$, for $n = 10^s$, $s = 1, 2, \ldots, 4$, with the initial vector $x = [0, 0, \ldots, 0]^T \in \mathbb{R}^n$ and assess to what extent the convergence statement for the CG method is sharp.

29.8 Sparse Matrices and Preconditioning

Problem 29.8.1 Show that if $A \in \mathbb{R}^{n \times n}$ is a band matrix with bandwidth $w \in \mathbb{N}$, i.e. $a_{ij} = 0$ for $|i - j| > w$, then the factors of the LU and Cholesky decompositions are also band matrices with bandwidth w, provided they exist.

Problem 29.8.2

(i) Let $A, B \in \mathbb{R}^{n \times n}$ be sparse matrices and $b \in \mathbb{R}^n$. Construct as efficient as possible algorithms for the calculation of AB and Ab and determine their complexity.
(ii) Show that the product of two sparse matrices is generally not sparse.

Problem 29.8.3 Let $A \in \mathbb{R}^{n \times n}$ with $n = w^2$ for a $w \in \mathbb{N}$ be defined by

$$a_{ij} = \begin{cases} 8, & |i - j| = 0, \\ 1, & |i - j| \in \{1, w\}. \end{cases}$$

(i) Show that A is a sparse band matrix.
(ii) Show that A has a Cholesky decomposition, the factors of which are not sparse.

Problem 29.8.4 Represent the matrix

29.8 Sparse Matrices and Preconditioning

$$A = \begin{bmatrix} 1 & 0 & 0 & 3 & 4 \\ 0 & 2 & 5 & 0 & 1 \\ 4 & 0 & 0 & 1 & 3 \\ 2 & 0 & 1 & 0 & 0 \\ 0 & 0 & 7 & 6 & 0 \end{bmatrix}$$

in coordinate and CCS format and calculate Ax for $x = [1, 2, \ldots, 5]^\mathsf{T}$ using the coordinate vectors.

Problem 29.8.5 Let $A, C \in \mathbb{R}^{n \times n}$ be symmetric and positive definite.

(i) Show that the product CA is generally neither symmetric nor positive definite.
(ii) Show that CA is positive definite with respect to the scalar product $(x, y) \mapsto (Cx) \cdot y$.

Problem 29.8.6 Let $A \in \mathbb{R}^{n \times n}$ be defined by $a_{ii} = 2$ for $i = 1, 2, \ldots, n$ and $a_{ij} = -1$ for $i, j = 1, 2, \ldots, n$ with $|i - j| = 1$ and let $b = [1, 1, \ldots, 1]^\mathsf{T} \in \mathbb{R}^n$. For $n = 5$, perform as many iterations of the Gauss-Seidel method until the first two decimal places of the entries of the solution vector no longer change. Use the sparsity of the matrix A to perform the matrix-vector multiplications as efficiently as possible.

Problem 29.8.7 Determine the condition numbers with respect to row sum norm of the matrix

$$A = \begin{bmatrix} 1 & 2 \\ 3 & 4 \end{bmatrix}$$

and its row equilibration CA.

Problem 29.8.8 Investigate whether the row equilibration CA of a matrix A can be formulated in the form $L^\mathsf{T} A L$ and whether this also leads to a reduction of the condition number.

Problem 29.8.9

(i) Let $A \in \mathbb{R}^{n \times n}$ be symmetric and positive definite. Determine a Cholesky decomposition $C_{SGS} = VV^\mathsf{T}$ of the symmetric Gauss-Seidel preconditioning matrix $C_{SGS} = [(L + D)D^{-1}(D + L)]^{-1}$ with the decomposition $A = L + D + L^\mathsf{T}$ of A into diagonal and lower and upper parts.
(ii) Show that $A - C_{SGS}^{-1} = -LD^{-1}L^\mathsf{T}$ holds.
(iii) Calculate the difference $A - C_{SGS}^{-1}$ for

$$A = \begin{bmatrix} 2 & -1 & 0 & 0 \\ -1 & 2 & -1 & 0 \\ 0 & -1 & 2 & -1 \\ 0 & 0 & -1 & 2 \end{bmatrix}.$$

```
1 U = zeros(m+2,m+2); U(2:m+1,2:m+1) = reshape(x,m,m)';
2 dx = 1/(m+1); mesh(0:dx:1,0:dx:1,U);
```

Fig. 29.2 Plotting a function defined by a matrix U

Problem 29.8.10 Let $A, M \in \mathbb{R}^{n \times n}$ be regular with the property that $\|I - MA\| = \delta < 1$ with respect to a suitable operator norm on $\mathbb{R}^{n \times n}$. Show that the estimates $\|MA\| \leq 1 + \delta$ and $\|(MA)^{-1}\| \leq 1/(1 - \delta)$ hold and deduce $\text{cond}(MA) \leq (1 + \delta)/(1 - \delta)$.

Project 29.8.1 Implement the preconditioned CG method and test it for the linear system $Ax = b$, where $A \in \mathbb{R}^{n \times n}$ with $n = m^2$ and $T_m \in \mathbb{R}^{m \times m}$ are defined by

$$A = \begin{bmatrix} T_m & -I_m & & \\ -I_m & \ddots & \ddots & \\ & \ddots & \ddots & -I_m \\ & & -I_m & T_m \end{bmatrix}, \quad T_m = \begin{bmatrix} 4 & -1 & & \\ -1 & \ddots & \ddots & \\ & \ddots & \ddots & -1 \\ & & -1 & 4 \end{bmatrix},$$

and $b \in \mathbb{R}^n$ is given by $b = (m + 1)^{-2}[1, 1, \ldots, 1]^\mathsf{T}$. Use the preconditioning by row equilibration, incomplete Cholesky decompositions of different bandwidths and the symmetric Gauss-Seidel preconditioning. Compare the iteration numbers for $m = 2^s \cdot 10$, $s = 0, 1, \ldots, 4$, and the termination parameter $\varepsilon_{stop} = (m + 1)^{-2}/10$. Visualise the solution $x \in \mathbb{R}^{m^2}$ of the equation system using the commands shown in Fig. 29.2. A smooth function in the domain $(0, 1)^2$ should be displayed, which vanishes on the boundary.

Project 29.8.2

(i) Define the matrices A = eye(n) and B = speye(n) in MATLAB and calculate A*x and B*x for x = ones(n,1). Measure the time required for the dimensions $n = 10^s$, $s = 1, 2, \ldots, 5$. Explain any differences.
(ii) Construct using the MATLAB commands sparse and spdiags the band matrix $A \in \mathbb{R}^{n \times n}$ with $n = w^2$ for $w \in \mathbb{N}$ and $a_{ii} = 8$ and $a_{ij} = 1$ for $|i - j| \in \{1, w\}$. Check the occupancy structure of the matrix using the command spy(A) for various numbers w. Solve the linear system $Ax = b$ with $b = [1, 1, \ldots, 1]^\mathsf{T}$ for $w = 10^2$ and repeat this after executing the command A = full(A). Comment on your observations.

29.9 Multidimensional Approximation

Problem 29.9.1 Show that the simplex $T = \text{conv}\{z_0, z_1, \ldots, z_d\}$ defined by the corners $z_0, z_1, \ldots, z_d \in \mathbb{R}^d$ is non-degenerate if and only if the vectors $z_i - z_0$ for

29.9 Multidimensional Approximation

$i = 1, 2, \ldots, d$ are linearly independent and in this case, the volume is given by the absolute value of $\det[z_1 - z_0, z_2 - z_0, \ldots, z_d - z_0]/d!$.

Problem 29.9.2 Show that with the reference triangle $\widehat{T} = \text{conv}\{0, e_1, e_2\}$ for $j, k \geq 0$ we have

$$\int_{\widehat{T}} s^j t^k \, d(s, t) = \frac{j! k!}{(j + k + 2)!}$$

and deduce the exactness of the partial degree 2 of the quadrature formula defined by

$$\widehat{\xi} = \frac{1}{6} \begin{bmatrix} 1 & 4 & 1 \\ 1 & 1 & 4 \end{bmatrix}^\mathsf{T}, \quad \widehat{w} = \frac{1}{6}[1, 1, 1]^\mathsf{T}$$

Problem 29.9.3 For a non-degenerate simplex $T \subset \mathbb{R}^d$ and functions $f, g \in C^1(T)$, let $\widehat{f}, \widehat{g} \in C^1(\widehat{T})$ be defined on the reference simplex $\widehat{T} \subset \mathbb{R}^d$ by $\widehat{f} = f \circ \Phi_T$ and $\widehat{g} = g \circ \Phi_T$ with an affine-linear diffeomorphism $\Phi_T : \widehat{T} \to T$. Show that

$$\int_T \nabla f \cdot \nabla g \, dx = \det D\Phi_T \int_{\widehat{T}} \nabla \widehat{f} \cdot \left(D\Phi^\mathsf{T} D\Phi\right)^{-1} \nabla \widehat{g} \, d\widehat{x}.$$

Problem 29.9.4 For $n \in \mathbb{N}$, let $\omega^k = (e^{ijk2\pi/n})_{j=0,\ldots,n-1} \in \mathbb{C}^n$ for $k = 0, 1, \ldots, n$ and $T_n = (e^{ijk2\pi/n})_{j,k=0,\ldots,n-1} \in \mathbb{C}^{n \times n}$.

(i) Show that an orthogonal basis is defined by $E^{k\ell} = \omega^k(\omega^\ell)^\mathsf{T}$ for $k, \ell = 0, 1, \ldots, n - 1$ with respect to the matrix scalar product defined by $E : F = \sum_{j_1, j_2=0}^{n-1} E_{j_1 j_2} \overline{F}_{j_1 j_2}$, by proving $E : F = \text{tr}(E \overline{F}^\mathsf{T})$.

(ii) Show that for any matrix $Y \in \mathbb{C}^{n \times n}$ and $B = (b_{k\ell})_{k,\ell=0,\ldots,n-1} = T_n Y T_n$ we have

$$Y = \sum_{k,\ell=0}^{n-1} b_{k\ell} E^{k\ell}.$$

Problem 29.9.5 For the matrix $F = (f_{jk})_{j,k=0,1} \in \mathbb{R}^{2 \times 2}$, let the vector $f \in \mathbb{R}^4$ be defined by $f = [f_{00}, f_{01}, f_{10}, f_{11}]^\mathsf{T} \in \mathbb{R}^4$. Show that the two-dimensional Fourier transform of F and the one-dimensional Fourier transform of f lead to different results.

Problem 29.9.6 For a non-degenerate simplex $T = \text{conv}\{z_0, z_1, \ldots, z_d\}$ and $i = 0, 1, \ldots, d$ let $\varphi_i : T \to \mathbb{R}$ be the affine-linear function with the property $\varphi_i(z_j) = $

δ_{ij}, $j = 0, 1, \ldots, d$. Show that

$$\varphi_i(x) = \frac{\det \begin{bmatrix} 1 & \ldots & 1 & 1 & 1 & \ldots & 1 \\ z_0 & \ldots & z_{i-1} & x & z_{i+1} & \ldots & z_n \end{bmatrix}}{\det \begin{bmatrix} 1 & \ldots & 1 & 1 & 1 & \ldots & 1 \\ z_0 & \ldots & z_{i-1} & z_i & z_{i+1} & \ldots & z_n \end{bmatrix}}.$$

Problem 29.9.7 Let $(\mathscr{T}_n)$ be a sequence of triangulations of the domain $\Omega \subset \mathbb{R}^2$ with maximum mesh widths $h_n > 0$, for which $h_n \to 0$ holds as $n \to \infty$. Furthermore, assume that for all interior angles α of the triangles in $\mathscr{T}_n$ the estimate $\alpha \geq \alpha_0 > 0$ holds with a constant α_0 independent of n. Show that a constant $K \in \mathbb{N}$ exists, independent of n, such that each triangle in $\mathscr{T}_n$ has at most K neighbours.

Problem 29.9.8 Let $\mathscr{T}_h$ be a triangulation of $\Omega \subset \mathbb{R}^d$, $f \in C^1(\overline{\Omega})$ and $\mathscr{I}_h f \in \mathscr{S}^{1,0}(\mathscr{T}_h)$ the nodal interpolant of f. Show that

$$\|f - \mathscr{I}_h f\|_{C^0(\overline{\Omega})} \leq h \|\nabla f\|_{C^0(\overline{\Omega})}.$$

Problem 29.9.9 Let $Q : C^0([0,1]) \to \mathbb{R}$ be a quadrature formula with non-negative weights and points $(w_i, t_i)_{i=0,\ldots,n}$ and degree of exactness $k \geq 0$, and let $Q^d : C^0([0,1]^d) \to \mathbb{R}$ be defined by

$$Q^d(f) = \sum_{i_1=0}^{n} \sum_{i_2=0}^{n} \cdots \sum_{i_d=0}^{n} w_{i_1} w_{i_2} \ldots w_{i_d} f(t_{i_1}, t_{i_2}, \ldots, t_{i_d}).$$

Show for the case $d = 3$, that

$$\left| I^d(f) - Q^d(f) \right| \leq \sum_{i=1}^{d} \sup_{\widehat{x}_i \in [0,1]^{d-1}} \left| I f_{\widehat{x}_i} - Q f_{\widehat{x}_i} \right|,$$

where $f_{\widehat{x}_i}$ for $\widehat{x}_i = (x_1, \ldots, x_{i-1}, x_{i+1}, \ldots, x_d) \in [0,1]^{d-1}$ denotes the mapping

$$t \mapsto f(x_1, \ldots, x_{i-1}, t, x_{i+1}, \ldots, x_d).$$

Problem 29.9.10 Let $T = \text{conv}\{z_0, z_1, \ldots, z_d\} \subset \mathbb{R}^d$, $d \in \{2, 3\}$, be a non-degenerate simplex and let $\varphi_0 : T \to \mathbb{R}$ be the hat function associated with the vertex z_0. Further, let S_0 be the side of the triangle or tetrahedron opposite the node z_0 and let n_0 be the outer unit normal to T on S_0. Show that

$$\nabla \varphi_0 = \frac{-|S_0|}{d|T|} n_0$$

holds with the area or volume $|T|$ of T and the length or area $|S_0|$ of S_0.

29.9 Multidimensional Approximation

Project 29.9.1 For $d \in \mathbb{N}$ and a function $f \in C^0([0, 1]^d)$, its integral on the cube $[0, 1]^d$ should be determined numerically.

(i) Write a routine that implements the iterated trapezoidal rule Q_{Trap}^d and test it for $d = 5$ and the functions

$$f_1(x) = \prod_{i=1}^{d} x_i^2, \quad f_2(x) = \sin(x_1 x_2 \ldots x_d).$$

Verify the order of convergence in the case of f_1 and determine the computational effort.

(ii) For uniformly and independently distributed random variables $\xi^1, \xi^2, \ldots, \xi^N \in [0, 1]^d$ a *Monte-Carlo quadrature formula* is defined by

$$Q_{MC}^N(f) = \frac{1}{N} \sum_{i=1}^{N} f(\xi^i).$$

It can be shown that the expected value of $|I^d(f) - Q_{MC}^N(f)|$ is of the order $\mathcal{O}(N^{-1/2})$. Verify this convergence behaviour with the above examples and determine the computational effort of $Q_{MC}^N(f)$. Realisations of suitable pseudo-random variables can be generated with the MATLAB command `rand(d,1)`.

(iii) Discuss in which situations the use of an iterated or a Monte Carlo quadrature formula is advantageous.

Project 29.9.2 A common format for storing triangulations consists of a list $Z \in \mathbb{R}^{N \times d}$ with the coordinates of the nodes $z_1, z_2, \ldots, z_N \in \mathbb{R}^d$, which also defines a numbering of the nodes, and a list $T \in \mathbb{R}^{L \times (d+1)}$, which contains the numbers of the nodes of the individual triangles or tetrahedra $T_1, T_2, \ldots, T_L$.

(i) Write a program that performs a uniform refinement of a given triangulation of a two-dimensional domain in the above format. Each triangle should be divided into four congruent sub-triangles by bisecting its sides, as shown in Fig. 29.3. Test your routine on two simple examples. You can visualise triangulations in MATLAB with the command `trimesh(T,Z(:,1),Z(:,2))`.

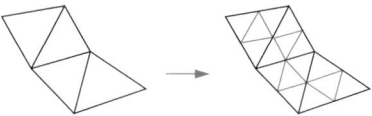

Fig. 29.3 Refinement of a triangulation by bisecting the edges

(ii) Implement a composite quadrature formula for a given triangulation of a domain $\Omega \subset \mathbb{R}^2$ that uses a Gaussian 5-point quadrature formula on each triangle. Experimentally verify the exactness and convergence properties of the formula using a sequence of uniformly refined triangulations and the function $f(x_1, x_2) = \sin(2\pi x_1)\sin(2\pi x_2)$ in the domain $\Omega = (-1, 1)^2 \setminus (0, 1)^2$.

Chapter 30
Problems on Numerics for Differential Equations

30.1 Ordinary Differential Equations

Problem 30.1.1

(i) Show that the function $y(t) = G^{-1}(F(t) + c)$ resulting from the formal equivalence

$$\frac{dy}{dt} = f(t)g(y) \iff \frac{1}{g(y)}dy = f(t)dt \iff \int \frac{1}{g(y)} = \int f(t)$$

with antiderivatives $G(y)$ of $1/g(y)$ and $F(t)+c$ of $f(t)$ solves the differential equation $y' = f(t)g(y)$ and discuss sufficient conditions for the well-posedness of this representation.

(ii) How can initial conditions be taken into account and to what extent is the solution unique?

(iii) Construct a non-trivial solution to the initial value problem $y' = y^{2/3}$, $y(0) = 0$.

Problem 30.1.2 Justify that the two-body problem for describing the flight altitude z of a body near the Earth's surface is described by the equation $z'' = -g$, where $g \approx 9.812 \, \text{m/s}^2$ is the acceleration due to gravity. Use the sizes $m_{Earth} \approx 5.974 \cdot 10^{24}$ kg and $r_{Earth} \approx 6.371 \cdot 10^6$ m.

Problem 30.1.3 We consider the string pendulum of length $\ell > 0$, sketched in Fig. 30.1, at the end of which a weight of mass m is attached. Determine the tangential acceleration a_{tan} to show that the deflection angle $\phi : [0, T] \to \mathbb{R}$ can be described by the differential equation $\phi'' = -(g/\ell)\sin(\phi)$ with the acceleration due to gravity g if friction effects are neglected. Simplify the differential equation for small angles and derive the solution of the resulting equation.

© The Author(s), under exclusive license to Springer-Verlag GmbH, DE, part of Springer Nature 2025
S. Bartels, *Numerical Mathematics 3x9*, La Matematica per il 3+2 160,
https://doi.org/10.1007/978-3-662-70890-3_30

Fig. 30.1 Mathematical description of a string pendulum

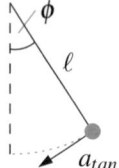

Problem 30.1.4 Sketch the phase diagram of the predator-prey model

$$y'_1 = \alpha(1 - y_2)y_1, \quad y'_2 = \beta(y_1 - 1)y_2,$$

in the range $[0, 5]^2$ for the parameters $\alpha = 1$ and $\beta = 1$. Use this to deduce the occurrence of periodic solutions and the positivity of solutions for suitable initial data.

Problem 30.1.5 Sketch the phase diagram for the equation of the undamped string pendulum $\phi'' = -(g/\ell)\sin(\phi)$, by writing the differential equation as a first order system. Draw various solution curves into the diagram and interpret them physically.

Problem 30.1.6 For a natural number $n \geq 2$ let y be a solution of the differential equation $y' = f(t)y + g(t)y^n$. Show that the function $z = y^{1-n}$ satisfies a differential equation that can be solved using the method of variation of constants.

Problem 30.1.7 Let $\widehat{y}$ be a solution of the differential equation $y' = f(t)y + g(t)y^2 + h$. Show that with every solution z of the differential equation $z' = -(f(t) + 2\widehat{y}(t)g(t))z - g(t)$ and the formula $z = 1/(y - \widehat{y})$ further solutions of the first differential equation can be obtained. To what extent is this observation useful?

Problem 30.1.8 Construct the solution of the initial value problem $my'' + ry' + D(y - \ell) = 0$, $y(0) = \ell$, $y'(0) = v_0$, which describes the deflection of a spring pendulum of length ℓ. Use the approach $y(t) = cz(t) + \ell$, where $z(t) = e^{\lambda t}$ for a $\lambda \in \mathbb{C}$. Discuss qualitative properties of solutions for different ratios of D and r.

Problem 30.1.9

(i) Let $A \in \mathbb{R}^{n \times n}$ be diagonalisable, i.e. there exist a diagonal matrix $D \in \mathbb{R}^{n \times n}$ and a regular matrix $R \in \mathbb{R}^{n \times n}$, such that $A = R^{-1}DR$ holds. Determine the solution of the system of differential equations $y' = Ay$ with initial condition $y(0) = y_0$.
(ii) How can the procedure be generalised if the Jordan normal form $J = RAR^{-1}$ of the matrix A is given?

Problem 30.1.10 Determine non-trivial solutions of the differential equations $y' = ty$, $y' = \sin(t)y$, and $y' = \cos(t)e^y$.

Project 30.1.1 Differential equations can be solved approximately in MATLAB with the routine ode45. In the case of the system $y' = f(t, y)$ in the interval $[0, T]$

30.1 Ordinary Differential Equations

```
function test_ode
T = 1; y_0 = [1,2];
[t_vec,y_vec] = ode45(@f,[0,T],y_0);
plot(t_vec,y_vec(:,1),'-r'); hold on;
plot(t_vec,y_vec(:,2),'-b'); hold off;

function dy = f(t,y)
A = [-2,0;0,-5]; dy = A*y;
```

Fig. 30.2 Numerically solving an initial value problem with MATLAB routines

with initial condition $y(0) = y_0$ this is realised for the mapping $f(t, y) = Ay$ in the MATLAB program shown in Fig. 30.2. The routine ode45 provides a list t_vec of time points $0 = t_0 < t_1 < \cdots < t_N = T$ and a matrix y_vec with corresponding approximations $\tilde{y}(t_i)$ of the exact solution values $y(t_i)$ at the time points t_i, $i = 0, 1, \ldots, N$. Modify the program test_ode.m to solve the following initial value problems approximately and to graphically display the approximate solutions:

(i) the initial value problem of the predator-prey model

$$y'_1 = \alpha y_1(1 - y_2), \quad y'_2 = \beta y_2(y_1 - 1)$$

in the interval $[0, T]$ with $T = 10$ and $\alpha = 2$, $\beta = 1$ and the initial conditions $y_1(0) = 3$ and $y_2(0) = 1$;

(ii) the initial value problem of the spring pendulum

$$my'' + ry' + D(y - \ell) = 0$$

in the interval $[0, T]$ with $T = 1$ and $m = 1$, $D = 1$, $\ell = 1$ and various values $r \in \{0, 1, 5\}$ and the initial conditions $y(0) = \ell$ and $y'(0) = 1$;

(iii) the initial value problem of the undamped pendulum

$$y'' = -(g/\ell)\sin(y)$$

with $g = 1$, $\ell = 1$ and the initial conditions $y(0) = 0$ and $y'(0) \in \{1, 2, 4, 8\}$;

(iv) the initial value problem

$$y'' - Ny' - (N+1)y = 0$$

in the interval $[0, 1]$ with initial conditions $y(0) = 1$, $y'(0) = -1$, whose exact solution is given by $y(t) = e^{-t}$, for $N = 1, 2, 10$ and small perturbations of the initial condition $y(0) = 1$.

Project 30.1.2 A point grid (x_i, y_i), $i = 1, 2, \ldots, N$ on a rectangular set $[a, b] \times [c, d] \subset \mathbb{R}^2$ for the step sizes $d_x, d_y > 0$ in x- respectively y-direction is

```
1  function test_phase_diagram
2  a = 1; b = 4; dx = 1/4;
3  c = -3; d = 3; dy = 1/6;
4  [x,y] = meshgrid(a:dx:b,c:dy:d);
5  r = (x.^2+y.^2).^(1/2);
6  v = sin(r); w = cos(r);
7  quiver(x,y,v,w,'c'); hold on;
8  v0 = 1.5; w0 = 2;
9  streamline(x,y,v,w,v0,w0); hold off;
```

Fig. 30.3 Representation of a phase diagram

generated in MATLAB by [x,y] = meshgrid(a:dx:b,c:dy:d). Here, x and y are matrices that contain the x and y coordinates of the grid points. A discrete vector field, defined by matrices v and w, by associating the vector (v_i, w_i) with each grid point (x_i, y_i), can be visualised using quiver(x,y,v,w). An integral curve of the discrete vector field starting at a point (v_0, w_0) is represented by streamline(x,y,v,w,v0,w0). The MATLAB program shown in Fig. 30.3 implements this for a simple example. Modify the program to display the phase diagrams of the following differential equations and plot two corresponding integral curves in each case:

(i) the initial value problem of the predator-prey model

$$y_1' = \alpha(1 - y_2)y_1, \quad y_2' = \beta(y_1 - 1)y_2$$

with $\alpha = \beta = 1$;

(ii) the initial value problem of the spring pendulum

$$my'' + ry' + D(y - \ell) = 0$$

with $m = r = D = \ell = 1$;

(iii) the initial value problem of the undamped string pendulum

$$y'' = -(g/\ell)\sin(y)$$

with $g = 1$, $\ell = 1$.

30.2 Existence, Uniqueness and Stability

Problem 30.2.1 Let $L, T > 0$. Show that the space $C^0([0, T])$ is complete with respect to the norm $\|u\|_L = \sup_{t \in [0,T]} e^{-2Lt} |u(t)|$.

Problem 30.2.2 Solve the initial value problem $y' = y^3$, $y(0) = y_0$, sketch the solution and discuss the applicability of the Picard-Lindelöf theorem.

30.2 Existence, Uniqueness and Stability

Problem 30.2.3 Construct infinitely many solutions of the initial value problem $y' = y^{1/3}$, $y(0) = 0$, sketch some of them and discuss the applicability of the Picard-Lindelöf theorem.

Problem 30.2.4 Determine and sketch the iterates y^k, $k = 0, 1, \ldots, 4$, of the Banach fixed point iteration

$$y^{k+1}(t) = y_0 + \int_0^t f(s, y^k(s))\,ds,$$

using the starting function $y^0(t) = y_0$ for the cases $f(t, y) = ay$ and $y_0 = 1$ as well as $f(t, y) = 1 + y^2$ and $y_0 = 0$.

Problem 30.2.5 Assume that the function $y : [0, T) \to \mathbb{R}$ is a solution of the initial value problem $y' = f(y)$, $y(0) = y_0$. Show that y is unique, provided $f \in C^1(\mathbb{R})$.

Problem 30.2.6 Let $f \in C^m([0, T] \times \mathbb{R})$ and $y \in C^1([0, T])$ be a solution of the differential equation $y' = f(t, y)$. Show that $y \in C^{m+1}([0, T])$ holds.

Problem 30.2.7 Generalise the existence and uniqueness statements for differential equations with delay effect, which for a delay parameter $t_v > 0$ seek a solution of the differential equation $y'(t) = f(t, y(t), y(t - t_v))$ for $t \in (0, T)$ and the initial condition $y(t) = y_0$ for $t \in (-t_1, t_0)$.

Problem 30.2.8 Consider the differential equation $y'' + t^{-1}y' + 4ty = 0$ with initial data $y'(0) = 0$ and $y(0) = y_0$. Use the power series approach $y(t) = \sum_{n=0}^{\infty}(a_n/n!)t^n$, to represent the solution of the equation as a series. Discuss the convergence of this series.

Problem 30.2.9 For a continuous mapping $A : [0, T] \to \mathbb{R}^{n \times n}$ we consider the system of differential equations $y' = A(t)y$.

(i) Modify the proof of the Picard-Lindelöf theorem to show the existence of a unique solution with the initial condition $y(0) = y_0$ for $y_0 \in \mathbb{R}^n$.
(ii) Show that the set L of all solutions of the system $y' = A(t)y$ defines a vector space.
(iii) Consider the mapping $E_0 : L \to \mathbb{R}^n$, $y \mapsto y(0)$, and infer that $\dim L = n$ holds.

Problem 30.2.10 Let $g : \mathbb{R}^n \to \mathbb{R}$ be a continuously differentiable, non-negative mapping and let $f : \mathbb{R}^n \to \mathbb{R}^n$ be defined by $f = -\nabla g$. Show that every solution $y : [0, T] \to \mathbb{R}^n$ of the initial value problem $y' = f(y)$, $y(0) = y_0$, satisfies the identity

$$\int_0^t |y'(s)|^2\,ds + g(y(t)) = g(y_0)$$

for every $t \in [0, T]$.

Project 30.2.1 The altitude of a body in the Earth's gravitational field, taking into account frictional forces, is described at high speeds by the equation

$$my''(t) + \eta \operatorname{sign}(y'(t))|y'(t)|^2 = -mg$$

where $\eta \geq 0$ is a friction coefficient, for example, depending on the shape of the body. Determine experimentally with the MATLAB routine ode45 values for η to describe the free and braked fall of a parachutist, so that the free fall from a height of 4 km to a height of 1 km takes about 60 s and the subsequent parachute flight to landing takes about 180 s. Simulate with the parameters found different jump heights and heights for triggering the parachute. What maximum average speeds do you observe for the free fall and the parachute flight?

Project 30.2.2 Use the MATLAB routine ode45 to approximate and display the two-body problem

$$m_1 y_1'' = \gamma \frac{m_1 m_2}{\|y_1 - y_2\|^2} \frac{y_2 - y_1}{\|y_1 - y_2\|},$$

$$m_2 y_2'' = \gamma \frac{m_1 m_2}{\|y_1 - y_2\|^2} \frac{y_1 - y_2}{\|y_1 - y_2\|}$$

for different initial data and mass ratios $m_1/m_2 \in \{1, 2, 10\}$. Construct both initial data that lead to the existence of a solution defined for all positive times, as well as initial data, for which the solution only exists in a finite interval.

30.3 Single-Step Methods

Problem 30.3.1 Let $y \in C^2(\mathbb{R}_{\geq 0})$ and $\tau > 0$. For $k \in \mathbb{N}_0$ define $t_k = k\tau$ and set $y^k = y(t_k)$. Show that for the quantities

$$d_t^- y^k = \frac{y^k - y^{k-1}}{\tau}, \quad d_t^+ y^k = \frac{y^{k+1} - y^k}{\tau},$$

$k = 1, 2, \ldots, K-1$, the estimates

$$|d_t^\pm y^k - y'(t_k)| \leq \frac{\tau}{2} \sup_{t \in t_k \pm [0,\tau]} |y''(t)|$$

hold. What estimate can be proven for the difference $|\widehat{d_t} y^k - y'(t_k)|$ with the quantity

$$\widehat{d_t} y^k = \frac{y^{k+1} - y^{k-1}}{2\tau},$$

30.3 Single-Step Methods

$k = 1, 2, \ldots, K - 1$?

Problem 30.3.2 Let $(y_\ell)_{\ell=0,\ldots,K}$ be a non-negative sequence of numbers and $\alpha, \beta \geq 0$, such that for $\ell = 0, 1, \ldots, K$ the estimate

$$y_\ell \leq \alpha + \sum_{k=0}^{\ell-1} \beta y_k$$

holds. Show that $y_\ell \leq \alpha(1+\beta)^\ell \leq \alpha \exp(K\beta)$ for $\ell = 0, 1, \ldots, K$. Infer the discrete version of Gronwall's lemma.

Problem 30.3.3 Use the explicit and implicit Euler method for the differential equation $y'(t) = 2\alpha t y(t)$ with step sizes $\tau = 1/2^\ell$, $\ell = 1, 2, 3$, as well as the initial value $y_0 = 1$ and $\alpha = \pm 3$, to determine the approximate solutions of both methods at time $T = 1$ and compare these with the exact solution. Comment on your results.

Problem 30.3.4 For an increment function Φ and $z_k \in \mathbb{R}$, let $z : [t_k, t_{k+1}] \to \mathbb{R}$ be the solution of the initial value problem $z'(t) = f(t, z(t))$, $z(t_k) = z_k$, and $z_{k+1} = z_k + \tau \Phi(t_k, z_k, z_{k+1}, \tau)$. With this, the consistency quantities $\mathscr{C}$ and $\widetilde{\mathscr{C}}$ are defined by

$$\mathscr{C}(t_k, z_k, \tau) = \frac{z(t_{k+1}) - z_k}{\tau} - \Phi(t_k, z_k, z_{k+1}, \tau),$$

$$\widetilde{\mathscr{C}}(t_k, z_k, \tau) = \frac{z(t_{k+1}) - z_k}{\tau} - \Phi(t_k, z_k, z(t_{k+1}), \tau).$$

Assume that the increment function Φ is uniformly Lipschitz continuous in the third argument with Lipschitz constant L. Show that for $\tau \leq 1/(2L)$ the equivalence

$$c^{-1} |\widetilde{\mathscr{C}}(t_k, z_k, \tau)| \leq |\mathscr{C}(t_k, z_k, \tau)| \leq c |\widetilde{\mathscr{C}}(t_k, z_k, \tau)|$$

holds. Specify the constant c which depends only on L.

Problem 30.3.5 Let f be a Lipschitz continuous function. Show that the implicit Euler method

$$y_{k+1} = y_k + \tau f(t_{k+1}, y_{k+1})$$

is consistent of order $p = 1$, that is, that $|\mathscr{C}(t_k, z_k, \tau)| \leq c\tau$ with a suitable constant $c \geq 0$.

Problem 30.3.6 Let $f \in C^2([0, T] \times \mathbb{R})$. Show that the method

$$y_{k+1} = y_k + \tau \left[f(t_k, y_k) + \frac{\tau}{2} \left(\partial_t f(t_k, y_k) + \partial_y f(t_k, y_k) f(t_k, y_k) \right) \right]$$

is consistent of order $p = 2$.

Problem 30.3.7 Show that the step size condition $\tau \leq 1/(2M)$ of the error estimate for single-step methods cannot be avoided in general for the implicit and explicit Euler methods. Consider for this the equation $y' = \lambda y$ with appropriately chosen numbers $\lambda \in \mathbb{R}$.

Problem 30.3.8 Provide a formula for the approximations $(y_k)_{k=0,1,\ldots}$ of the initial value problem $y' = \lambda y$, $y(0) = y_0$ defined by the implicit and explicit Euler methods. Sketch the exact and numerical solutions for three different values of λ and different time step sizes. Discuss in the case $\lambda < 0$ the boundedness of the approximations and the exact solution.

Problem 30.3.9 Determine numbers $a, b, c, d \in \mathbb{R}$, for which the explicit single-step method defined by the increment function

$$\Phi(t_k, y_k, \tau) = af(t_k, y_k) + bf\big(t_k + c\tau, y_k + \tau df(t_k, y_k)\big)$$

possesses the consistency order $p = 2$.
Hint: Justify and use the approximation $f(t + c\tau, y + d\tau f(t, y)) = f(t, y) + \partial_t f(t, y)c\tau + \partial_y f(t, y)d\tau f(t, y) + \mathcal{O}(\tau^2)$ and differentiate the differential equation.

Problem 30.3.10 Use the implicit function theorem to ensure the existence of a unique solution y_{k+1} of the equation

$$y_{k+1} = y_k + \tau \Phi(t_k, y_k, y_{k+1}, \tau)$$

under suitable conditions on the function Φ and the step size τ.

Project 30.3.1 The MATLAB program shown in Fig. 30.4 implements the explicit Euler–Collatz method defined by the increment function $\Phi(t_k, y_k, \tau) = f(t_k + \tau/2, y_k + \tau f(t_k, y_k)/2)$ for the spring pendulum equation

$$y'' + ry' + D(y - \ell) = 0$$

with the initial data $y(0) = y_0$ and $y'(0) = v_0$.

(i) Experimentally investigate the dependence of the approximate solutions on the parameters r and D.
(ii) Use the exact solution $y(t) = (v_0/\omega)e^{-rt/2}\sin(\omega t)$ with $\omega = (D - r^2/4)^{1/2}$ of the initial value problem for the special case $r = 1/10$, $D = 1$, $y_0 = \ell = 0$, $v_0 = 1$ and determine the approximation error $|y_K - y(t_K)|$ for the step sizes $\tau = 2^{-s}$, $s = 1, 2, \ldots, 7$, at the time $t_K = 100$.
(iii) Modify the program to implement the explicit and implicit Euler methods as well as Heun's method. Compare the qualitative behaviour of the different approximate solutions for the time horizon $T = 1000$.

30.3 Single-Step Methods

```
function spring_pendulum
T = 10; y_0 = 0; v_0 = 1;
s = 5; tau = 2^(-s); K = floor(T/tau);
y = zeros(K+1,2);
y(1,:) = [y_0,v_0];
for k = 1:K
    y(k+1,:) = y(k,:)+tau*Phi((k-1)*tau,y(k,:),tau);
    plot(tau*(0:k),y(1:k+1,1),'r');
    axis([0,T,-5,5]); drawnow;
end
D = 1; r = 1/10; omega = sqrt(D-r^2/4); t = K*tau;
% y_ex = ...
% abs(y_ex-y(K+1))

function val = Phi(t,y,tau)
val = (f(t,y)+f(t+tau,y+tau*f(t,y)))/2;

function vec = f(t,y)
r = 1/10; D = 1; ell = 0;
vec = [y(2),-r*y(2)-D*(y(1)-ell)];
```

Fig. 30.4 Numerical solution of the initial value problem for the spring pendulum

```
function predator_prey
T = 10; tau = 1/100; K = floor(T/tau);
alpha = 2; beta = 1;
y = zeros(K+1,2);
y(1,:) = [3,1];
for k = 1:K
    y(k+1,1) = y(k,1)+tau*alpha*y(k,1)*(1-y(k,2));
    y(k+1,2) = y(k,2)+tau*beta*y(k,2)*(y(k,1)-1);
    plot(tau*(0:k),y(1:k+1,1),'b'); hold on;
    plot(tau*(0:k),y(1:k+1,2),'r'); hold off;
    axis([0,T,0,4]); drawnow;
end
```

Fig. 30.5 Numerical solution of the initial value problem for the predator-prey model

Project 30.3.2 The MATLAB program shown in Fig. 30.5 calculates an approximate solution of the predator-prey model.

(i) Comment on each line of the program and identify the realised numerical method.
(ii) Test various step sizes and observe the qualitative behaviour of the approximate solutions. For which step sizes do meaningful results emerge?
(iii) Modify a line of the program to obtain an implicit method. How does the qualitative behaviour of the numerical solutions change?

30.4 Runge-Kutta Methods

Problem 30.4.1 Determine the iterates $(y_k)_{k=0,1,\ldots}$ of the explicit Euler method, the Euler–Collatz method and the classical Runge-Kutta method in the approximation of the initial value problem $y' = \lambda y$, $y(0) = y_0$, by constructing an expression $g(\tau\lambda)$ such that $y_{k+1} = g(\tau\lambda)y_k$, $k = 0, 1, 2, \ldots$. Determine the order of convergence of the approximation errors $|y(t_k) - y_k|$, by using the identity $y(t_{k+1}) = e^{\tau\lambda}y(t_k)$ and considering the difference $e^{\tau\lambda} - g(\tau\lambda)$ for the three methods.

Problem 30.4.2 Derive sufficient conditions for the third order consistency of a Runge-Kutta method in the case of autonomous differential equations.

Problem 30.4.3 Show, by constructing polynomial solutions of suitable initial value problems, that the conditions $\sum_{\ell=1}^{m} \gamma_\ell = 1$, $\sum_{\ell=1}^{m} \gamma_\ell \alpha_\ell = 1/2$ and $\sum_{\ell=1}^{m} \sum_{j=1}^{m} \gamma_\ell \beta_{\ell j} = 1/2$ are necessary for the consistency order $p = 2$ of a Runge-Kutta method.

Problem 30.4.4 Determine a two-stage Runge-Kutta method of consistency order $p = 4$, based on the Gaussian quadrature formula with the quadrature points $x_0, x_1 = 1/2 \pm 1/(2\sqrt{3})$ and corresponding weights $w_0 = w_1 = 1/2$.

Problem 30.4.5 Determine the Butcher tableau of the Runge-Kutta method defined by the increment function

$$\Phi(t, y, \tau) = \frac{1}{6}(\eta_1 + 4\eta_2 + \eta_3),$$

$$\eta_1 = f(t, y), \quad \eta_2 = f(t + \tau/2, y + \tau\eta_1/2),$$

$$\eta_3 = f(t + \tau, y + \tau(-\eta_1 + 2\eta_2))$$

and show that it has the consistency order $p = 3$.

Problem 30.4.6 Which quadrature formulas underlie the classical Runge-Kutta method, the 3/8 rule and the Radau-3 method, and what degrees of exactness do they possess?

Problem 30.4.7 Assume that the autonomous system $z' = F(z)$, $z(0) = z_0$, is the equivalent formulation of the differential equation $y' = f(t, y)$, $y(0) = y_0$, obtained by introducing the auxiliary variable w with $w' = 1$ and $w(0) = 0$. Show that Runge-Kutta methods in both cases provide identical approximations of y.

Problem 30.4.8 Construct a Runge-Kutta method of consistency order $p = 4$ based on the Simpson rule.

Problem 30.4.9 Show for the case of autonomous differential equations that the classical Runge-Kutta method defined by $\alpha = [0, 1/2, 1/2, 1]^T$, $\gamma =$

$[1/6, 1/3, 1/3, 1/6]^T$ and $\beta \in \mathbb{R}^{4\times 4}$ with the non-trivial entries $\beta_{21} = 1/2$, $\beta_{32} = 1/2$, $\beta_{43} = 1$ is consistent of order $p = 4$.

Problem 30.4.10

(i) Let $A \in \mathbb{R}^{m\times m}$ be such that $\|A\| < 1$ with respect to an operator norm. Show that the matrix $I_m - A$ is invertible with $(I_m - A)^{-1} = \sum_{n=0}^{\infty} A^n$.

(ii) Formulate the Newton method for solving the fixed point equation $\eta = \Psi(\eta)$ for determining a coefficient vector $\eta \in \mathbb{R}^m$ in a Runge-Kutta method and discuss its well-posedness.

Project 30.4.1 The MATLAB program shown in Fig. 30.6 implements an explicit Runge-Kutta method for solving a scalar differential equation $y' = f(t, y)$, $y(0) = y_0$.

(i) Document each line of the program.
(ii) Verify that the exact solution for the case $f(t, y) = -2y + 5\cos(t)$ and $y_0 = 2$ is given by $y(t) = 2\cos(t) + \sin(t)$. Determine for the step sizes $\tau = 2^{-s}$, $s = 0, 1, \ldots, 5$, the approximation error $|y(T) - y_K|$ with $T = t_K = 10$.
(iii) Modify the program to implement the explicit Euler method, the Euler–Collatz method, the classic Runge-Kutta method and the 3/8 rule.
(iv) Determine for all methods the approximation errors $|y(T) - y_K|$ at time $T = 10$ with the step sizes $\tau = 2^{-s}$, $s = 0, 1, \ldots, 5$. Present these comparatively

```
function runge_kutta_expl
T = 10; s = 2; tau = 2^(-s); K = floor(T/tau);
y = zeros(K+1,1); y(1) = 2;
for k = 1:K
    y(k+1) = y(k)+tau*Phi((k-1)*tau,y(k),tau);
end
plot(tau*(0:K),y(1:K+1),'b-o'); hold on;

function val = Phi(t,y,tau)
m = 2; alpha = [0,1/2]; beta = [0,0;1/2,0]; gamma = [0,1];
eta = zeros(m,1);
val = 0;
for ell = 1:m
    dy = 0;
    for j = 1:ell-1
        dy = dy+beta(ell,j)*eta(j);
    end
    eta(ell) = f(t+tau*alpha(ell),y+tau*dy);
    val = val+gamma(ell)*eta(ell);
end

function val = f(t,y)
val = -2*y+5*cos(t);
```

Fig. 30.6 Implementation of an explicit Runge-Kutta method

as polygonal chains in a graph with logarithmic axis scaling, which can be implemented in MATLAB with the command loglog.

Project 30.4.2 Write two MATLAB routines for the numerical approximation of ordinary differential equations with general implicit Runge-Kutta methods. Use both a fixed point iteration and the Newton method with an appropriate termination criterion. Investigate the respective iteration numbers in the time steps for the Radau-3 method using the example $y' = (1 + y^2)^{1/2}$, $y(0) = 0$, in the interval $[0, T]$ with $T = 4$, whose exact solution is given by $y(t) = \sinh(t)$.

30.5 Multistep Methods

Problem 30.5.1 For a step size $\tau > 0$ and time steps $t_k = k\tau$, $k \in \mathbb{N}_0$, let values $w_k \in \mathbb{R}$ be given.

(i) Construct the interpolation polynomial $q \in \mathscr{P}_2$ defined by the interpolation pairs $(t_{k+\ell}, w_{k+\ell})_{\ell=0,1,2}$ and integrate this over the interval $[t_{k+2}, t_{k+3}]$ to obtain coefficients $(\beta_\ell)_{\ell=0,1,2}$ such that

$$\int_{t_{k+2}}^{t_{k+3}} q(t)\,dt = \tau \sum_{\ell=0}^{2} \beta_\ell w_{k+\ell}.$$

(ii) Construct the interpolation polynomial $q \in \mathscr{P}_2$ defined by the interpolation pairs $(t_{k+\ell}, w_{k+\ell})_{\ell=0,1,2}$ and integrate this over the interval $[t_{k+1}, t_{k+2}]$ to obtain coefficients $(\beta_\ell)_{\ell=0,1,2}$ such that

$$\int_{t_{k+1}}^{t_{k+2}} q(t)\,dt = \tau \sum_{\ell=0}^{2} \beta_\ell w_{k+\ell}.$$

Problem 30.5.2 Determine the maximum number $p \in \mathbb{N}$ such that the identities

$$\sum_{\ell=0}^{m} \alpha_\ell = 0, \quad \sum_{\ell=0}^{m} \left(\alpha_\ell \ell^q - \beta_\ell q \ell^{q-1}\right) = 0, \quad q = 1, 2, \ldots, p,$$

hold for the Adams-Bashforth and the Adams-Moulton method with $m = 3$ and $m = 2$ respectively.

Problem 30.5.3 Show that the Adams-Moulton method is well-defined under the condition $\tau \|\beta\|_1 L < 1$, where L is the uniform Lipschitz constant of the function f associated with the differential equation.

30.5 Multistep Methods

Problem 30.5.4 Determine the consistency order of the *leap-frog* method on the one hand directly with an error estimate for the approximation of the time derivative using $y'(t_k) \approx (y(t_{k+1}) - y(t_{k-1}))/(2\tau)$ and on the other hand by checking the general consistency criterion for multistep methods.

Problem 30.5.5 For a step size $\tau > 0$ and time steps $t_k = k\tau$, $k \in \mathbb{N}_0$, let values $w_k \in \mathbb{R}$ be given. Determine the derivative $p'(t_{k+m})$ of the interpolation polynomial $p \in \mathscr{P}_m$ for the interpolation pairs $(t_{k+\ell}, w_{k+\ell})_{\ell=0,\ldots,m}$ with $m = 1, 2, 3$. Discuss how a multistep method can be constructed with this.

Problem 30.5.6 Construct a multistep method by approximating the integral in the representation

$$y(t_{k+2}) = y(t_k) + \int_{t_k}^{t_{k+2}} f(s, y(s))\, ds$$

with the Simpson rule and determine the consistency order of the method obtained in this way.

Problem 30.5.7 Show, by constructing suitable initial value problems, that the sufficient consistency criterion for linear multistep methods

$$\sum_{\ell=0}^{m} \alpha_\ell = 0, \quad \sum_{\ell=0}^{m} \left(\alpha_\ell \ell^q - \beta_\ell q \ell^{q-1}\right) = 0, \quad q = 1, 2, \ldots, p,$$

is necessary.

Problem 30.5.8 Show that for each $m \geq 1$ there is exactly one linear m-step method of consistency order $2m$ and none of consistency order $2m + 1$. Use the normalisation $\beta_0 = 1$ for this purpose and formulate the general consistency criterion as a system of linear equations $A[\widehat{\alpha}, \widehat{\beta}]^\mathsf{T} = b$ with $\widehat{\alpha} = [\alpha_1, \ldots, \alpha_m]^\mathsf{T}$ and $\widehat{\beta} = [\beta_1, \ldots, \beta_m]^\mathsf{T}$. Use the fundamental theorem of algebra to investigate the matrix A^T.

Problem 30.5.9 Let a linear, explicit multistep method be defined by $(\widehat{\alpha}_\ell, \widehat{\beta}_\ell)_{\ell=0,\ldots,m}$ and a linear, implicit multistep method be defined by $(\alpha_\ell, \beta_\ell)_{\ell=0,\ldots,m}$. The approximation y_{k+m} is defined by $y_{k+m} = y_{k+m}^{(\nu)}$, where $y_{k+m}^{(\nu)}$ is calculated by the iteration rule

$$\widetilde{y}_{k+m}^{(i+1)} = -\sum_{\ell=0}^{m-1} \alpha_\ell y_{k+\ell} + \tau \sum_{\ell=0}^{m-1} \beta_\ell f(t_{k+\ell}, y_{k+\ell}) + \tau \beta_m f(t_{k+m}, y_{k+m}^{(i)})$$

with the initialisation $y_{k+m}^{(0)} = \widetilde{y}_{k+m}$ for

$$\widetilde{y}_{k+m} = -\sum_{\ell=0}^{m-1} \widehat{\alpha}_\ell y_{k+\ell} + \tau \sum_{\ell=0}^{m-1} \widehat{\beta}_\ell f(t_{k+\ell}, y_{k+\ell}).$$

Show that this defines an explicit multistep method of consistency order $p = \min\{p_{expl} + \nu, p_{impl}\}$, where p_{expl} and p_{impl} denote the consistency orders of the explicit and implicit methods, respectively.

Problem 30.5.10 Investigate for which values $z = \tau\lambda \in \mathbb{R}$ you obtain bounded approximations of the initial value problem $y' = \lambda y$ in $(0, \infty)$, $y(0) = 1$ with the two-step methods defined by:

α_2	α_1	α_0	β_2	β_1	β_0
1	-1	0	0	$3/2$	$-1/2$
1	-1	0	$5/12$	$8/12$	$-1/12$
1	$-4/3$	$1/3$	$2/3$	0	0

Write the methods in the form $Y_{k+1} = BY_k$ and investigate the matrix $B \in \mathbb{R}^{2\times 2}$.

Project 30.5.1 We consider the initial value problem $y' = f(t, y)$ for $t \in (0, T]$, $y(0) = y_0$, with $f(t, y) = (1 + y^2)^{1/2}$, $y_0 = 0$ and $T = 1$. The exact solution is given by $y(t) = \sinh(t)$.

(i) Implement the Adams-Bashforth method.
(ii) Use a fixed point iteration with a suitable termination criterion to implement the Adams-Moulton method.
(iii) Realize the Adams-Bashforth–Moulton method.
(iv) Compare the errors $|y(T) - y_K|$ at the final time $t_K = T$ of the three methods for $m = 2, 3, 4$ and step sizes $\tau = 2^{-\ell}$, $\ell = 2, 3, \ldots, 6$, in three tables. As initial values, you can use the function values of the exact solution.

Project 30.5.2 Write a short program for the algorithmic determination of the consistency order of a given multistep method. Test it for the Adams methods with $m = 1, 2, 3, 4$ steps as well as for the method with $m = 6$ and

$$[\alpha_6, \alpha_5, \ldots, \alpha_0] = \frac{1}{147}[147, -360, 450, -400, 225, -72, 10],$$

$$[\beta_6, \beta_5, \ldots, \beta_0] = \frac{1}{147}[60, 0, 0, 0, 0, 0, 0].$$

30.6 Convergence of Multistep Methods

Problem 30.6.1 Let $A \in \mathbb{R}^{m\times m}$ be the diagonalisable companion matrix of the difference equation defined by $(\alpha_\ell)_{\ell=0,\ldots,m}$ with linearly independent eigenvectors $v_1, v_2, \ldots, v_m$. Show that the sequence $(y_k)_{k\geq 0}$ is a solution of the homogeneous difference equation if and only if for the vectors $Y_k = [y_k, y_{k+1}, \ldots, y_{k+m-1}]^T$ we

30.6 Convergence of Multistep Methods

have that $Y_k = \sum_{j=1}^m \lambda_j^k \gamma_j v_j$, $k \geq 0$, with suitable numbers $\gamma_j \in \mathbb{R}$ and the roots λ_j of the polynomial $q(\lambda) = \lambda^m + \alpha_{m-1}\lambda^{m-1} + \cdots + \lambda\alpha_1 + \alpha_0$.

Problem 30.6.2 Investigate the zero stability and consistency of the multistep method $y_{k+2} - 4y_{k+1} + 3y_k = -2\tau f(t_k, y_k)$.

Problem 30.6.3 For $(\alpha_\ell)_{\ell=0,\ldots,m}$ with $\alpha_m = 1$ we consider the linear homogeneous difference equation

$$\sum_{\ell=0}^m \alpha_\ell y_{k+\ell} = 0.$$

(i) Show that for m initial values $y_0, y_1, \ldots, y_{m-1} \in \mathbb{R}$ exactly one sequence $(y_k)_{k\geq 0}$ exists, which solves the homogeneous difference equation.
(ii) Show that the homogeneous difference equation has m linearly independent solutions $(y_k)_{k\geq 0}$.

Problem 30.6.4 Let $\lambda \in \mathbb{C}$ be an s-fold root of the polynomial $q(z) = z^m + \alpha_{m-1}z^{m-1} + \cdots + \alpha_1 z + \alpha_0$ and let $(y_k)_{k\geq 0}$ be defined by $y_k = k^r \lambda^k$ with $r \in \mathbb{N}$, $r < s$. Furthermore, for $f \in C^1(\mathbb{R})$ and $x \in \mathbb{R}$ the function $Af \in C^0(\mathbb{R})$ is defined by $Af(x) = xf'(x)$.

(i) Prove the identity

$$\sum_{\ell=0}^m \alpha_\ell y_{k+\ell} = \lambda^k \sum_{\nu=0}^r \binom{r}{\nu} k^\nu \sum_{\ell=0}^m \alpha_\ell \ell^{r-\nu} \lambda^\ell = \lambda^k \sum_{\nu=0}^r \binom{r}{\nu} k^\nu A^{r-\nu} q(\lambda).$$

(ii) Let x_0 be an $(r+1)$-fold root of $f \in C^r(\mathbb{R})$, that is, we have $f(x_0) = f'(x_0) = \cdots = f^{(r)}(x_0) = 0$. Show that $A^i f(x_0) = 0$ for $i = 0, 1, \ldots, r$.
(iii) Conclude that $(y_k)_{k\geq 0}$ is a solution of the linear homogeneous difference equation $\sum_{\ell=0}^m \alpha_\ell y_{k+\ell} = 0$ and discuss the boundedness of this sequence.

Problem 30.6.5 Let $R \in \mathbb{C}^{m\times m}$ be regular and let $\|\cdot\|$ be a norm on $\mathbb{C}^m$. Show that by $A \mapsto \sup_{x\in\mathbb{R}^m\setminus\{0\}} \|RAx\|/\|x\|$ an operator norm on $\mathbb{R}^{m\times m}$ is defined.

Problem 30.6.6 Investigate the zero stability of the Fibonacci sequence $y_{k+2} = y_{k+1} + y_k$ and the Chebyshev recursion $T_{k+2}(x) = 2xT_{k+1}(x) - T_k(x)$.

Problem 30.6.7 Assume that the Jordan normal form of the companion matrix $A \in \mathbb{R}^{m\times m}$ of a difference equation is real and the Dahlquist root condition is violated. Show that then $\varrho(A) > 1$ holds.

Problem 30.6.8 Specify the constants C_0, C_1, C_2 in the general convergence statement for multistep methods and discuss in which situations the error estimation is of practical use.

Problem 30.6.9 Let $f \in C^1([0, T] \times \mathbb{R})$ with $|\partial_z f(t, z)| \leq C$ for all $(t, z) \in [0, T] \times \mathbb{R}$. Show that the Adams-Moulton, Adams-Bashforth, and Adams-Bashforth-Moulton methods satisfy the conditions of the general convergence statement for multistep methods.

Problem 30.6.10 Let $J = T^{-1}AT$ be the Jordan normal form of the matrix $A \in \mathbb{R}^{m \times m}$ with Jordan blocks J_i, $i = 1, 2, \ldots, r$. For $\varepsilon \geq 0$, let $D \in \mathbb{R}^{m \times m}$ be the diagonal matrix with entries $d_{kk} = \varepsilon^{k-1}$ for $k = 1, 2, \ldots, m$. Show that the matrix $\widetilde{J} = D^{-1}JD$ is given by the blocks

$$\widetilde{J}_i = \begin{bmatrix} \lambda_i & \varepsilon & & \\ & \ddots & \ddots & \\ & & \ddots & \varepsilon \\ & & & \lambda_i \end{bmatrix},$$

$i = 1, 2, \ldots, r$.

Project 30.6.1 Formulate algorithms for the systematic experimental analysis of the zero stability of a difference equation, on the one hand by testing randomly selected initial values and on the other hand by solving a suitable eigenvalue problem. Discuss the reliability of the assessment determined in this way and test your algorithms with the coefficients

$$[\alpha_2, \alpha_1, \alpha_0] = [1, 4, -5],$$
$$[\alpha_2, \alpha_1, \alpha_0] = [1, -4, 3],$$
$$[\alpha_2, \alpha_1, \alpha_0] = [1, 0, -1],$$
$$[\alpha_4, \alpha_3, \alpha_2, \alpha_1, \alpha_0] = [1, -48/25, 36/25, -16/25, 3/25].$$

Project 30.6.2 The BDF methods *(backward differentiation formulas)* are given for $m \geq 1$ by

$$\sum_{\ell=0}^{m} \widehat{\alpha}_\ell y_{k+\ell} = \tau f(t_{k+m}, y_{k+m})$$

with the coefficients $\widehat{\alpha}_m = \sum_{j=1}^{m} 1/j$ and

$$\widehat{\alpha}_\ell = (-1)^{m-\ell} \sum_{j=m-\ell}^{m} \frac{1}{j}\binom{j}{m-\ell},$$

$\ell = 0, 1, \ldots, m - 1$. Use the BDF methods with $m = 1, 2, \ldots, 7$ for the numerical approximation of the initial value problem $y' = f(t, y)$ in $(0, T]$, $y(0) = y_0$, with $f(t, y) = -2y + 5\cos(t)$, $y_0 = 1$ and $T = 1$, whose exact solution is given by

$y(t) = 2\cos(t) + \sin(t)$. Determine the experimental convergence rates at time $T = 1$ with suitable sequences of time step sizes and the approach $e_\tau \approx c\tau^\gamma$, so that for two different step sizes it follows

$$\gamma \approx \log(e_\tau/e_{\tau'})/\log(\tau/\tau').$$

30.7 Stiff Differential Equations

Problem 30.7.1 Let $A \in \mathbb{R}^{2\times 2}$ with eigenvalues $\lambda_1, \lambda_2 \in \mathbb{C}$. Draw the phase diagrams of the differential equation $z' = Az$ in a neighbourhood of the origin for four typical situations characterised by

(i) $\lambda_1, \lambda_2 \in \mathbb{R}_{>0}$, (ii) $\lambda_1, \lambda_2 \in \mathbb{R}_{<0}$, (iii) $\lambda_1, \lambda_2 \in \mathbb{R}$, $\lambda_1 \lambda_2 < 0$, (iv) $\lambda_1 = \overline{\lambda_2}$.

Problem 30.7.2 Assume that a numerical method leads to bounded approximations of the scalar differential equation $y' = \lambda y$ for every step size $\tau > 0$, provided $\mathrm{Re}(\lambda) \leq 0$ holds. Furthermore, let $A \in \mathbb{R}^{n\times n}$ be complex diagonalisable and the eigenvalues of A have exclusively negative real parts. Show that the method is A-stable.

Problem 30.7.3

(i) Show that the application of a linear multistep method to the differential equation $y' = \lambda y$ leads to a homogeneous difference equation.
(ii) Define the concept of A-stability for linear multistep methods, so that it is consistent with the definition for single-step methods in the case of the implicit Euler method.
(iii) Investigate the A-stability of the methods defined by $m = 2$ and

$$[\alpha_2, \alpha_1, \alpha_0] = [1, -4/3, 1/3], \quad [\beta_2, \beta_1, \beta_0] = [2/3, 0, 0]$$

or $m = 3$ and

$$[\alpha_3, \alpha_2, \alpha_1, \alpha_0] = [1, -18/11, 9/11, -2/11],$$
$$[\beta_3, \beta_2, \beta_1, \beta_0] = [6/11, 0, 0, 0]$$

respectively.

Problem 30.7.4 Let $A \in \mathbb{R}^{n\times n}$ be negative definite, i.e. there exists a number $\alpha > 0$, such that $z^\top A z \leq -\alpha \|z\|^2$ for all $z \in \mathbb{R}^n$. Show that the solution of the initial value problem $y' = Ay$ converges exponentially fast to 0 for every initial value $y_0 \in \mathbb{R}^n$.

Problem 30.7.5 Let $\alpha \in \mathbb{R}^m$, $\beta \in \mathbb{R}^{m \times m}$ and $\gamma \in \mathbb{R}^m$ be the coefficients of a Runge–Kutta method. Show that the associated stability function is a polynomial or rational function.

Problem 30.7.6 Investigate the Runge–Kutta method defined by the Butcher table

1/3	5/12	−1/12
1	3/4	1/4
	3/4	1/4

for A- and L-stability.

Problem 30.7.7 Let the function $f : \mathbb{R}^n \to \mathbb{R}^n$ be one-sided Lipschitz continuous, i.e. for all $z, w \in \mathbb{R}^n$ we have

$$\langle f(z) - f(w), z - w \rangle \leq L\|z - w\|^2.$$

Show that the differential equation $y' = f(y)$ has at most one solution for every initial value y_0 and discuss the well-posedness of the differential equations $y' = -y^3$ and $y' = y^3$.

Problem 30.7.8

(i) Let $G \in C^1(\mathbb{R}^n)$. Show that G is convex if and only if

$$\nabla G(z) \cdot (w - z) + G(z) \leq G(w)$$

holds for all $z, w \in \mathbb{R}^n$.

(ii) Let $G \in C^2(\mathbb{R}^n)$. Show that G is convex if and only if $D^2 g(x)$ is positive semi-definite for all $x \in \mathbb{R}^n$.

Problem 30.7.9 Peano's theorem states that every initial value problem $y' = f(y)$, $y(0) = y_0$, with a continuous function $f : \mathbb{R}^n \to \mathbb{R}^n$ has a solution in an interval $(0, \varepsilon)$ and $\varepsilon > 0$ can be chosen arbitrarily, provided the solution remains bounded. Show that the initial value problem $y' = -\nabla G(y)$ with a coercive function $G \in C^1(\mathbb{R}^n)$ for every initial value $y_0 \in \mathbb{R}^n$ has a solution defined on all $\mathbb{R}_{\geq 0}$ and discuss the applicability to the differential equation $y' = -y^3$.

Problem 30.7.10

(i) Let $G : \mathbb{R} \to \mathbb{R}$ be defined by $G(z) = (1 - z^2)^2$. Sketch the function G and show that G is μ-convex.
(ii) Assume $G(x) \geq -c_1 + c_2 |x|^p$ with $p \geq 1$. Show that G is coercive and, provided G is also continuous, has a minimum.

Project 30.7.1 We consider the initial value problem $y' = -\alpha(y - \cos(t))$, $y(0) = 0$ in the interval $[0, T]$ with $T = 1$ and $\alpha = 50$.

(i) Verify that the solution of the problem is given by

30.8 Step Size Control

$$y(t) = \frac{\alpha}{1+\alpha^2}\bigl(\sin(t) + \alpha\cos(t) - \alpha e^{-\alpha t}\bigr).$$

(ii) Solve the problem approximately with the explicit and implicit Euler method, the trapezoidal method, and the classical Runge–Kutta method with the step sizes $\tau = 2^{-\ell}/10$, $\ell = 0, 1, 2, 3$. Present the errors at time T comparatively in a table.

(iii) Present the approximations for some step sizes and the exact solution in a graph and discuss the results.

Project 30.7.2 We consider the initial value problem $y' = -\alpha y^3$, $y(0) = 1$, in the interval $[0, T]$ with $T = 1$ and $\alpha = 200$.

(i) Show that the initial value problem defines a gradient flow for a suitable function G and determine the exact solution.

(ii) Test the explicit and implicit Euler method for the approximate solution of the problem. Use the Newton method to solve nonlinear equations approximately. Document your observations.

(iii) Test the semi-implicit Euler method

$$y_{k+1} = y_k - \tau \alpha y_k^2 y_{k+1}$$

as well as the linearised implicit Euler method

$$y_{k+1} = y_k - \tau \alpha \bigl(f(y_k) + f'(y_k)(y_{k+1} - y_k)\bigr),$$

where $f(y) = y^3$, and document your observations.

(iv) Experimentally determine for each of the above methods step sizes for which the sequence $(G(y_k))_{k=0,\ldots,K}$ is monotonically decreasing.

30.8 Step Size Control

Problem 30.8.1 Let $\widehat{y}_\tau : [0, T] \to \mathbb{R}$ be the affine-linear interpolant of the approximations calculated with the implicit Euler method for the initial value problem $y' = f(y)$, $y(0) = y_0$, and let $y \in C^1([0, T])$. Show that $\widehat{y}_\tau$ converges uniformly to y on $[0, T]$ as $\tau_{max} = \max_{k=1,\ldots,K} \tau_k \to 0$.

Problem 30.8.2 Let $\widehat{y}_\tau, \overline{y}_\tau : [0, T] \to \mathbb{R}$ be the interpolants of a sequence $(y_k)_{k=0,\ldots,K}$ with maximum step size $\tau = \max_{k=1,\ldots,K} \tau_k$. Show that for $k = 1, 2, \ldots, K$ we have that

$$\sup_{t\in[t_{k-1},t_k]} |\widehat{y}_\tau(t) - \overline{y}_\tau(t)| \le \tau \sup_{t\in[t_{k-1},t_k]} |\widehat{y}'_\tau(t)|.$$

Problem 30.8.3 Let $\bar{y}_\tau, \widehat{y}_\tau : [0, T] \to \mathbb{R}$ be the piecewise constant and piecewise affine interpolant of the sequence $(y_k)_{k=0,\ldots,K}$ for the uniform step size $\tau > 0$. Show that for every function $v \in C^1([0, T])$ the identities

$$\int_0^T v' \bar{y}_\tau \, dt = -\sum_{k=0}^{K-1}(y_{k+1} - y_k)v(t_k) + y_K v(T) - y_0 v(0)$$

and

$$\int_0^T v' \widehat{y}_\tau \, dt = -\int_0^T v \widehat{y}'_\tau \, dt + y_K v(T) - y_0 v(0)$$

hold.

Problem 30.8.4 Let $y \in C^0([0, T])$. Determine conditions under which, for a given parameter $\delta > 0$, there exists a number $\tau > 0$ such that

$$|y(t + \tau) - y(t)| \leq \delta$$

for all $t \in [0, T-\tau]$. Show with an example that this does not generally hold without additional assumptions on y.

Problem 30.8.5 Derive an a posteriori error estimate for the explicit Euler method.

Problem 30.8.6 Show for the case of an autonomous differential equation $y' = f(y)$ with a Lipschitz-continuous function $f : \mathbb{R} \to \mathbb{R}$, that the adaptive method based on the a posteriori error estimate always terminates, i.e. the final time is reached.

Problem 30.8.7 Let $\widehat{y}'_\tau = f(\widehat{y}_\tau) + R_\tau$ and $y' = f(y)$ in the interval $(0, T)$ and $\widehat{y}_\tau(0) = y(0)$. Show that for the error $e(t) = y(t) - \widehat{y}_\tau(t)$ we have that

$$\sup_{t \in [0,T]} |e(t)| \leq \max_{t \in [0,T]} |R_\tau(t)| \exp(LT),$$

and compare this estimate with other a posteriori error estimates.

Problem 30.8.8 Consider a numerical method of consistency order p. Construct by extrapolation of approximations to the step sizes τ, $\tau/2$ and $\tau/4$ a method of consistency order $p + 2$. Discuss the total effort of the obtained method compared to the use of the original method with the step size τ^3 in the case $p = 1$.

Problem 30.8.9 Determine the Butcher table of the method obtained by extrapolation of the implicit Euler method with step sizes τ and $\tau/2$ and discuss its consistency order.

Problem 30.8.10 Let $f \in C^1([0, T])$ and $g \in C^0([0, T])$ with $f, g \geq 0$ and $c_0 \geq 0$, such that $f'(t) \leq c_0 + \bigl(g(t)f(t)\bigr)^{1/2}$ holds for all $t \in [0, T]$.

30.9 Symplectic, Shooting and dG Methods

(i) Show that for all $a, b \in \mathbb{R}$ and $\gamma > 0$ we have $ab \leq \gamma a^2/2 + b^2/(2\gamma)$.

(ii) Show using the Gronwall Lemma that for every $\delta > 0$ we have

$$\max_{t \in [0,T]} f(t) \leq \left(f(0) + c_0 T + (\delta T/2) \max_{t \in [0,T]} g(t)\right) \exp\left(T/(2\delta)\right).$$

(iii) Prove without using the Gronwall Lemma that

$$\max_{t \in [0,T]} f(t) \leq 2f(0) + 2c_0 T + T^2 \max_{t \in [0,T]} g(t).$$

(iv) Discuss the advantages and disadvantages of the estimates from (ii) and (iii).

Project 30.8.1 Implement the adaptive algorithm for step size control and test it with the implicit Euler method for the initial value problems

$$y'(t) = -(y(t) - 100\cos(t)), \quad t \in (0, 1], \quad y(0) = 0,$$

and

$$y''(t) = 20\left(1 - y(t)^2\right)y'(t) - y(t), \quad t \in [0, 100], \quad y(0) = 1/10, \quad y'(0) = 0.$$

Use different parameters $\delta > 0$ for the condition $|y_{k+1} - y_k| \leq \delta$ and display the variable step sizes as a function of time. Compare the effort and accuracy of the adaptive method for calculating the approximations on a uniform grid. Use the fact that the solution of the first initial value problem is given by $y(t) = 50(\sin(t) + \cos(t) - e^{-t})$.

Project 30.8.2 Implement the extrapolation of the trapezoidal method with step sizes τ and $\tau/2$ and verify the improved consistency order using the example of the initial value problem $y'(t) = -y(t) + \cos(t)$, $y(0) = 0$ in the interval $[0, T]$ with $T = 1$, whose exact solution is given by $y(t) = (\sin(t) + \cos(t) - e^{-t})/2$. Display the approximations $(y_k^\tau)_{k=0,\ldots,K}$, $(y_k^{\tau/2})_{k=0,\ldots,2K}$ and $(\widetilde{y}_k)_{k=0,\ldots,K}$ comparatively in a graph.

30.9 Symplectic, Shooting and dG Methods

Problem 30.9.1 Formulate the kinetic energy $mv^2/2$ and the potential energy mgh in suitable polar coordinates to derive a Hamilton function for the pendulum.

Problem 30.9.2 Show that Newton's law of inertia can be interpreted as a Hamiltonian system. Assume that the acting force is given as the negative gradient of a potential.

Problem 30.9.3 With a function $V \in C^1(\mathbb{R})$, a Hamiltonian system is given by the function $H : \mathbb{R}^{N \times 3} \times \mathbb{R}^{N \times 3} \to \mathbb{R}$,

$$H(q, p) = \sum_{i=1}^{N} \frac{\|p_i\|^2}{2m_i} + \frac{1}{2} \sum_{\substack{i,j=1 \\ i \neq j}}^{N} V(\|q_i - q_j\|).$$

Show that the total momentum P and the total angular momentum L, which are defined with the three-dimensional cross product as

$$P = \sum_{i=1}^{N} p_i, \quad L = \sum_{i=1}^{N} q_i \times p_i,$$

of the system are conserved.

Problem 30.9.4 Let $J \in \mathbb{R}^{2n \times 2n}$ be defined by

$$J = \begin{bmatrix} & I_n \\ -I_n & \end{bmatrix}.$$

(i) Show that $\omega : \mathbb{R}^{2n} \times \mathbb{R}^{2n} \to \mathbb{R}$, $\omega(z_1, z_2) = z_1^\top J z_2$, defines a skew-symmetric bilinear form.
(ii) Let P be a parallelogram in $\mathbb{R}^2$, spanned by the vectors z_1 and z_2. Show that the area of P is given by $|\omega(z_1, z_2)|$. How is the sign of $\omega(z_1, z_2)$ to be interpreted?
(iii) Construct a nonlinear mapping $\Psi : \mathbb{R}^2 \to \mathbb{R}^2$, which is symplectic.

Problem 30.9.5 Determine all symplectic matrices $A \in \mathbb{R}^{2 \times 2}$.

Problem 30.9.6 To describe the path of a planet of mass m in the gravitational field of a stationary sun of mass $M \gg m$ we use the Hamilton function

$$H(q, p) = \frac{\|p\|^2}{2m} - \gamma \frac{mM}{\|q\|}.$$

(i) Assume that the motion of the body takes place in a plane and is described by the function $q : [0, T] \to \mathbb{R}^2$. Furthermore, let $p = mq'$. Use polar coordinates (r, ϕ) to show that

$$H(q, p) = \frac{m}{2}\left((r')^2 + (r\phi')^2\right) - \frac{\gamma mM}{r}.$$

(ii) Use the constancy of the angular momentum $\widehat{L} = q \times p$, whose length is given by $L = mr^2\phi'$, and the total energy $H(q(t), p(t)) = H_0$, to show that for the radius as a function of the angle we have

30.9 Symplectic, Shooting and dG Methods

$$\left(\frac{dr}{d\phi}\right)^2 = \frac{2mr^4}{L^2}\left(H_0 + \frac{\gamma Mm}{r} - \frac{L^2}{2mr^2}\right).$$

(iii) Prove that every ellipse $\{(x, y) \in \mathbb{R}^2 : (x/a)^2 + (y/b)^2 = c^2\}$ in polar coordinates with respect to a focus can be represented by $r(\phi) = s/(1 + \varepsilon \cos(\phi))$, $\phi \in [0, 2\pi]$, and show that

$$\left(\frac{dr}{d\phi}\right)^2 = \frac{r^4}{s^2}\left(\varepsilon^2 - 1 + \frac{2s}{r} - \frac{s^2}{r^2}\right).$$

(iv) Conclude that the path of the planet is described by an ellipse.

Problem 30.9.7 Show that the midpoint method is symplectic, but in the case of the Hamiltonian

$$H(q, p) = \sum_{i=1}^{N} \frac{\|p_i\|^2}{2m_i} + \frac{1}{2} \sum_{i,j=1\ i\neq j}^{N} V(\|q_i - q_j\|)$$

requires the solution of nonlinear systems of equations.

Problem 30.9.8 Show that the implicit Euler method is not symplectic.

Problem 30.9.9

(i) Show that the method

$$\begin{bmatrix} q_{k+1} \\ p_{k+1} \end{bmatrix} = \begin{bmatrix} q_k \\ p_k \end{bmatrix} + \tau \begin{bmatrix} \partial_p H(q_{k+1}, p_k) \\ -\partial_q H(q_{k+1}, p_k) \end{bmatrix}$$

is symplectic.
(ii) What disadvantages arise compared to the partitioned Euler method, where on the right-hand side the expressions $\partial_p H(q_k, p_{k+1})$ and $-\partial_q H(q_k, p_{k+1})$ are used?

Problem 30.9.10 Show that the discontinuous Galerkin method for $\ell = 1$ leads to a variant of the midpoint method.

Project 30.9.1 A ball of mass $m = 10\,\text{g}$ is to be shot vertically upwards so that it reaches the ground again exactly after 10 seconds. Taking into account air resistance, an initial velocity s is sought such that for the solution of the initial value problem

$$my'' + \eta \operatorname{sign}(y')|y'|^2 = -mg, \quad t \in (0, 10], \quad y(0) = 0, \quad y'(0) = v_0$$

we have $y(10) = 0$. Here $g = 9.81\,\text{m/s}^2$ and $\eta = 2 \cdot 10^{-4}\,\text{kg/m}$. Use the bisection method and the Newton method to solve the problem approximately. To define a suitable starting value, you can first solve the problem neglecting friction effects. Test other friction coefficients and discuss the convergence behaviour of the methods. Check the plausibility of your results.

Project 30.9.2 Use the explicit and implicit Euler method, the midpoint method and the partitioned Euler method to simulate the pendulum described by the Hamiltonian

$$H(\phi, \psi) = \frac{1}{2}\psi^2 - \cos(\phi)$$

in the time interval $[0, T]$ with $T = 10$. Display the trajectories $t \mapsto (\phi(t), \psi(t))$ in the phase diagram and plot the total energy $t \mapsto H(\phi(t), \psi(t))$ as well as the kinetic and potential energy comparatively for the methods and various step sizes. Solve nonlinear systems of equations with the Newton method.

Part V
Supplementary Material

Chapter 31
Results from Linear Algebra

31.1 Scalar Product of Vectors

On the vector space $\mathbb{R}^n$, the mapping

$$\cdot : \mathbb{R}^n \times \mathbb{R}^n \to \mathbb{R}, \quad (v,w) \mapsto v \cdot w = v^\mathsf{T} w = \sum_{i=1}^n v_i w_i$$

defines a bilinear mapping, which is referred to as *scalar product*. The Euclidean length of a vector is thus given by

$$\|v\|_2 = (v \cdot v)^{1/2} = \Big(\sum_{i=1}^n v_i^2\Big)^{1/2}.$$

Two linearly independent vectors $v, w \in \mathbb{R}^n$ span a plane and with the angle α between these vectors within the plane, we have

$$v \cdot w = \cos(\alpha)\|v\|_2\|w\|_2.$$

Two vectors $v, w \in \mathbb{R}^n$ are called *orthogonal*, denoted by $v \perp w$, if $v \cdot w = 0$.

31.2 Determinant of Square Matrices

In the case $n = 2$, an oriented area of the parallelogram spanned by two vectors $v, w \in \mathbb{R}^2$ is defined by

$$\det[v,w] = v_1 w_2 - v_2 w_1.$$

More generally, the oriented volume of a parallelepiped spanned by the vectors $v_1, v_2, \ldots, v_n \in \mathbb{R}^n$ is given by the *determinant* $\det V$ of the matrix V, whose columns are the vectors $v_1, v_2, \ldots, v_n$. The sign of the determinant defines an equivalence relation on the set of bases of $\mathbb{R}^n$ and thus allows the definition of a positive and negative orientation. The value of a determinant can be calculated recursively with the Laplace expansion theorem, which states that for each $i = 1, 2, \ldots, n$ the identity

$$\det V = \sum_{j=1}^n v_{ij}(-1)^{i+j} \det \widehat{V}_{ij}$$

holds, where $\widehat{V}_{ij} \in \mathbb{R}^{(n-1)\times(n-1)}$ is obtained from V by deleting the i-th row and j-th column and for every real number $s \in \mathbb{R}$ the identity $\det s = s$ holds. For triangular matrices $R \in \mathbb{R}^{n \times n}$, that is $r_{ij} = 0$ for all $i > j$ or for all $i < j$, one deduces $\det R = r_{11} r_{22} \ldots r_{nn}$.

31.3 Image and Kernel of Linear Mappings

For a matrix $A \in \mathbb{R}^{m \times n}$, or the linear mapping $x \mapsto Ax$, $x \in \mathbb{R}^n$, identified with it, its *image* and *kernel* are defined by

$$\operatorname{Im} A = \{w \in \mathbb{R}^m : \exists v \in \mathbb{R}^n, w = Av\},$$
$$\ker A = \{v \in \mathbb{R}^n : Av = 0\}.$$

With this, the identities

$$\mathbb{R}^m = \operatorname{Im} A + \ker A^\mathsf{T}, \quad \mathbb{R}^n = \operatorname{Im} A^\mathsf{T} + \ker A,$$

hold, where the decompositions are even orthogonal, that is for $w = Av \in \operatorname{Im} A$ and $u \in \ker A^\mathsf{T}$ we have

$$w \cdot u = w^\mathsf{T} u = (Av)^\mathsf{T} u = (v^\mathsf{T} A^\mathsf{T}) u = v^\mathsf{T} (A^\mathsf{T} u) = 0.$$

Thus, $\operatorname{Im} A$ is the orthogonal complement of $\ker A^\mathsf{T}$, that is $\operatorname{Im} A = (\ker A^\mathsf{T})^\perp$. The rank of a matrix A is the dimension of the image of the induced linear mapping, that is

$$\operatorname{rank} A = \dim \operatorname{Im} A.$$

The *rank* of a matrix corresponds to the number of linearly independent column vectors. Elementary arguments reveal that the rank of a matrix matches the rank of the transposed matrix. From the orthogonality of the above decompositions, the formulas

$$m = \operatorname{rank} A + \dim \ker A^\mathsf{T}, \quad n = \operatorname{rank} A + \dim \ker A,$$

in particular, $\operatorname{rank} A = \operatorname{rank} A^\mathsf{T}$. For an endomorphism or a square matrix $A \in \mathbb{R}^{n \times n}$, it follows that it is bijective if and only if it is surjective, that is $\operatorname{Im} A = \mathbb{R}^n$, or injective, that is $\ker A = \{0\}$. In this case, A is regular or invertible and it holds that $\det A \neq 0$.

31.4 Eigenvalues and Diagonalisability

Characteristic information about a matrix A and the associated linear mapping are contained in the *eigenvalues*, which are the roots of the *characteristic polynomial* of degree n

$$p_A(t) = \det(A - t I_n).$$

A number $\lambda \in \mathbb{R}$ is an eigenvalue of A if and only if an associated eigenvector $v \in \mathbb{R}^n \setminus \{0\}$ with $Av = \lambda v$ exists. The set of eigenvalues is also referred to as the *spectrum*. Every triangular matrix $R \in \mathbb{R}^{n \times n}$ has, taking into account multiplicities, n eigenvalues, which are given by the diagonal entries of R. A matrix $A \in \mathbb{R}^{n \times n}$ is called *diagonalisable*, if an invertible matrix $V \in \mathbb{R}^{n \times n}$ and a diagonal matrix $D \in \mathbb{R}^{n \times n}$ exist, such that $V^{-1} A V = D$ holds. In this case, A and D have the same eigenvalues and these are given by the diagonal entries of D. Furthermore, the column vectors of V are associated eigenvectors, since

$$[A v_1, \ldots, A v_n] = A [v_1, \ldots, v_n] = A V = V D = [v_1, \ldots, v_n] D$$
$$= [\lambda_1 v_1, \ldots, \lambda_n v_n].$$

This implies that A is diagonalisable if and only if there is a basis consisting of eigenvectors of A. An example of a non-diagonalisable matrix is

$$A = \begin{bmatrix} 0 & 1 \\ 0 & 0 \end{bmatrix},$$

because the characteristic polynomial of A has the double root $\lambda = 0$ and if A were diagonalisable, there would be an invertible matrix $V \in \mathbb{R}^{2 \times 2}$ with $V^{-1} A V = 0$, which would imply $A = 0$. Symmetric matrices are always diagonalisable and there exists an orthonormal basis consisting of eigenvectors, that is there exist linearly independent eigenvectors $v_1, \ldots, v_n$ with $\|v_j\|_2 = 1$ and $v_j \cdot v_k = 0$ for $1 \leq j, k \leq n$ with $j \neq k$.

31.5 Jordan Normal Form

The characteristic polynomial of a matrix $A \in \mathbb{R}^{n \times n}$ always has n complex roots, however, the mapping defined by $A : \mathbb{C}^n \to \mathbb{C}^n$, $z \mapsto Az$ is generally not diagonalisable. Every matrix $A \in \mathbb{R}^{n \times n}$ is however complex triangularisable, that is there exist an invertible matrix $T \in \mathbb{C}^{n \times n}$ and an upper triangular matrix $J \in \mathbb{C}^{n \times n}$, whose diagonal entries are the complex eigenvalues of A, such that $A = T^{-1}JT$ holds. The existence of the *Jordan normal form* states that J can be chosen so that

$$J = \begin{bmatrix} J_1 & & & \\ & J_2 & & \\ & & \ddots & \\ & & & J_r \end{bmatrix}$$

with block matrices $J_i \in \mathbb{R}^{s_i \times s_i}$, $i = 1, 2, \ldots, r$, the so-called Jordan blocks, which are associated with eigenvalues λ_{ℓ_i}, $i = 1, 2, \ldots, r$, through

$$J_i = \begin{bmatrix} \lambda_{\ell_i} & 1 & & \\ & \lambda_{\ell_i} & \ddots & \\ & & \ddots & 1 \\ & & & \lambda_{\ell_i} \end{bmatrix}.$$

Here, the number of Jordan blocks associated with an eigenvalue λ corresponds to its geometric multiplicity, that is the dimension of $\ker(A - \lambda I_n)$. The sum of the sizes of the Jordan blocks of an eigenvalue λ corresponds to its algebraic multiplicity, that is the multiplicity of the root λ of the characteristic polynomial $p_A(t)$.

Chapter 32
Results from Analysis

32.1 Continuous and Differentiable Functions

The *intermediate value theorem* guarantees for every continuous function $f \in C^0([a, b])$ with the property $f(a)f(b) \le 0$ the existence of a $\xi \in [a, b]$, such that $f(\xi) = 0$ holds. The *Bolzano–Weierstrass theorem* states that every function $f \in C^0([a, b])$ attains its maximum and minimum, that is, there exist $\xi_{max}, \xi_{min} \in [a, b]$ with $f(\xi_{max}) \ge f(x) \ge f(\xi_{min})$ for all $x \in [a, b]$. A function $f : [a, b] \to \mathbb{R}$ is called differentiable at $x_0 \in [a, b]$, if a number $L \in \mathbb{R}$, a number $\delta > 0$ and a function $\varphi : [0, \delta) \to \mathbb{R}$ exist, such that

$$f(x) = f(x_0) + L(x - x_0) + \varphi(x - x_0)$$

for all $x \in [a, b]$ with $|x - x_0| < \delta$ and $\lim_{s \to 0} \varphi(s)/|s| \to 0$ holds. In this case, L is called the derivative of f at x_0 and we write $f'(x_0) = L$. If f is differentiable at every point $x_0 \in [a, b]$, and if the induced function $x_0 \mapsto f'(x_0)$ is continuous, then f is called continuously differentiable and we write $f \in C^1([a, b])$. Inductively, k-times continuously differentiable functions $f \in C^k([a, b])$ can be defined. For each $k \in \mathbb{N}_0$ the set $C^k([a, b])$ is a vector space, on which

$$\|f\|_{C^k([a,b])} = \max_{i=0,\dots,k} \sup_{x \in [a,b]} |f^{(i)}(x)|$$

the so-called supremum norm is defined. If a function f or one of its derivatives up to order k is only continuous in the open interval (a, b), we write $f \in C^k(a, b)$. In this case, f or a derivative of f can be unbounded and the norm $\|f\|_{C^k([a,b])}$ may not be defined.

32.2 Mean Value Theorem and Taylor Polynomials

If $f \in C^0([a,b]) \cap C^1(a,b)$, the mean value theorem states, or in the special case $f(a) = f(b)$ *Rolle's theorem*, that there exists a $\xi \in (a,b)$ with

$$\frac{f(a) - f(b)}{a - b} = f'(\xi).$$

According to the *fundamental theorem of differential and integral calculus* the identity

$$\int_a^b f'(x)\,dx = f(b) - f(a),$$

holds, and with the *mean value theorem* it follows

$$f(b) - f(a) = \int_a^b f'(x)\,dx = f'(\xi)(b-a)$$

for a $\xi \in (a,b)$. More generally, for a function $f \in C^{k+1}([a,b])$ and $x_0 \in [a,b]$ *Taylor's formula* states that

$$f(x) = f(x_0) + f'(x_0)(x - x_0) + \cdots + \frac{1}{k!}f^{(k)}(x_0)(x - x_0)^k + R_{k+1}(x_0)$$

$$= \sum_{j=0}^k \frac{1}{j!}f^{(j)}(x_0)(x - x_0)^j + R_{k+1}(x_0),$$

with a *remainder term* $R_{k+1}(x_0)$, such that the *Lagrange representation*

$$R_{k+1}(x_0) = \frac{1}{k!}\int_{x_0}^x (x-t)^k f^{(k+1)}(t)\,dt = \frac{1}{(k+1)!}f^{(k+1)}(\xi)(x-x_0)^{k+1}$$

holds with a number $\xi \in [x_0, x]$. The Taylor formula thus defines an approximating polynomial $T_{k,x_0}f$ of degree k with the property

$$\|f - T_{k,x_0}f\|_{C^0([a,b])} \leq \frac{\|f^{(k+1)}\|_{C^0([a,b])}}{(k+1)!}(b-a)^{k+1}.$$

32.3 Landau Symbols

The approximation property of the Taylor polynomial can be written more concisely for a function $f \in C^{k+1}([a, b])$ as

$$f(x) - T_{k,x_0} f(x) = \mathcal{O}(|x - x_0|^{k+1}), \quad x \to x_0.$$

Here, the so-called *Landau symbol* $\mathcal{O}(|x-x_0|^{k+1})$ stands for an expression $\varphi(x-x_0)$ with a function $\varphi : \mathbb{R} \to \mathbb{R}$, for which numbers $\delta > 0$ and $c \geq 0$ exist, such that $|\varphi(s)|/|s|^{k+1} \leq c$ for all $s \in \mathbb{R} \setminus \{0\}$ with $|s| \leq \delta$. For all $x_0 \in [a, b]$ and $x \in [a, b]$ with $|x - x_0| < \delta$ it therefore holds that

$$|f(x) - T_{k,x_0} f(x)| \leq c|x - x_0|^{k+1}.$$

More generally, for $f \in C^k([a, b])$ the property holds that

$$f(x) - T_{k,x_0} f(x) = o(|x - x_0|^k), \quad x \to x_0,$$

where the Landau symbol $o(|x - x_0|^k)$ represents an expression $\varphi(x - x_0)$ with a function $\varphi : \mathbb{R} \to \mathbb{R}$ that has the property $\lim_{s \to 0} \varphi(s)/|s|^k \to 0$. Based on the Taylor formula, the Weierstrass approximation theorem can be proven, which states that every function $f \in C^0([a, b])$ can be uniformly approximated by polynomials. In contrast to the notation $\mathcal{O}(n^p)$ used, for example, in complexity analysis of algorithms, here the limit $s \to 0$ is considered. The most important cases of the Landau symbols can be summarised as follows:

$$g(n) = \mathcal{O}(n^p), \; n \to \infty \iff \exists c \geq 0 \; \forall n \in \mathbb{N}, \; |g(n)| \leq cn^p,$$

$$\psi(s) = \mathcal{O}(|s|^p), \; s \to 0 \iff \exists c \geq 0 \limsup_{s \to 0} |\psi(s)|/|s|^p \leq c,$$

$$\psi(s) = o(|s|^p), \; s \to 0 \iff \lim_{s \to 0} |\psi(s)|/|s|^p = 0.$$

Usually, it is clear from the context which limit is meant, so the addition $n \to \infty$, $s \to 0$ or $x \to x_0$ is often omitted.

32.4 Fundamental Theorem of Algebra

A point $x_0 \in [a, b]$ is referred to as an ℓ-fold root of a function $f \in C^r([a, b])$ if $r \geq \ell - 1$ and $f^{(j)}(x_0) = 0$ for $j = 0, 1, \ldots, \ell - 1$. In the case of a polynomial p of degree $k \geq \ell$ it follows that

$$p(x) = (x - x_0)^\ell r(x)$$

with a polynomial r of degree $k - \ell$. If one identifies a polynomial $p(x) = a_k x^k + a_{k-1} x^{k-1} + \cdots + a_0$ of degree k with real coefficients with a mapping $f : \mathbb{C} \to \mathbb{C}$, $f(z) = a_k z^k + a_{k-1} z^{k-1} + \ldots a_0$, then according to the *fundamental theorem of algebra* the function f always has k roots $z_1, z_2, \ldots, z_k \in \mathbb{C}$, which are not generally pairwise different. If $z_j \in \mathbb{C} \setminus \mathbb{R}$ is a strictly complex root, then the complex conjugate number $\bar{z}_j$ is also a root. If polynomials p and q are given, then there exist polynomials s and r such that

$$p(x) = s(x)q(x) + r(x)$$

holds for all $x \in \mathbb{R}$. With the condition that the degree of the remainder r is truly smaller than that of s, s and r are uniquely determined, provided p or q is not identically zero.

32.5 Multidimensional Calculus

A continuous mapping $f : U \to \mathbb{R}^m$ defined on an open set $U \subset \mathbb{R}^n$ is called (totally) differentiable at the point $x_0 \in U$, if a linear mapping $L : \mathbb{R}^n \to \mathbb{R}^m$ exists, such that

$$f(x) - f(x_0) = L(x - x_0) + o(\|x - x_0\|_2)$$

holds. In this case, the differential $Df(x_0)$ of f at the point x_0 is defined as the linear mapping L and is identified with the representing, so-called *Jacobian* or functional matrix. This matrix is also denoted by $Df(x_0) \in \mathbb{R}^{m \times n}$ and its entries are for $i = 1, 2, \ldots, m$ and $j = 1, 2, \ldots, n$ given by the partial derivatives

$$\partial_j f_i(x_0) = \frac{\partial f_i}{\partial x_j}(x_0) = \lim_{x \to x_0} \frac{f_i(x_0 + h e_j) - f(x_0)}{h}$$

If f is differentiable at every point $x_0 \in U$, we write $f \in C^1(U; \mathbb{R}^m)$. In the case $m = 1$, that is $f : U \to \mathbb{R}$, the *gradient* of f is defined by $\nabla f(x) = (Df(x))^\mathsf{T} \in \mathbb{R}^n$. With this definition, we have

$$Df(x)[s] = \nabla f(x) \cdot s$$

for all $s \in \mathbb{R}^n$. If all partial derivatives of ∇f are continuously differentiable, then the symmetric *Hessian matrix* $D^2 f$ corresponds to the functional matrix of ∇f. The multidimensional Taylor formula implies that for $f \in C^2(U)$, U convex, and $x, x_0 \in U$ a $\xi = t x_0 + (1-t)x \in U$, $t \in [0, 1]$, exists, such that

$$f(x) = f(x_0) + \nabla f(x) \cdot (x - x_0) + \frac{1}{2} D^2 f(\xi)[x - x_0, x - x_0].$$

Here, $D^2 f(\xi)[d, d]$ stands for the expression $d^\mathsf{T} D^2 f(\xi) d$. A necessary condition for an extremum of a function $f \in C^1(U)$ at the point $x_0 \in U$ is that $\nabla f(x_0) = 0$ holds. If additionally $f \in C^2(U)$ is fulfilled and $D^2 f(x_0)$ is positive definite, it follows that x_0 is a local isolated minimum. In the case of a convex function, x_0 is even a global, unique minimum.

Chapter 33
Introduction to C++

33.1 Structure

The programming language C++ is a compiler-based language, which means that programs created with a text editor like *emacs* or *kate* are translated into machine-readable code with the help of the compiler. For this to work flawlessly, programs must be written within a predefined framework. A C++ language program begins with the integration of required predefined routines, which are provided in libraries and classes, such as mathematical functions or input and output functions. This is optionally followed by self-defined functions and at the end is the main program beginning with `main()`. In the main program are variable definitions and commands such as value assignments and function calls. The program on the left in Fig. 33.1 shows a simple example, in which the square of a number is calculated in a subroutine. This is called from the main program with an argument. If the program is saved as a text file under the name `comp_square.cc`, it can be compiled in the same directory with the command

```
$ g++ comp_square.cc -o comp_square.out
```

To start the program, use the command

```
$ ./comp_square.out
```

33.2 Classes

The `iostream` class provides routines for input and output, while the `cmath` class implements elementary mathematical functions. The basic arithmetic operations `+,-,*,/` can be used without the inclusion of libraries. To output text and numbers, the command `std::cout << "text \n"` or `std::cout << x` is used,

```
// comp_square.cc
#include <iostream>
#include <cmath>
double square(double x){
  return pow(x,2.0);
}
int main(){
  double x, y;
  x = 3.8;
  y = square(x);
  std::cout << "square is ";
  std::cout << y << "\n";
}
```

```
// simple_loop.cc
#include <iostream>
int main(){
  int i;
  for (i=0; i<5; i=i+1){
    std::cout << i << "\n";
  }
  std::cout << "\n";
  if (i==5){
    std::cout << "i is 5 \n";
  }
}
//
```

Fig. 33.1 Elementary programs in C++

Table 33.1 Input and output functions, comments and elementary mathematical functions

cout, cin	Output and input of text and variables
\n, endl	Creation of a line break
/*...*/, //	Multi-line and single-line comments
cos, sin, tan	Trigonometric functions
exp, log, log10	Exponential function and logarithms
pow, sqrt	Power and square root
floor, ceil, fabs	Rounding to integers and absolute function

where \n causes a line break. The reading of values for a variable is done with std::cin >> x. By the instruction

```
using namespace std;
```

the additions std:: can be avoided. Further commands are listed in Table 33.1.

33.3 Types

Every variable must be declared in C++, that is, before its use it is determined whether it will hold an integer or a floating-point number, i.e. values of type *integer* or *double* to be stored, see Table 33.2. The use of numbers and arithmetic operations is also associated with types, for example, the expression 2 is interpreted as a variable of type *integer*, while 2. is used as a variable of type *double*. The operation 2/3 is executed in C++ as a binary operation of the higher-value variable type, which means

$$2/3 = 0, \quad 2./3. \approx 0.\overline{6}, \quad 2/3. \approx 0.\overline{6}, \quad 2./3 \approx 0.\overline{6}.$$

Variables can, for example, be converted with x = (double)a. Lists and matrices of fixed size are referred to as *arrays* and can be initialised via double x[n]

Table 33.2 Variable types in C++

`int`	Integer machine numbers
`float, double`	Floating point numbers of single and double precision
`bool`	Boolean variables with values 0 and 1

```
if (condA) statementA else if (condB) statementB else statementC
while (cond) statement
for (init; cond; step) statement
```

Fig. 33.2 Case distinction, repetition and enumeration in C++

or `double A[m][n]`. The indexing of array entries starts with 0. Incorrect indexing of arrays generally does not lead to an error message and must be excluded in the program. When declaring a variable, a value can already be assigned, provided it is not an array whose size is defined by a variable.

33.4 Control Statements

In C++, case distinctions, repetitions and enumerations can be implemented in the formats shown in Fig. 33.2. Here, `cond` stands for a logical condition that can be defined via a logical operation like a<b, while `statement` stands for a list of commands enclosed by curly brackets. The expressions `init` and `step` stand for an initialisation like `i=0` and a statement of the kind `i=i+1`, whose execution is repeated as long as the condition `cond`, for example `i<5`, is evaluated as true. First, `init` is executed, then `cond` is checked, then the command block `statement` is processed and finally `step` is executed, before the condition `cond` is evaluated again and this process is repeated until `cond` is false. Occasionally, the use of the `do while` loop is also useful, in which the condition is checked after rather than before the execution of the commands.

33.5 Logical Expressions and Increments

Boolean variables, binary operations for comparing machine numbers, the logical conjunctions *and*, *or* and the negation are available for formulating logical conditions, see Table 33.3. Comparisons are in brackets, for example (a<b). Floating point numbers are only compared to machine accuracy and due to possible disturbances, a test for exact equality of two floating point numbers is not very meaningful. The command `i=i+1` can be replaced in C++ by `i++` or `++i` and used in arithmetic or logical expressions. In the case of `++i`, the variable is first increased and then the expression is evaluated, whereas when using `i++`, the variable is first increased. The logical expressions (`++i==1`) or (`i++==1`) thus lead to different truth values.

Table 33.3 Logical operations as well as increment- and decrement functions

==, !=, >, >=, <, <=	Logical comparison of machine numbers
&&, ||, !	Logical conjunctions *and*, *or* as well as negation
i++, ++i, i--, --i	Pre- and post-increment as well as -decrement
b+=x	Short form for b=b+x

33.6 Functions

Functions can either return one variable or none. If a value is returned, the type of the function value is placed before the function name, otherwise `void` is used. Following the function name is a list of arguments in brackets. Arguments of simple types like `double` and `int` are handled by *call by value*, meaning they are copied into a local variable. The corresponding calling variable in the main program remains unchanged. Arrays are not allowed as a return value of a function and are therefore passed to functions as arguments by *call by reference* and are changed as global variables by the subroutine. In the program shown on the left in Fig. 33.3, the array `x` is passed to the function `mod_vector` and used there under the name `vec`. After the function has run, the values of the array `x` are changed. The use of the asterisk in the declaration of the function's argument is important here.

```
1  // static_array.cc
2  #include <iostream>
3  using namespace std;
4  void mod_vector(double* vec){
5    vec[0] = 2.0;
6    vec[1] = 1.0;
7  }
8  int main(){
9    const int n = 2;
10   double x[n];
11   x[0] = 1.0;
12   x[1] = 2.0;
13   mod_vector(x);
14   cout << "x[0] = " << x[0];
15   cout << "\n";
16   cout << "x[1] = " << x[1];
17   cout << "\n";
18 }
```

```
1  // functions.cc
2  #include <iostream>
3  using namespace std;
4  void fun_1(double z){
5    z = z+1.0;
6  }
7  void fun_2(double* z){
8    *z = *z+1.0;
9  }
10 int main(){
11   double x = 1.0;
12   fun_1(x);
13   cout << "x = " << x << "\n";
14   fun_2(&x);
15   cout << "x = " << x << "\n";
16 }
17 //
18 //
```

Fig. 33.3 Passing of arrays as well as simple variables and pointers to functions

33.7 Pointers

A pointer is a variable that contains the address of a section in the computer's memory. The pointer allows for reading or changing the content of the corresponding memory section. The size of the section depends on whether a floating point number or integer machine number is to be stored there. If `ptr` is a pointer, then the value of the variable that is found at the address contained in `ptr` is given by the ordinary variable `*ptr`. Conversely, for an ordinary variable `var`, `&var` defines a pointer that contains the address of the corresponding memory location. A pointer is declared, for example, using `double* ptr`. When passing the address of a variable, i.e. the corresponding pointer, to a function, then this variable is treated by *call by reference*, so that the content of the variable is manipulated with global effects. The above-described passing of arrays to functions follows this principle. In the program shown on the right in Fig. 33.3, the function `fun_1` does not change the value of the variable `x` of the main program, while the function `fun_2` increases its value. A pointer defined in a function can be used as the return value of the function. In this case, the function or its value must be declared using `double*`. The passing of variables to functions via pointers avoids, for example, the copying of large data.

33.8 Dynamic Arrays

The above-described use of arrays reaches its limits when the dimension of the arrays is only determined during the course of the program run. The `vector` class provides tools for declaring vectors with entries of certain types and their manipulation. The length of such vectors can then be changed arbitrarily. To be able to use vectors with floating-point number entries concisely in programs, a new variable type should be defined, such as the type `doubleVec`:

`typedef typename std::vector<double> doubleVec;`

In a corresponding variable declaration, the length of the vector and entries can then optionally be specified. By

`doubleVec x(5,1.);`

a vector `x` of length 5 with floating-point number entries, which are initially set to one, is defined. The length and the entries can be changed with the methods shown in Table 33.4. Figure 33.4 shows as an example of application the input and output of variable length vectors.

Table 33.4 Methods for manipulating a variable of a vector type

`x.size()`	Returns the length of the vector `x`
`x.resize(n)`	Changes the length of a vector
`x.push_back(val)`	Appends the entry `val` to a vector
`x.pop_back()`	Deletes the last element of a vector

```cpp
// dynamic_vectors.cc
#include <iostream>
#include <vector>
typedef typename std::vector<double> doubleVec;
doubleVec scan_vector(doubleVec x){
   for (int i=0; i<x.size(); ++i){
      std::cout << "x[" << i << "] = ";
      std::cin >> x[i];
   }
   std::cout << "\n";
   return(x);
}
void print_vector(doubleVec x){
   for (int i=0; i<x.size(); ++i){
      std::cout << "x[" << i << "] = " << x[i] << "\n";
   }
   std::cout << "\n";
}
int main(){
   int dim = 5;
   doubleVec y(dim,1.);
   print_vector(y);
   std::cout << "dim = \n";
   std::cin >> dim;
   y.resize(dim);
   y = scan_vector(y);
   print_vector(y);
}
```

Fig. 33.4 Input and output of vectors of arbitrary length

33.9 Working with Matrices

A matrix $A \in \mathbb{R}^{m \times n}$ can be identified with a vector $\widehat{A} \in \mathbb{R}^{mn}$ by writing the columns of A one below the other into a vector. We have, when numbering the entries with the indices $i = 0, 1, \ldots, m - 1$ and $j = 0, 1, \ldots, n - 1$, that

$$A_{ij} = \widehat{A}_{i+jm}.$$

With this identification, matrices in C++ can be treated like vectors. Occasionally, it is preferable to treat matrices as two-dimensional arrays. If the size is known in advance, they can be used like simple variables. The passage to functions is then done by *call by reference*, as shown in Fig. 33.5.

33.10 Time Measurement, Saving and Packages

The time.h library provides the variable type clock_t, the command clock(), and the constant CLOCKS_PER_SEC for performing runtime measurements. Their use is illustrated in Fig. 33.6.

33.10 Time Measurement, Saving and Packages

```cpp
// matrix.cc
#include <iostream>
const int m = 3;
const int n = 2;
void mod_matrix(double mat[m][n]){
  mat[0][0] = 7.0;
  mat[2][1] = 8.0;
}
int main(){
  double A[m][n] = {{1.,2.},{3.,4.},{5.,6.}};
  mod_matrix(A);
  std::cout << "A[0][0] = " << A[0][0] << "\n";
  std::cout << "A[2][1] = " << A[2][1] << "\n";
}
```

Fig. 33.5 Use of two-dimensional arrays

```cpp
// runtime.cc
#include <iostream>
#include <time.h>
int main(){
  double diff, dt, x = 0.33;
  clock_t t1, t2;
  t1 = clock();
  for (int i=0; i<100000; i++){
    x*x;
  }
  t2 = clock();
  diff = double(t2-t1);
  dt = diff/CLOCKS_PER_SEC;
  std::cout << "runtime = ";
  std::cout << dt << "\n";
}
```

```cpp
// save_data.cc
#include <fstream>
int main(){
  std::fstream f;
  double x[3] = {0.,1.,3.};
  f.open("var.dat",
      std::ios::out);
  if (f.good()){
    for (int i=0; i<3; i++){
      f << x[i] << "\n";
    }
  }
  f.close();
}
//
//
```

Fig. 33.6 Runtime measurement and data storage

To save data in files, methods from the `fstream` class can be used. With the variable type `fstream` and the `open` method, a pointer to a file can be defined, into which the redirection operator `<<` is then written. Once the writing is finished, the file must be closed with `close`. The example program shown on the right in Fig. 33.6 stores a vector in the file `var.dat`. This can be read by MATLAB with the command `load var.dat` and assigns the values of the vector to the variable `var`.

The packages BLAS and LAPACK provide implementations of numerical methods, for example for solving systems of linear equations and eigenvalue problems. The package Eigen contains, for example, methods for efficiently solving sparse systems of linear equations.

Chapter 34
Introduction to MATLAB

34.1 Structure

MATLAB stands for *Matrix Laboratory* and is a commercial software package, which provides implementations of a multitude of numerical methods and allows the creation of your own programs. It is an interpreter language, meaning programs are sequences of commands that are executed without compilation. The user interface essentially consists of the *command window*, in which commands are entered, and an editor, in which programs can be created. These are stored in the format `prog.m`, and can then be started in a command line or from other programs using the command `prog`. A command is terminated with a semicolon. If this is not done, the result of the operation is displayed. Variables are by default defined as type *double*, but they can easily be used like variables of type *integer*, for example when indexing arrays. As a rule, variables are treated by MATLAB as matrices.

34.2 Lists and Arrays

Central objects in MATLAB are matrices or arrays and lists. These are defined using square brackets. Entries of a row are separated by commas and different rows by semicolons. Access to the entries of an array starts with index 1. Submatrices like $A_{IJ} = (a_{ij})_{i \in I, j \in J}$ can be extracted via index lists; boolean lists can also be used instead of index lists. Table 34.1 shows some important operations.

Table 34.1 Creation of lists and arrays

[a,b,...;x,y,...]	Definition of an array (commas optional)
[a,b,...],[x;y;...]	Definition of a row or column vector
A(i,j), I(j)	Access to the entries of an array
a:b, a:step:b	List from a to b with step size 1 or *step*
A(i,:), A(:,j)	i-th row and j-th column of A
A(I,J)	Submatrix defined by lists I and J
ones(n,m)	Array with entries 1
zeros(n,m)	Array with entries 0
accumarray(I,X)	Construction of an array by summation

Table 34.2 Elementary matrix operations

A'	Transposed matrix
A+B, A-B, A*B	Addition, subtraction and product of matrices
inv(A), det(A)	Inverse and determinant of a matrix
x = A \ b	Solution of the linear system $Ax = b$
eye(n)	Identity matrix of dimension n
A.*B, A./B	component-wise multiplication and division
diag(A)	Extraction of the diagonal elements
[L,U] = lu(A)	LU decomposition of a matrix
L = chol(A)	Cholesky decomposition of a matrix
[Q,R] = qr(A)	QR decomposition of a matrix
[V,D] = eig(A)	Approximation of eigenvectors and eigenvalues

34.3 Matrix Operations

The basic matrix operations are defined in MATLAB and can be used in a canonical way, whereby the well-posedness of the operation should be ensured. Matrix factorisations and approximations of eigenvectors and eigenvalues are also available. Some standard routines are listed in Table 34.2.

34.4 Manipulation of Arrays

Various set-theoretic operations and rearrangements of arrays are available in routines. These usually allow further arguments and output values, with which the execution can be specified such as the formation of the row or column-wise maximum. Table 34.3 shows some useful commands.

Table 34.3 Manipulation of arrays

A(:)	Rearrangement of an array into a column vector
reshape(A,m,n)	Rearrangement of an array as an $m \times n$ array
repmat(A,m,n)	Repeated arrangement of an array
unique(A)	Extraction of the elements of an array
setdiff(A,B)	Complement of A and B
sort(A)	Sorting of the entries of an array
sum(A,1), sum(A,2)	Column and row-wise summation
max(A), min(A)	Column-wise extreme values of an array
size(A), length(I)	Dimensions of an array and length of a list

Table 34.4 Elementary functions

sqrt(x), x^y	Square root and powers
exp(x), ln(x)	Exponential function and logarithm
sin(x), cos(x), pi	Trigonometric functions and constant π
norm(x,p)	p-norm of a vector

34.5 Elementary Functions

Numerical realisations of some functions are available under their respective names. They can be applied to arrays, which usually realises the component-wise execution. In exceptions like A^n, the component-wise execution is generated by A.^n. A brief overview can be found in Table 34.4.

34.6 Loops and Control Statements

Loops can be realised over lists or control statements in an obvious way. The comparison of variables can be applied to arrays. Figure 34.1 shows the structure of the most important case distinctions and Table 34.5 shows some important commands.

```
if (condA) statementA elseif (condB) statementB
    else statementC end
while (cond) statement end
for i = I statement
```

Fig. 34.1 Case distinction, repetition and enumeration in MATLAB

Table 34.5 Logical operations

a==b, a~=b	Logical test for equality or inequality
a<b, a<=b	Logical comparison of two numbers
E&&F, E||F	Logical *and* and *or*
tic ... toc	Measurement of CPU time

Table 34.6 Displaying of objects

disp(A)	Display of the variable *A*
plot(X,Y,'-*')	Polygonal chain through points $(X(k), Y(k))$ in $\mathbb{R}^2$
hold on, hold off	Display of multiple objects in one graphic
mesh(X,Y,Z)	Display of a two-dimensional graph
meshgrid	Generation of a grid
axis([x1,x2,...])	Limitation of the displayed region
xlabel, ylabel	Labelling of the axes
legend	Insertion of a legend
figure(k)	Selection of a graphic window
subplot(n,m,j)	Display of multiple plots in one window
quiver, quiver3	Visualisation of vector fields
trisurf	Graph of a function on a triangular grid
tetramesh	Display of a decomposition into tetrahedra

34.7 Text and Graphic Output

If a program is started via a command line, intermediate results can be output in the command window. Functions or other objects can be displayed in graphic windows called *figures*. A selection of corresponding MATLAB commands can be found in Table 34.6.

34.8 Creating New Functions

New functions with multiple inputs and outputs can be defined using the framework shown in Fig. 34.2. The concluding `end` is optional. Functions should be saved as a file with the name of the function, for example `new_function.m`. A file can contain multiple function definitions, but only the first can be called from outside via the file name. For this one has to be in the directory of the file or the path must have been set up as a search path.

```
function [y1,y2,..] = new_function(x1,x2,..)
..
end
```

Fig. 34.2 Framework for a newly created function

34.11 Examples

Table 34.7 Various commands

whos, clear	Display and deletion of all variables
clc, clf	Clearing of the command or graphic window
addpath	Addition of a search path for functions
save, load	Loading and saving of variables
Ctrl-C	Termination of a program
fopen	Opening of a file
printf	Formatted output
strcat	Concatenation of strings

34.9 Various Commands

In addition to some Unix commands such as cd and ls, various commands for managing the used files and directories as well as variables are available, which are shown in Table 34.7.

34.10 Sparse Matrices

For matrices with many vanishing entries, the effort of solving associated linear systems can be reduced, provided the matrices are defined using the matrix type *sparse*. For index lists $I \subset \{1, 2, \ldots, m\}$ and $J \subset \{1, 2, \ldots, n\}$ and a vector X of the same length, a matrix $A \in \mathbb{R}^{m \times n}$ is defined by

$$a_{ij} = \sum_{k : I(k)=i, J(k)=j} X(k),$$

that is, for multiple occurring index pairs, the associated entries are summed. Access to individual entries of a sparse matrix is generally inefficient. Some important commands are listed in Table 34.8.

34.11 Examples

In Fig. 34.3, the input of various commands in the command window of MATLAB is shown. The calculation of the determinant of a matrix according to the Laplace expansion theorem leads to a recursion, the MATLAB implementation of which is

Table 34.8 Generation of sparse matrices

sparse(I,J,X,m,n)	Composition of a sparse matrix
speye(n)	Identity matrix as a sparse matrix

```
>> A = [2,1;1,2]; b = [1;1];          >> x = [pi/2,0,1];
>> x = A\b                            >> sin(x)

 x =                                   ans =

    0.3333                                1.0000         0    0.8415
    0.3333
                                      >> sqrt(-1)
>> x'
                                       ans =
 ans =
                                          0.0000 + 1.0000i
    0.3333    0.3333
                                      >>
>>                                    >>
```

Fig. 34.3 Execution of commands in the command window

```
 1  function val = laplace(A)          1  function x = bisect(a,b)
 2  n = size(A,1);                     2  x = a; z = b;
 3  val = 0;                           3  tol = 1e-4;
 4  if n == 1                          4  while z-x > tol
 5      val = A(1,1);                  5      c = (x+z)/2;
 6  else                               6      if f(x)*f(c)<0
 7      for j = 1:n                    7          z = c;
 8          I = 2:n;                   8      else
 9          J = [1:j-1,j+1:n];         9          x = c;
10          val = val+(-1)^(1+j)...   10      end
11              *A(1,j)...            11  end
12              *laplace(A(I,J));     12
13      end                           13  function y = f(x)
14  end                               14  y = x^3+cos((pi/2)*x);
```

Fig. 34.4 Calculation of the determinant according to Laplace (left) and implementation of the bisection method (right)

shown in Fig. 34.4. An implementation of the bisection method and its application to a function $f(x)$ is also shown in Fig. 34.4.

The graphical representation of various functions in a graphics window is illustrated by the program `several_plots.m` shown in Fig. 34.5. The function `plot_bubble.m` shown next to it evaluates a function $f(x)$ defined on $\mathbb{R}^2$ and displays it graphically. The graphics generated by the functions are shown in Fig. 34.6.

34.12 Free Alternative

Octave is a freely available software package, which is largely compatible with MATLAB. However, some solution routines for ordinary differential equations are not available in Octave.

34.12 Free Alternative

```
function several_plots
dx = .1;
X = 0:dx:pi;
Y1 = sin(X); plot(X,Y1,'-r');
hold on;
Y2 = cos(X); plot(X,Y2,':k');
hold off;
legend('sin','cos');
disp('press key'); pause; clf
Z1 = exp(X); plot(X,Z1,'-+');
hold on;
Z2 = log(X); plot(X,Z2,'-*');
hold off;
legend('exp','log');
```

```
function plot_bubble
dx = .1;
dy = .1;
[X,Y] = ...
    meshgrid(-2:dx:2,-2:dy:2);
W = f([X(:),Y(:)]);
Z = reshape(W,size(X));
mesh(X,Y,Z);

function y = f(x)
y = zeros(size(x,1),1);
r = sum(x.^2,2).^(1/2);
I = r<1;
y(I) = exp(-1./(1-r(I).^2));
```

Fig. 34.5 Representation of one-dimensional functions (left) and a function defined on $\mathbb{R}^2$ (right)

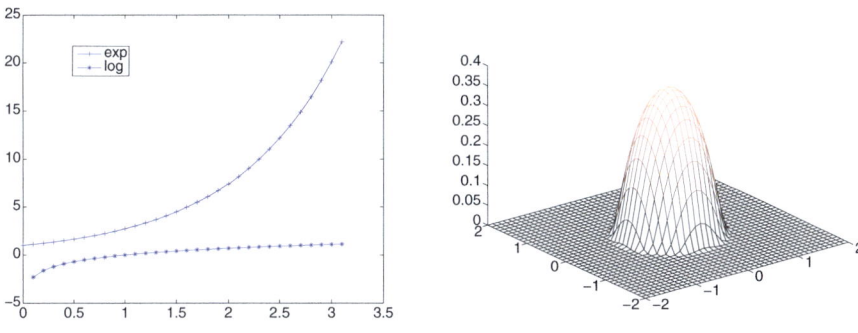

Fig. 34.6 Graphic outputs of the functions several_plots.m (left) and plot_bubble.m (right)

Chapter 35
Introduction to Python

35.1 Structure

The programming language Python is an interpreter language that allows both the interactive input of commands in a console as well as the execution of programs or scripts, which are saved as a sequence of commands in a file. Python is well-known for its good readability of programs, the implementation in common operating systems and the availability of extensive libraries and modules. Variables in Python do not need to be declared and can be used directly. It is important to maintain block structures by indenting commands and marking their beginning with a colon. In Fig. 35.1 two simple Python programs are shown, which are saved as text files with the specified filenames. They are for example started with the command

```
$ python3 comp_square.py
```

from a console.

Typically, each instruction is written into a separate line. A longer command can be continued into further lines with a backslash. To write multiple commands in one line, they are separated with a semicolon. Programs usually start with the inclusion of modules and the definition of functions.

```
# comp_square.py
def square(x):
    x_sq = x**2
    return x_sq
x = 3.8
y = square(x)
print(y)
```

```
# simple_loop.py
import math
n = 10
delta_x = math.pi/n
for i in range(n):
    x = math.sin(i*delta_x)
    print(i,x)
```

Fig. 35.1 Elementary programs in Python; indentations are essential components for marking program blocks

35.2 Elementary Commands

The basic arithmetic operations and the input and output of variables are directly available in Python. The assignment of values is done with an equals sign. The command `print` displays variables and text flexibly and always ends automatically with a line break. Single-line comments are marked with a hash, multi-line comments with triple quotation marks. Further mathematical functions are provided by the modules `math` and `numpy`, which are included via the command `import`. To avoid cumbersome commands, a library can be renamed when included using `as`. The output of floating point numbers can be formatted, for example by:

`print("x = {:>7.4f}".format(x))`

Here, space for seven characters is reserved right-aligned and four decimal places of the variable x are displayed. Table 35.1 provides an overview of some commands.

Table 35.1 Input and output functions, comments, importing of modules as well as mathematical functions

print, input	Output and input of text and variables
\n	Generation of a line break
"""..."""`, #	Multi-line and single-line comments
**, %	Power and remainder in division
a+=b	Short form for a=a+b
import [as]	Inclusion of libraries
e, pi	Euler's number and π
cos, sin, tan	Trigonometric functions
exp, log, log10	Exponential function and logarithms
pow, sqrt	Power and square root
floor, ceil, abs	Rounding to whole numbers and absolute function

Table 35.2 Immediately available variable types in Python

int	Integer machine numbers
float	Double precision floating point numbers
bool	Boolean variables with values `false` and `true`
str	Strings
list	Lists

```
if (condA): statementA elif (condB): statementB else: statementC
while (cond): statement
for i in <enumerating object>: statement
```

Fig. 35.2 Control structures in Python: conditional, repetition and enumeration

35.3 Types

Python provides common data types for working with variables, see Table 35.2. The conversion of a variable or object is done, for example, using `i = int(str)`.

Although Python automatically performs type conversions, so for example 1/2 yields the result `0.5`, Python relies on various C++ libraries where this is usually not the case. Floating point numbers should therefore be specified with a point for safety, such as `1./2.` or `1.0/2.0`.

35.4 Control Statements

Conditional statements and loops in Python have a special structure. The blocks to be executed are introduced with a colon. For `for`-loops, the specification of an enumerating object is required, over which the iteration is performed. For this, the command `range(start,stop,step)` is often used, which generates a sequence of numbers. The specifications of the start value and the step size are optional, if they are omitted, then 0 and 1 are used. The stop value is not included in the list, for example

$$\text{range(n)} \equiv (0, 1, \ldots, n-1)$$

is an enumeration of the integers from 0 to $n-1$. The syntactic structures of the most important control statements are shown in Fig. 35.2.

35.5 Logical Expressions

To formulate conditions in conditional statements and repetitions, logical expressions are needed, which can be constructed using the comparisons and conjunctions shown in Table 35.3.

35.6 Functions

Functions begin with the command `def` and can have no or several arguments. They usually return one or more values with the command `return`. This command can be omitted if a variable to be changed is passed by *call by reference* and directly modified by the function. Lists are always passed in this way, so the explicit return can and should be omitted. If a function does not contain a return command, the value `None` is automatically returned. The exemplary use of functions is shown in Fig. 35.3.

35.7 Lists

Python allows the direct definition of lists or vectors without special commands for declaration or changing their length. The elements of an array are separated by commas and enclosed in square brackets:

x = [1.0,2.3,0.7]

Access to one or more elements of a list is also done using square brackets, the indexing of lists starts with index zero. The formal addition of two lists using x+y results in their combination into a larger list. The most important list operations are

Table 35.3 Logical operations for formulating conditions

==, !=, >, >=, <, <=	Logical comparison of machine numbers
in	Logical query for containment
and, or, not	Logical conjunctions *and*, *or* and negation

```
1 # example_function.py
2 def two_values(x,y):
3     u = x*y; v = x+y
4     return u, v
5 a = 2.0; b = 1.0
6 [c,d] = two_values(a,b)
7 print(c,d)
```

```
1 # function_list.py
2 def several_squares(v):
3     for i in range(len(v)):
4         v[i] = v[i]**2
5 x = [1,2,3,4];
6 several_squares(x)
7 print(x)
```

Fig. 35.3 Definition and call of functions with and without explicit return of values

35.8 Timing, Saving and Plotting

Table 35.4 Methods for editing lists

[]	Empty list
[0]*n	Create a list with n zero entries
x[i], x[-m], x[i:s:j]	Access to individual or multiple elements
len(x)	Returns the length of the list x
x.append(a)	Appends an element at the end of the list
x.sort()	Sorts a list
max(x), min(x)	Maximum and minimum of a list
del x[i:j]	Deletes a range of a list

```
# read_vector.py
def scan_vector(x,n):
    for i in range(n):
        str = input("x[i] = ")
        x.append(float(str))
x = []
str = input("dim = ")
n = int(str)
scan_vector(x,n)
print(x)
```

```
# numpy_matrices.py
import numpy as np
m = 5; n = 3
A = np.zeros((m,n))
x = np.ones(n)
for i in range(m):
    for j in range(n):
        A[i][j] = i*n+j
y = np.matmul(A,x)
print(A,x,y)
```

Fig. 35.4 Reading a vector and matrix-vector multiplication

shown in Table 35.4. The program shown in Fig. 35.4 uses the list commands for interactive input of a vector.

For working with matrices, the numpy module with the multidimensional fields and matrix operations defined therein is recommended, a program example is shown in Fig. 35.4.

35.8 Timing, Saving and Plotting

To determine the runtime of a program, the time module provides the function t = time.time(), which returns an absolute time in seconds. To access a file, the command open('file.txt','r') is used when data is to be read. If a file is to be created or overwritten, the argument 'w' is to be used, when appending data the argument 'a'. The addition of the command with ... as serves for error handling and ensures the correct closing of the file. The matplotlib module provides routines for displaying graphs. Examples of the use of the commands and modules can be found in Figs. 35.5 and 35.6. The module scipy contains methods for the numerical solution of differential equations.

```
1  # runtime.py
2  import time
3  t1 = time.time()
4  for i in range(10**6):
5      x = i*i # i**2
6  t2 = time.time()
7  print("runtime = ",t2-t1)
```

```
1  # save_data.py
2  x = [1.0,3.0,6.0]
3  with open('var.dat','w') as f:
4      for i in range(len(x)):
5          f.write("{:>10.4f}\n".\
6              format(x[i]))
7  #
```

Fig. 35.5 Runtime measurement and data saving in Python

```
1   # plot_function.py
2   import math
3   import matplotlib.pyplot as plt
4   n = 100; x = [0]*n;
5   y = [0]*n; z = [0]*n;
6   for i in range(n):
7       x[i] = i*2.0*math.pi/n
8       y[i] = math.sin(x[i])
9       z[i] = math.cos(x[i])
10  plt.plot(x,y,"-b",label="sin")
11  plt.plot(x,z,":k",label="cos")
12  plt.legend(loc="lower left")
13  plt.show()
```

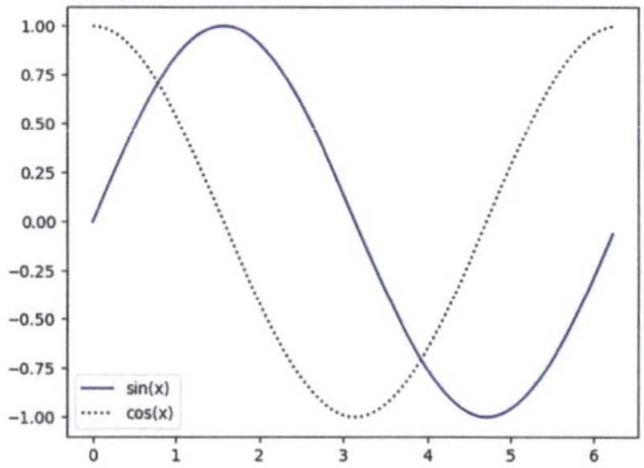

Fig. 35.6 Plotting of functions in Python

Chapter 36
Example Programs in MATLAB, C++ and Python

36.1 *LU* Decomposition and Solving Triangular Systems

The calculation of the *LU* decomposition is pre-implemented in MATLAB and the following commands determine the solution of a system of linear equations. Alternatively, this can be done with the command x = A\beta .

```
>> A = [2,-1,0;-1,2,-1;0,-1,2]; b = [1;1;1];
>> [L,U] = lu(A);
>> y = L\b; x = U\y;
```

The MATLAB program `lu_solution.m` shown in Fig. 36.1 calculates the *LU* decomposition of a given matrix using the Crout algorithm, which is based on the identities

$$u_{ik} = a_{ik} - \sum_{j=1}^{i-1} \ell_{ij} u_{jk}, \quad \ell_{ki} = \left(a_{ki} - \sum_{j=1}^{i-1} \ell_{kj} u_{ji}\right)/u_{ii}$$

The given matrix A is overwritten with the entries of the factors L and U, which is possible due to the normalisation of L, i.e. the condition $\ell_{ii} = 1, i = 1, 2, \ldots, n$.

Electronic Supplementary Material The online version of this chapter (https://doi.org/10.1007/978-3-662-70890-3_36) contains supplementary material, which is available to authorized users.

© The Author(s), under exclusive license to Springer-Verlag GmbH,
DE, part of Springer Nature 2025
S. Bartels, *Numerical Mathematics 3x9*, La Matematica per il 3+2 160,
https://doi.org/10.1007/978-3-662-70890-3_36

```
 1  function lu_solution
 2  n = 3;
 3  A = [2,-1,0;-1,2,-1;0,-1,2];
 4  b = [1;1;1];
 5  A = lu_crout(n,A);
 6  y = solve_lower_normalized(n,A,b);
 7  x = solve_upper(n,A,y);
 8  disp(x);
 9
10  function A = lu_crout(n,A)
11  for i = 1:n
12      for k = i:n
13          s = 0;
14          for j = 1:i-1
15              s = s+A(i,j)*A(j,k);
16          end
17          A(i,k) = A(i,k)-s;
18      end
19      for k = i+1:n
20          s = 0;
21          for j=1:i-1
22              s = s+A(k,j)*A(j,i);
23          end
24          A(k,i) = (A(k,i)-s)/A(i,i);
25      end
26  end
27
28  function y = solve_lower_normalized(n,L,b)
29  y = zeros(n,1);
30  for j = 1:n
31      sum = 0;
32      for k = 1:j-1
33          sum = sum+L(j,k)*y(k);
34      end
35      y(j) = b(j)-sum;
36  end
37
38  function x = solve_upper(n,U,y)
39  x = zeros(n,1);
40  for j = n:-1:1
41      sum = 0;
42      for k = j+1:n
43          sum = sum+U(j,k)*x(k);
44      end
45      x(j) = (y(j)-sum)/U(j,j);
46  end
```

Fig. 36.1 Solution of a system of linear equations using an explicitly calculated LU decomposition in MATLAB (program available at https://doi.org/10.1007/978-3-662-70890-3_36)

With the help of the decomposition, the system of equations $Ax = b$ is then solved by explicitly solving two systems of equations with triangular matrices, i.e.

$$Ax = b \iff Ly = b, \; Ux = y.$$

The forward substitution used in this process takes advantage of the fact that the matrix L is normalised. In the subroutines, only the upper triangular part or the strict lower triangular part of the passed matrix is used.

An implementation in the programming language C++ is shown in Fig. 36.2. It is particularly important to note that the indexing of vectors starts at zero. The compilation and execution of the program is done with the commands:

```
$ g++ lu_solution.c -o lu_solution.out
$ ./lu_solution.out
```

The analogous implementation of the LU method in the programming language Python is shown in Fig. 36.3. The program is started using:

```
$ python3 lu_solution.py
```

Here too, the indexing of lists starts with index 0.

36.2 Polynomial Interpolation and Neville's Scheme

Neville's scheme allows the evaluation of an interpolation polynomial p defined by nodes $x_0, x_1, \ldots, x_n$ and values $y_0, y_1, \ldots, y_n$ at a point z via the formula

$$p_{i,j}(z) = \frac{(z - x_i)p_{i+1,j-1}(z) - (z - x_{i+j})p_{i,j-1}(z)}{x_{i+j} - x_i}$$

for $j = 1, 2, \ldots, n$ and $i = 0, 1, \ldots, n - j$ with the initialisation $p_{i,0}(z) = y_i$ for $i = 0, 1, \ldots, n$. Then $p(z) = p_{0,n}(z)$ holds. In Fig. 36.4, the MATLAB program `neville_scheme.m` is shown, which calculates the values of the interpolation polynomial at the points z_k, $k = 0, 1, \ldots, N$, using the Neville scheme and subsequently approximates the interpolation polynomial graphically by a polygonal line through the points $(z_k, p(z_k))$, $k = 0, 1, \ldots, N$. The calculation is done recursively with the subroutine `neville_recursive` and also by successive evaluation of the above formula in the subroutine `neville_forward`. When accessing arrays, the indices are always increased by the value 1, as the index 0 is not allowed in MATLAB. The final result is thus given by the entry P(1,n+1). Instead of using the array P, the local variable y could also be overwritten in each step of the loop over variable j to save memory.

In MATLAB, various interpolation methods are available. The function values of a cubic spline interpolant can for example be calculated with the following commands:

```cpp
// lu_solution.cc
#include <iostream>
const int n = 3;
void lu_crout(double A[n][n]){
  double s;
  for (int i=0; i<n; i++){
    for (int k=i; k<n; k++){
      s = 0.0;
      for (int j=0; j<=i-1; j++){s = s+A[i][j]*A[j][k];}
      A[i][k] = A[i][k]-s;
    }
    for (int k=i+1; k<n; k++){
      s = 0.0;
      for (int j=0;j<=i-1;j++){s = s+A[k][j]*A[j][i];}
      A[k][i] = (A[k][i]-s)/A[i][i];
    }
  }
}
void solve_lower_normalized(double L[n][n], double b[n],
                double y[n]){
  double s;
  for (int j=0; j<n; j++){
    s = 0.0;
    for (int k=0; k<=j-1; k++){s = s+L[j][k]*y[k];}
    y[j] = b[j]-s;
  }
}
void solve_upper(double U[n][n], double y[n], double x[n]){
  double s;
  for (int j=n-1; j>=0; j--){
    s = 0.0;
    for (int k=j+1; k<n; k++){s = s+U[j][k]*x[k];}
    x[j] = (y[j]-s)/U[j][j];
  }
}
int main(){
  double x[n], y[n];
  double A[n][n] = {{2.,-1.,0.},{-1.,2.,-1.},{0.,-1.,2.}};
  double b[n] = {1.,1.,1.};
  lu_crout(A);
  solve_lower_normalized(A,b,y);
  solve_upper(A,y,x);
  for (int i=0; i<n; i++){
    std::cout << "x[" << i << "] = " << x[i] << "\n";
  }
}
```

Fig. 36.2 Calculation of the *LU* decomposition and subsequent solving of a system of equations in C++ (program available at https://doi.org/10.1007/978-3-662-70890-3_36)

36.2 Polynomial Interpolation and Neville's Scheme

```python
# lu_solution.py
import numpy as np
def lu_crout(A,n):
    for i in range(n):
        for k in range(i,n):
            s = 0.0
            for j in range(i):
                s = s+A[i][j]*A[j][k]
            A[i][k] = A[i][k]-s
        for k in range(i+1,n):
            s = 0.0
            for j in range(0,i):
                s = s+A[k][j]*A[j][i]
            A[k][i] = (A[k][i]-s)/A[i][i]
def solve_lower_normalized(L,b,y):
    for j in range(n):
        s = 0.0
        for k in range(0,j):
            s = s+L[j][k]*y[k]
        y[j] = b[j]-s
def solve_upper(U,y,x):
    for j in range(n-1,-1,-1):
        s = 0.0
        for k in range(j+1,n):
            s = s+U[j][k]*x[k]
        x[j] = (y[j]-s)/U[j][j]
""" MAIN PROG """
n = 3;
A = np.array([[2.0,-1.0,0.0],[-1.0,2.0,-1.0],[0.0,-1.0,2.0]])
b = np.ones(n); y = np.zeros(n); x = np.zeros(n)
lu_crout(A,n)
solve_lower_normalized(A,b,y)
solve_upper(A,y,x);
print(x)
```

Fig. 36.3 Calculation of the LU decomposition and subsequent solving of a system of equations in Python (program available at https://doi.org/10.1007/978-3-662-70890-3_36)

```
>> x = [-1,-1/3,1/3,1]; y = [-1,1,-1,1];
>> N = 100; z = -1+2*[0:N]/N;
>> w = interp1(x,y,z,,spline,);
>> plot(z,w);
```

Here, the interpolation pairs $(x_j, y_j)_{j=0,\ldots,n}$ are used to calculate an interpolating cubic spline, this is evaluated at the points $(z_k)_{k=0,\ldots,N}$ and finally graphically displayed.

A C++ implementation analogous to the program `neville_scheme.m` is shown in Fig. 36.5. Its compilation and execution is similar to the above C++ program. The program uses dynamic lists for the nodes and values as well as evaluation points and associated function values. A Python implementation is shown in Fig. 36.6.

```matlab
function neville_scheme
n = 3;
x = [-1,-1/3,1/3,1];
y = [-1,1,-1,1];
N = 20; z = zeros(N+1,1);
w_rec = zeros(N+1,1);
w_for = zeros(N+1,1);
for k = 0:N
    z(k+1) = -1+2*k/N;
    w_rec(k+1) = neville_recursive(z(k+1),x,y,0,n);
    w_for(k+1) = neville_forward(z(k+1),x,y,n);
end
plot(z,w_rec,'b-o'); hold on;
plot(z,w_for,'r-x'); hold off;

function val = neville_recursive(z,x,y,i,j)
if j == 0
    val = y(i+1);
else
    val = ((z-x(i+1))*neville_recursive(z,x,y,i+1,j-1)...
        -(z-x(i+j+1))*neville_recursive(z,x,y,i,j-1))/...
        (x(i+j+1)-x(i+1));
end

function val = neville_forward(z,x,y,n)
P = zeros(n+1,n+1);
for i = 0:n
    P(i+1,1) = y(i+1);
end
for j = 1:n
    for i = 0:n-j
        P(i+1,j+1) = ((z-x(i+1))*P(i+2,j)...
            -(z-x(i+j+1))*P(i+1,j))/(x(i+j+1)-x(i+1));
    end
end
val = P(1,n+1);
```

Fig. 36.4 Recursive and direct implementation of the Neville scheme for evaluating the Lagrange interpolation polynomial through the interpolation pairs (x_i, y_i), $i = 0, 1, \ldots, n$, in MATLAB (Program available at https://doi.org/10.1007/978-3-662-70890-3_36)

36.3 Numerical Solution of Ordinary Differential Equations

The implicit Euler method approximates the solution of an initial value problem $y' = f(t, y)$, $y(0) = y_0$, by the recursively defined sequence

$$y_{k+1} = y_k + \tau f(t_{k+1}, y_{k+1}) = y_k + \tau \Phi(t_k, y_k, y_{k+1}, \tau).$$

This generally requires the solution of a nonlinear system of equations at each time step, which under suitable conditions is approximated with the fixed point iteration

$$z_{i+1} = \Psi(z_i) = y_k + \tau \Phi(t_k, y_k, z_i, \tau)$$

36.3 Numerical Solution of Ordinary Differential Equations

```
// neville_scheme.cc
#include <iostream>
#include <vector>
typedef typename std::vector<double> doubleVec;
const int n = 3;
double neville_recursive(double z, doubleVec x, doubleVec y,
              int i, int j){
  if (j==0){
    return y[i];
  }
  else{
    return ((z-x[i])*neville_recursive(z,x,y,i+1,j-1)
      -(z-x[i+j])*neville_recursive(z,x,y,i,j-1))/(x[i+j]-x[i]);
  }
}
double neville_forward(double z, doubleVec x, doubleVec y){
  double P[n+1][n+1];
  for (int i=0; i<=n; i++){
    P[i][0] = y[i];
  }
  for (int j=1; j<=n; j++){
    for (int i=0; i<=n-j; i++){
      P[i][j] = ((z-x[i])*P[i+1][j-1]-(z-x[i+j])*P[i][j-1])/
      (x[i+j]-x[i]);
    }
  }
  return P[0][n];
}
int main(){
  int N = 20;
  doubleVec x(n+1), y(n+1);
  doubleVec z(N+1), w_rec(N+1), w_for(N+1);
  x[0] =-1.0; x[1] =-1.0/3; x[2] = 1.0/3; x[3] = 1.0;
  y[0] =-1.0; y[1] = 1.0;   y[2] =-1.0;   y[3] = 1.0;
  for (int k=0; k<=N; k++){
    z[k] = -1.0+2.0*(double)k/N;
    w_rec[k] = neville_recursive(z[k],x,y,0,n);
    w_for[k] = neville_forward(z[k],x,y);
    std::cout << "w_rec = " << w_rec[k] << ", ";
    std::cout << "w_for = " << w_for[k] << "\n";
  }
}
```

Fig. 36.5 Recursive and direct implementation of the Neville scheme for evaluating an interpolation polynomial at various points in C++ (Program available at https://doi.org/10.1007/978-3-662-70890-3_36)

or a Newton method for the equation

$$F(z) = z - y_k - \tau \Phi(t_k, y_k, z, \tau) = 0,$$

that is the iteration

$$z_{i+1} = z_i - F(z_i)/F'(z_i).$$

```python
# neville_scheme.py
import numpy as np
def neville_recursive(z,x,y,i,j):
    if (j==0):
        return y[i]
    else:
        return ((z-x[i])*neville_recursive(z,x,y,i+1,j-1) \
                -(z-x[i+j])*neville_recursive(z,x,y,i,j-1)) \
                /(x[i+j]-x[i])
def neville_forward(z,x,y,n):
    P = np.zeros((n+1,n+1))
    for i in range(n+1):
        P[i][0] = y[i]
    for j in range(1,n+1):
        for i in range(n-j+1):
            P[i][j] = ((z-x[i])*P[i+1][j-1]- \
                       (z-x[i+j])*P[i][j-1])/ \
                       (x[i+j]-x[i])
    return P[0][n]
""" MAIN PROG """
n = 3; N = 20
x = np.array([-1,-1/3,1/3,1])
y = np.array([-1,1,-1,1])
z = np.zeros(N+1);
w_rec = np.zeros(N+1); w_for = np.zeros(N+1)
for k in range(N+1):
    z[k] = -1+2*k/N;
    w_rec[k] = neville_recursive(z[k],x,y,0,n);
    w_for[k] = neville_forward(z[k],x,y,n);
    print("w_rec = {:>7.4f}, w_for = {:>7.4f}" \
          .format(w_rec[k],w_for[k]))
```

Fig. 36.6 Recursive and direct implementation of the Neville scheme for evaluating an interpolation polynomial at various points in Python (Program available at https://doi.org/10.1007/978-3-662-70890-3_36)

As an initial value z_0, the solution from the previous time step is used. Both approaches are implemented in the MATLAB program shown in Fig. 36.7. To take into account the indexing of arrays in MATLAB starting with 1, a routine inc was defined in the program, which increases a given number by the value 1. This allows the iteration rule to be very directly transferred from the theoretical algorithm.

Various methods for the numerical solution of differential equations are already pre-implemented in MATLAB routines, such as in the routine ode45, which returns a list of time points and associated approximations. The following lines show an example of the use of this routine. Other MATLAB routines for solving ordinary differential equations with different accuracy, effort and stability properties are the routines ode23, ode113, ode15s, ode23s, ode23t, ode23tb.

```
>> T = 10; y_0 = 1;
>> f = @(t,y)cos(2*t)*y^2;
>> [t_list,y_list] = ode45(f,[0,T],y_0);
>> plot(t_list,y_list)
```

36.3 Numerical Solution of Ordinary Differential Equations

```
function implicit_euler
y_0 = 1; T = 10;
tau = 1/100; K = floor(T/tau);
y(inc(0)) = y_0;
for k = 0:K-1
    t_k = k*tau;
    y(inc(k+1)) = fixed_point_iteration(t_k,y(inc(k)),tau);
    % y(inc(k+1)) = newton_iteration(t_k,y(inc(k)),tau);
end
plot(tau*(0:K),y);

function z = fixed_point_iteration(t,y_old,tau)
z = y_old; diff = 1; eps_stop = tau/10;
while diff > eps_stop
    z_new = y_old+tau*Phi(t,y_old,z,tau);
    diff = abs(z_new-z);
    z = z_new;
end

function z = newton_iteration(t,y_old,tau)
z = y_old; diff = 1; eps_stop = tau/10;
while diff > eps_stop
    F = z-y_old-tau*Phi(t,y_old,z,tau);
    dF = 1-tau*dPhi_y(t,y_old,z,tau);
    z_new = z-F/dF;
    diff = abs(z_new-z);
    z = z_new;
end

function val = Phi(t,y_old,y_new,tau)
val = f(t+tau,y_new);

function val = dPhi_y(t,y_old,y_new,tau)
val = df_y(t+tau,y_new);

function val = f(t,y)
val = cos(2*t)*y^2;

function val = df_y(t,y)
val = cos(2*t)*2*y;

function val = inc(k)
val = k+1;
```

Fig. 36.7 Two implementations of the implicit Euler method for the numerical solution of an ordinary differential equation in MATLAB; the solution of the nonlinear equation at each time step is done via a fixed point or Newton iteration (program available at https://doi.org/10.1007/978-3-662-70890-3_36)

An implementation in C++ of the implicit Euler method is shown in Fig. 36.8. A corresponding Python program can be found in Fig. 36.9.

```cpp
// implicit_euler.cc
#include <fstream>
#include <cmath>
#include <vector>
typedef typename std::vector<double> doubleVec;
double f(double t, double y){
  return cos(2.0*t)*pow(y,2.0);
}
double Phi(double t, double y_old, double y_new, double tau){
  return f(t+tau,y_new);
}
void save_solution(doubleVec y, int K){
  std::fstream f;
  f.open("sol.dat",std::ios::out);
  if (f.good()){
    for (int k=0; k<=K; k++){
      f << y[k] << "\n";
    }
  }
  f.close();
}
int main(){
  double y_0 = 1.0, T = 10.0, tau = 1.0/100.0, t_k;
  double z, z_new, diff, eps_stop = tau/10;
  int k, K = floor(T/tau);
  doubleVec y(K+1);
  y[0] = y_0;
  for (k=0; k<K; k++){
    t_k = k*tau;
    z = y[k];
    diff = 1.0;
    while (diff>eps_stop){
      z_new = y[k]+tau*Phi(t_k,y[k],z,tau);
      diff = fabs(z_new-z);
      z = z_new;
    }
    y[k+1] = z;
  }
  save_solution(y,K);
}
```

Fig. 36.8 Implementation of the implicit Euler method in C++; the nonlinear equations in the time steps are solved with a fixed point iteration (program available at https://doi.org/10.1007/978-3-662-70890-3_36)

36.3 Numerical Solution of Ordinary Differential Equations

```python
# implicit_euler.py
import numpy as np
import matplotlib.pyplot as plt
def f(t,y):
    return np.cos(2*t)*y**2
def Phi(t,y_old,y_new,tau):
  return f(t+tau,y_new)
def save_solution(y,K):
    with open('sol.dat','w') as f:
        for k in range(K):
            f.write("{:>10.4f} \n".format(y[k]))
""" MAIN PROG """
y_0 = 1.0; T = 10.0; tau = 1.0/100
eps_stop = tau/10; K = int(np.floor(T/tau))
y = np.zeros(K+1); t_list = np.zeros(K+1)
y[0] = y_0; t_list[0] = 0
for k in range(K):
    t_k = k*tau
    z = y[k]
    diff = 1
    while (diff>eps_stop):
        z_new = y[k]+tau*Phi(t_k,y[k],z,tau)
        diff = abs(z_new-z)
        z = z_new
    y[k+1] = z; t_list[k+1] = t_k+tau
save_solution(y,K);
plt.plot(t_list,y)
plt.show()
```

Fig. 36.9 Implementation of the implicit Euler method in Python; the nonlinear equations in the time steps are solved with a fixed point iteration (program available at https://doi.org/10.1007/978-3-662-70890-3_36)

Advanced Topics

Some important topics and concepts could not be included in this book. These are suitable as presentation topics for a seminar following a lecture on numerical methods.

Numerical Linear Algebra

- Convergence of the QR method for eigenvalue problems
- SOR method for the iterative solution of linear systems
- Stability properties of Gaussian elimination
- Perturbation results for eigenvalues of symmetric matrices
- Lanczos method for eigenvalue determination
- Aspects of the practical implementation of the Simplex algorithm

Numerical Analysis

- Lebesgue constant in numerical interpolation
- Barycentric Lagrange interpolation
- Polynomial approximation with respect to least squares
- GMRES method and Arnoldi process
- Euler–Maclaurin formula and Romberg quadrature
- Levenberg–Marquardt method
- Clenshaw–Curtis quadrature
- Chebyshev root finding method
- CAD methods

Numerics of Ordinary Differential Equations

- Splitting methods and exponential integrators
- Collocation, Gaussian and Radau methods
- Analysis of extrapolation methods
- Discussion of special Runge-Kutta methods
- Dahlquist's limit theorems

- Error constants in multistep methods
- Störmer–Verlet method for Hamiltonian systems
- Lagrange formulations and variational integrators
- Algebraic differential equations

Bibliography

The following textbooks and lecture notes on numerics were used in the preparation of the material presented:

1. Dahmen, W., Reusken, A.: Numerik für Ingenieure und Naturwissenschaftler. Springer-Lehrbuch. Springer, Berlin (2008)
2. Deiser, O., Lasser, C., Vogt, E., Werner, D.: 12×12 Schlüsselkonzepte zur Mathematik. Spektrum Akademischer Verlag, Heidelberg (2011)
3. Freund, R.W., Hoppe, R.H.W.: Stoer/Bulirsch: Numerische Mathematik 1. Springer-Lehrbuch. Springer, Berlin (2007)
4. Haftendorn, D., Riebesehl, D., Dammer, H.: Höhere Mathematik sehen und verstehen. Springer-Spektrum. Springer, Berlin (2024)
5. Hairer, E., Wanner, G.: Introduction à l'Analyse Numérique. Lecture Notes. Université de Genève, Geneva (2005)
6. Hämmerlin, G., Hoffmann, K.-H.: Numerische Mathematik. Springer-Lehrbuch. Springer, Berlin (1994)
7. Hanke-Bourgeois, M.: Grundlagen der Numerischen Mathematik und des Wissenschaftlichen Rechnens. Vieweg+Teubner, Berlin (2009)
8. Harbrecht, H.: Einführung in die Numerik und Numerik der Differentialgleichungen. Lecture Notes. Universität Basel, Basel (2014–2015)
9. Plato, R.: Numerische Mathematik kompakt. Springer, Berlin (2010)
10. Praetorius, D.: Numerische Mathematik. Lecture Notes. TU Wien, Vienna (2006)
11. Rannacher, R.: Einführung in die Numerische Mathematik. Lecture Notes. Universität Heidelberg, Heidelberg (2006). http://numerik.uni-hd.de/~lehre/notes/
12. Schaback, R., Wendland, H.: Numerische Mathematik. Springer-Lehrbuch. Springer, Berlin (2005)
13. Süli, E., Mayers, D.F.: An Introduction to Numerical Analysis. Cambridge University Press, Cambridge (2003)

Topics of numerical linear algebra, especially the efficient solution of large linear systems, can be found in the following books:

14. Golub, G.H., Van Loan, C.F.: Matrix Computations. Johns Hopkins Studies in the Mathematical Sciences. Johns Hopkins University Press, Baltimore (1996)
15. Hackbusch, W.: Iterative Solution of Large Sparse Systems of Equations. Applied Mathematical Sciences. Springer, Cham (2016)
16. Meister, A.: Numerik Linearer Gleichungssysteme. Springer, Berlin (2015)
17. Saad, Y.: Iterative Methods for Sparse Linear Systems. Society for Industrial and Applied Mathematics, Philadelphia (2003). http://www-users.cs.umn.edu/~saad/books.html
18. Strang, G.: Linear Algebra and Its Applications. Academic Press [Harcourt Brace Jovanovich, Publishers], New York-London (1980)
19. Trefethen, L.N., Bau, D., III: Numerical Linear Algebra. Society for Industrial and Applied Mathematics, Philadelphia (1997)

Detailed stability analyses and further topics of numerical analysis can be found in the following books:

20. Higham, N.J.: Accuracy and Stability of Numerical Algorithms. Society for Industrial and Applied Mathematics, Philadelphia (2002)
21. Krommer, A.R., Überhuber, C.W.: Computational Integration. Society for Industrial and Applied Mathematics, Philadelphia (1998)
22. Ortega, J.M., Rheinboldt, W.C.: Iterative Solution of Nonlinear Equations in Several Variables. Classics in Applied Mathematics. Society for Industrial and Applied Mathematics, Philadelphia (2000)
23. Overton, M.L.: Numerical Computing with IEEE Floating Point Arithmetic. Society for Industrial and Applied Mathematics, Philadelphia (2001)
24. Trefethen, L.N.: Six myths of polynomial interpolation and quadrature. Math. Today **47**(4), 184–188 (2011)

The numerical treatment of optimisation problems with constraints is covered in the following books:

25. Forst, W., Hoffmann, D.: Optimization - Theory and Practice. Springer Undergraduate Texts in Mathematics and Technology. Springer, New York (2010)
26. Nocedal, J., Wright, S.J.: Numerical Optimization. Springer Series in Operations Research and Financial Engineering. Springer, New York (2006)

The theory and numerics of ordinary differential equations are the subject of the following books:

27. Butcher, J.C.: Numerical Methods for Ordinary Differential Equations. Wiley, Chichester (2008)
28. Deuflhard, P., Bornemann, F.: Scientific Computing with Ordinary Differential Equations. Texts in Applied Mathematics. Springer, New York (2002)
29. Hairer, E., Norsett, S.P., Wanner, G.: Solving Ordinary Differential Equations I. Nonstiff Problems. Springer Series in Computational Mathematics. Springer, Berlin (1993)
30. Hairer, E., Wanner, G.: Solving Ordinary Differential Equations II. Stiff and Differential-Algebraic Problems. Springer Series in Computational Mathematics. Springer, Berlin (1996)
31. Iserles, A.: A First Course in the Numerical Analysis of Differential Equations. Cambridge Texts in Applied Mathematics. Cambridge University Press, Cambridge (1996)

32. Strehmel, K., Weiner, R., Podhaisky, H.: Numerik gewöhnlicher Differentialgleichungen. Vieweg+Teubner, Wiesbaden (2012)
33. Teschl, G.: Ordinary Differential Equations and Dynamical Systems. Graduate Studies in Mathematics. American Mathematical Society, Providence (2012). http://www.mat.univie.ac.at/~gerald/ftp/book-ode/
34. Walter, W.: Ordinary Differential Equations. Graduate Texts in Mathematics. Springer, New York (1998)

Introductions to MATLAB, C++ and Python are offered by the following books:

35. Flowers, C.H.: An Introduction to Numerical Methods in C++. Oxford University Press, Oxford (2000)
36. Higham, D.J., Higham, N.J.: MATLAB Guide. Society for Industrial and Applied Mathematics, Philadelphia (2005)
37. Kirsch, R., Schmitt, U.: Programmieren in C: Eine mathematikorientierte Einführung. Springer Masterclass. Springer, Berlin (2007)
38. Klein, B.: Numerical Programming with Python. Online Tutorial (2021). https://python-course.eu/numerical-programming/
39. Press, W.H., Teukolsky, S.A., Vetterling, W.T., Flannery, B.P.: Numerical Recipes 3rd Edition: The Art of Scientific Computing. Oxford University Press, Oxford (2007)
40. Quarteroni, A., Saleri, F.: Scientific Computing with MATLAB and Octave. Texts in Computational Science and Engineering. Springer, Berlin (2006)
41. The MathWorks Inc.: MATLAB Version: 9.13.0 (R2022b). The MathWorks, Natick (2022).
42. Mueller, J.P., Sizemor, J.: MATLAB for Dummies. Wiley, Hoboken (2021)

The necessary results of analysis and linear algebra can be found in the following books:

43. Lax, P.D., Terrell, M.S.: Multivariable Calculus with Applications. Undergraduate Texts in Mathematics. Springer, Cham (2018)
44. Liesen, J., Mehrmann, V.: Linear Algebra. Undergraduate Mathematics Series. Springer, Cham (2015)
45. Pedersen, S.: From Calculus to Analysis. Mathematics and Statistics. Springer, Cham (2015)

Index

A
A-conjugate, 138
Adams-Bashforth method, 202
Adams-Moulton method, 202
Adaptive algorithm, 230
Aitkin process, 124
Algorithm, 5
Approximate solving, 3
Approximation, 3
Armijo condition, 133
A-stable, 218
Asymptotic, 8
Asymptotic region, 128

B
Backward-differentiation-formulas, 202
Backward substitution, 17
Banach's fixed point theorem, 67
Band matrix, 147
Bandwidth, 76
Barycentric representations, 91
Bisection method, 129
Bolzano–Weierstrass theorem, 327
Boundary value problem, 239
Butcher tableau, 193

C
Cancellation, xxviii, 6
Characteristic polynomial, 325
Chebyshev nodes, 92
Chebyshev polynomial, 92
Cholesky decomposition, 21
Cholesky decomposition, incomplete, 151
Coercive, 222
Column sum norm, 13
Complexity, 8
Computational effort, 3
Conditioning, 5, 79
Condition number, 14, 79
Conjugate gradient method, 141
Conjugate vectors, 138
Consistency, 183, 203
Contraction, 67
Control procedure, 230
Convergence order, 128
Convex, 222
Corner, 46
Cycles, 49

D
Dahlquist's root condition, 211
Data error, 3
Degree of exactness, 116
Descent method, 133
Determinant, 324
Diagonalisable, 325
Diagonally dominant, 72
Difference quotient, 181
Differential equation, autonomous, 170
Discontinuous Galerkin method, 240
Discrete search, 132
Discretisation error, 184
Divided differences, 90
Double precision, 82

E
Eigenvalue, 325
Eigenvalue problem, 51
Elements, 156
Elimination method, 25
Euclidean norm, 11
Euler method, 182
Euler method, partitioned, 236
Exactness, 195
Experimental convergence order, 124
Explicit method, 182

F
Fast Fourier transform, 111
Feasible set, 45
Fill-in, 151
Fixed point iteration, xx
Floating point number, 81
Flow, 234
Fourier basis, 109
Fourier synthesis, 110
Fourier transform, 110
Frobenius norm, 13
Fundamental theorem, 328, 330

G
Gaussian normal equation, 33
Gauss quadrature, 120, 162
Gauss-Seidel method, 71
Gershgorin circles, 51
Givens rotation, 61
Global convergence, 69, 127
Gradient, 330
Gradient flow, 221
Gradient method, 133
Grid, 155
Gronwall's lemma, continuous, 176
Gronwall's lemma, discrete, 185

H
Hat function, 97, 160
Heat equation, 223
Hermite interpolation, 94
Heron's method, xx
Hessian matrix, 330
Horner scheme, 91
Householder transformation, 36

I
Image, 324
Implicit method, 182
Increment function, 182
Index set, 46
Initial condition, 167
Initial value problem, 167
Integral curve, 170
Intermediate value theorem, 327
Interpolant, 161, 227
Interpolation polynomial, 88
Interpolation task, 99
Interpolation, trigonometric, 107
Interval reduction, 132
Inverse iteration, 58
Irreducible, 72
Iteration, xxiv
Iterative method, 69

J
Jacobian matrix, 330
Jacobi method, 63, 71
Jordan normal form, 326

K
Kepler's barrel rule, 118
Kernel, 324
Krylov space, 140

L
Lagrange interpolation, 87
Lagrange polynomial, 87
Lagrange representation, 328
Landau notation, 8, 329
Leapfrog method, 202
Least squares problem, 33
Linear program, 45
Local convergence, 127
L-stable, 221
LU decomposition, 18

M
Machine number, 3, 81
Machine precision, 82
Mathematical operation, 3
Mean value theorem, 328
Mesh width, 156
Method error, xxxii
Method of least squares, 33
Midpoint method, 183
Midpoint rule, 117
Minimisation problem, 127
Model error, xxxii, 3

Moore–Penrose inverse, 43
Multi-body problem, 171, 233
Multistep method, 201

N
Neville's algorithm, 89, 357
Newton basis, 91
Newton-Cotes formula, 117
Newton method, 130
Newton's law of cooling, 167
Nodal basis, 160
Nodes, 87, 156
Norm, 11
Normalised LU decomposition, 18
Normalised triangular matrix, 17
Not-a-number, 83

O
Operator norm, 12
Order, 8
Order of a differential equation, 169
Order of convergence, 119, 186
Ordinary differential equation, 167
Orthogonal, 323
Orthogonal matrix, 35
Orthogonal polynomial, 121
Overflow, 83

P
Partial degree, 157
Peano's theorem, 175
Permutation matrix, 28
Phase diagram, 170
Picard-Lindelöf theorem, 174
Pivot search, 28
Polygonal chain method, 182
Positive definite, 21
Positive semidefinite, 21
Power method, 55
Preconditioning matrix, 148
Predator-prey model, 168
Predictor-corrector method, 206
Procedure, 5
Pseudoinverse, 43

Q
QR decomposition, 37
QR method, 58
Quadrature formula, 115, 158, 162
Quadrature formula, composite, 118

R
Rank, 324
Rayleigh quotient, 52
Reducible, 72
Regula-falsi method, 130
Relative error, 5
Relaxation method, 76
Remainder term, 328
Residual, 33, 139
Richardson method, 70
Rolle's theorem, 328
Root finding, 127
Rounding, 3, 83
Rounding error, xxxii
Row equilibration, 148
Row sum norm, 13
Runge-Kutta method, 192

S
Scalar product, 323
Secant method, 129
Self-stabilizing, 128
Separation of variables, 172
Shooting method, 239
Simplex, 156
Simplex method, 49
Simpson rule, 118
Single precision, 82
Single-step method, 182
Singular value decomposition, 42
Sparse, 147, 345
Spectral norm, 13
Spectral radius, 13, 69
Spectrum, 325
Spline, 97, 160
Stability, 6, 84, 177
Stability function, 219
Standard form, 45
Step size, 181
Step size control, 230
Stiff differential equation, 218
Symplectic method, 236

T
Taylor's formula, 328
Tensor grid, 155
Time steps, 181
Total degree, 157
Trapezoidal method, 193
Trapezoidal rule, 117
Triangular matrix, 17
Triangulation, 156
Two-body problem, 171

U
Unconditionally stable, 218
Underflow, 83
Uniform grid, 155
Uniform triangulation, 156
Unit root, 109

V
Values, 87
Vandermonde matrix, 88

Variables, 167
Variation of constants, 172

W
Weight function, 120

Z
Zero-stability, 211

MIX
Papier aus verantwortungsvollen Quellen
Paper from responsible sources
FSC® C105338

If you have any concerns about our products,
you can contact us on
ProductSafety@springernature.com

In case Publisher is established outside the EU,
the EU authorized representative is:
**Springer Nature Customer Service Center GmbH
Europaplatz 3, 69115 Heidelberg, Germany**

Printed by Libri Plureos GmbH
in Hamburg, Germany